T0171760

Basic Course in Race Car Technology

Lars Frömmig

Basic Course in Race Car Technology

Introduction to the Interaction of Tires, Chassis, Aerodynamics, Differential Locks and Frame

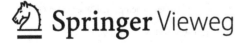 Springer Vieweg

Lars Frömmig
Volkswagen AG
Wolfsburg, Germany

ISBN 978-3-658-38472-2   ISBN 978-3-658-38470-8   (eBook)
https://doi.org/10.1007/978-3-658-38470-8

This Springer Vieweg imprint is published by the registered company Springer Fachmedien Wiesbaden GmbH, part of Springer Nature.
The registered company address is: Abraham-Lincoln-Str. 46, 65189 Wiesbaden, Germany

Series Foreword

Now the manual series is already in its second edition and the family of these books is growing. While there were already five volumes at the beginning, another volume has been added for this edition. The authoritative idea that individual special volumes can go into depth without space problems has thus proven itself. Since the publication of the first edition, further findings have been added, which have found their way into the corresponding chapters, or new chapters have been added.

That the contents nevertheless fit together and complement each other as if in a single book – one of the great strengths of the original book *Race Car Technology* – is ensured by the editor in a comparable way to how the project manager keeps an eye on the overall function in a major construction project.

The racing car technology handbook series is dedicated to the racing vehicle from conception, design and calculation to operation and its (further) development.

The first volume, *Basic Course in Race Car Technology*, thus offers not only current considerations but also a historical overview of motorsport, racing operations, such as the rescue chain, and a comprehensive overview of the technology used in racing cars as a general introduction to the subject. For more than 15 years, the author has been concerned with the driving dynamics and chassis tuning of production passenger cars.

Volume two, *Complete Vehicle*, starts with the chronological design process and therefore begins with concept considerations, considers safety aspects and the design of the driver's environment, describes aerodynamic influences, and then looks at the frame and outer skin design.

Volume three *Drive* deals with all forms of drive systems and their energy storage, continues in the sense of load flow via start-up elements and identification converters up to the side shafts. Electrical systems and electronic driving aids have also found their appropriate place in this volume.

Volume four, *Chassis*, is devoted exclusively to the decisive assembly and its components that determine driving behavior. Tires and wheels, wheel-guiding parts, springs and dampers, steering, and brakes are covered.

Volume five, *Data Analysis, Tuning and Development*, deals with the phase that follows once the vehicle has been designed and built. The development and tuning of a racing vehicle requires a much different approach than its construction, and key tools – such as data acquisition and analysis, simulation, and testing – are therefore presented. The subject of data acquisition and analysis is profoundly presented by an author who is confronted with this activity on a daily basis.

For volume six, *Practical Course in Vehicle Dynamics*, authors have been recruited who have decades of experience as race engineers and race drivers on the race track. In their work, they describe the practical tuning of racing vehicles, underpin what they present with examples of calculations and thus also build a bridge to the theoretical considerations in the other volumes.

I hope all readers will find "their" volume in the abundance offered and that they will get essential impulses for their studies, profession and/or leisure time from reading it, be it because they are designing a vehicle, building one, or operating and improving one, or because they are analyzing one with a thirst for knowledge.

Graz, Austria   Michael Trzesniowski
Spring 2019

Greeting

27.05.2019
EC Todsen

Second Edition: Racing Car Technology Handbook – Six Volumes

Motorsport continues to inspire. For as long as there have been cars, drivers have been pushing their racing cars to their technical and physical limits, engaging in gripping and exciting competitions. But the competition doesn't just take place on the race track. The foundation for success is laid in the development departments and design offices. In-depth knowledge of vehicle technology and development methodologies, along with thorough and timely project management, creative problem-solving skills and unconditional team play, determine victory or defeat.

Motorsport continues to be a model and guide for technological progress – be it in lightweight construction, material selection or aerodynamics. Chassis and tire technology also benefit immensely, and new safety concepts are often based on experience from the racetrack. However, the influence of motorsport is particularly evident in the powertrain: In addition to the impressive increase in performance and efficiency of the classic internal combustion engine drive system, the key future technologies of hybrid and purely electric drive have also successfully arrived in motorsport competition and are continuing this successfully and with public appeal in a partly completely new setting. It remains very exciting to observe which attractive innovations digital networking solutions and autonomous driving systems will generate in motorsport. I recommend that young engineers in particular acquire the tools for their future careers in motorsport. What you learn in motorsport sticks. Formula Student already offers an ideal environment to start with.

I am very pleased that the book series Handbuch Rennwagentechnik has been so well received and that the second edition has been published within 2 years. This shows that the competencies addressed are clearly presented in this work and conveyed in an understandable way.

This book series has deservedly become a well-known and valued reference work among experts. The work brings students closer to the fascination of motorsport and racing enthusiast laymen to a deeper technical understanding.

I wish you much success on and off the race track!

Deputy Chairman of the Executive Board   Peter Gutzmer
and Chief Technology Officer, Schaeffler AG
Herzogenaurach, Germany

Preface

On November 13, 1994, as a 15-year-old, I sit tensely in front of the television in the living room of my parents' house. It is shortly before 4 o'clock in the night. My parents and siblings are still asleep. The Australian Grand Prix is being broadcast, the last race of the 1994 Formula 1 season. The two remaining contenders for the Formula 1 world title, Michael Schumacher in a Benetton-Ford B194 and Damon Hill in a Williams-Renault FW16, are separated by just one point. The starting position is simple. Michael Schumacher will be world champion if he finishes ahead of Damon Hill or if Hill scores a maximum of one point more than Schumacher. After an eventful and tragic year, this is the 17th time and the 1st time since 1986 that the Formula One World Champion will be decided in the final race of the season. Nigel Mansell is on pole position with the second Williams, followed by Schumacher and Hill. Right after the start Mansell is overtaken by both of them. Damon Hill, who is often criticized by the media, drives the race of his life. He hounds Schumacher lap after lap. They turn into the pit lane together on lap 18. After the pit stop, the picture is the same: Schumacher leads the race ahead of Hill. The gap is at times less than a second. On lap 35, the cameras show a slow Benetton after turn 5. Schumacher has touched the boundary wall on the right in turn five after a driving mistake. Hill wants to pass Schumacher, who has returned to the track, on the left. Schumacher fends him off. Before the next right-hand bend, Hill pulls to the inside. Schumacher is half a nose in front and pulls his car sharply towards the apex. Hill's left front wheel hits the Benetton's right side pod, and shortly thereafter lifts the Benetton over its right rear wheel. The Benetton's sideways position is almost 90 ° for a fraction of a second before it falls back on its wheels and goes straight into the boundary wall. It's all over for Schumacher. Hill continues the race with a flat left front tire. The retired Schumacher stands at the boundary wall staring into space. The world title seems lost. Hill pulls into the pits. The Williams is inspected by the mechanics. The camera shows a close-up of the front upper wishbone, which is bent and cracked from the collision. Damon Hill also has to retire from the race. I'm going into ecstasies in front of the TV. Michael Schumacher becomes the first German Formula 1 World Champion.

The year 1994 sparked my interest in Formula 1 and motorsport in general, which continues to this day. In the course of my studies and my time as a research assistant at the Institute of Automotive Engineering at the TU Braunschweig, the technical aspects of this sport became increasingly important to me. Racing cars show what is technically possible when engineers – in contrast to the series production business – can act detached from economic constraints and from strict benefit-to-expense considerations. Driving performance and acceleration are achieved that otherwise only fighter pilots experience. Their consistent design, their "war paint", and the competition in a highly emotional and sometimes dangerous environment also contribute to their fascination. This fascination finally gave rise to the lecture "Racing Vehicles" in 2007, which I still hold annually at the Institute of Automotive Engineering at the TU Braunschweig and on whose contents this book is based.

The book *Grundkurs Rennwagentechnik* (*Basic Course in Race Car Technology*) is therefore primarily aimed at engineering students who are studying automotive engineering in depth or are involved in one of the numerous Formula Student projects, as well as engineers from the automotive and motorsport industries. In reading the book, a basic knowledge of engineering is advantageous, but not essential. In compiling the book, an attempt was made to achieve the most descriptive presentation of the technical content possible, so that hopefully motorsport enthusiasts of all kinds will gain interesting insights into the development of historical and modern racing vehicles.

Before it's said:

Ladies and gentlemen, start your engines!,

I would like to express my sincere thanks to the following for the provision of image and text material: ADAC, AP Racing, ATL Ltd, AUDI AG, ASME, BMW AG, BorgWarner Inc, Brembo S.p.A., Daimler AG, Dallara, DC Electronics Motorsport Specialist Ltd, Deutsches Zentrum für Luft- und Raumfahrt e.V., fka GmbH (Aachen), Institute for Aerodynamics and Gas Dynamics (University of Stuttgart), Simon McBeath (SM Design), McLaren Technology Group, Dr. Maria Mogg, MOOG, MoTeC, Nürburgring 1927 GmbH & Co. KG., Pirelli & C. S.p.A., Dr. Ing. h.c. F. Porsche AG, Racecar Engineering (The Chelsea Magazine Company), Red Bull Media House, Red Bull Racing, Sauber Motorsport AG, Steve Rendle (Haynes Publishing), Volkswagen AG, Volkswagen Motorsport GmbH, Williams Grand Prix Engineering Ltd, Prof. Xin Zhang (The Hong Kong University of Science and Technology), ZF Friedrichshafen AG.

I would also like to thank Springer Vieweg Verlag for their support during publication and Prof. Michael Trzesniowski for including the work in the successful book series Handbuch Rennwagentechnik.

This book is dedicated to my parents, my wife Alexandra and my two children, Lana and Nils.

Brunswick, Germany   Lars Frömmig

Contents

Symbols, Units and Abbreviations

Chapter 4

F_x	Longitudinal force [N]
$F_{x3,\,4}$	Longitudinal force of a rear wheel [N]
F_{xH}	Longitudinal force on rear axle [N]
F_{xy}	Resulting tire force in the road surface [N]
F_y	Lateral force [N]
f_R	Rolling resistance coefficient [N]
F_R	Rolling resistance force [N]
F_z	Normal load [N]
K_P	Rating index for air pressure setting [°]
K_γ	Rating index for camber setting [°]
m	Vehicle mass [kg]
M_{an}	Drive torque [Nm]
M_{br}	Braking torque [Nm]
n	Amount of substance [mol]
p	Tire inflation pressure [bar]
P_{an}	Drive power at the wheels [kW]
P_{Fzg}	Power on vehicle [kW]
$P_{V,S_{an}}$	Power loss due to traction slip [kW]
$P_{V,\alpha}$	Power loss due to side slip [kW]
r	Dynamic tire radius [m]
R_m	Universal gas constant [J/(mol-K)]
S_{an}	Traction slip
S_{br}	Brake slip
T	Tire temperature [°C]
v	Speed over ground [m/s]
v_R	Wheel peripheral speed [m/s]
V	Air volume in tire [m^3]

$\ddot{x}$	Longitudinal acceleration of the vehicle center of gravity [m/s^2]
$\ddot{y}$	Lateral acceleration of the vehicle center of gravity [m/s^2]
α	Side slip angle [°]
γ	Camber angle [°]
μ	Friction coefficient between tire and road surface
μ_y	adhesion potential for lateral forces
ω	Wheel speed [°/s]

Chapter 5

a_x	Longitudinal acceleration of the vehicle center of gravity [m/s^2], see $\ddot{x}$
a_y	Lateral acceleration of the vehicle center of gravity [m/s^2], see $\ddot{y}$
$a_{y,gr,V,H}$	Limiting lateral acceleration of the front axle, rear axle [m/s^2]
$c_{FV,H}$	Spring stiffness on front axle, rear axle [N/m]
$c_{StV,H}$	Stabilizer stiffness on front axle, rear axle [Nm/°]
c_W	Total roll stiffness [Nm/°]
F_{F1-4}	Spring force on wheel [N]
F_{Lx}	Drag force [N]
F_{x1-4}	Longitudinal force of a wheel [N]
$F_{xV,H}$	Longitudinal force on front axle, rear axle [N]
F_{y1-4}	Lateral force of a wheel [N]
$F_{yV,H}$	Lateral force on front axle, rear axle [N]
F_{z1-4}	Wheel load [N]
$F_{zV,H}$	Axle load on front axle, rear axle [N]
g	Gravitational acceleration, 9.81 m/s^2
h'	Distance between center of gravity of the body and rolling axis [m]
h''	Distance between center of gravity of body and pitch center [m].
h_A	Body center of gravity height [m]
i_L	Steering ratio
J_z	Yaw moment of inertia [kgm^2]
l	Wheelbase [m]
l_H	Distance between center of gravity and rear axle [m]
l''_H	Distance between pitch center and rear axle [m]
l_{HA}	Distance between center of gravity of the body and the rear axle [m]
l_V	Distance between center of gravity return and front axle [m]
l''_V	Distance between pitch center and front axle [m]
l_{VA}	Distance between center of gravity of the body and front axle [m]
m	Vehicle mass [kg]
m_A	Body mass [kg]

$m_{V,H}$	Unsprung mass of front axle, rear axle [kg]
$M_{FV,H}$	Rolling resistance moment of the body springs on the front axle, rear axle [Nm].
$M_{StV,H}$	Roll resistance moment of the stabilizer on the front axle, rear axle [Nm].
$M_{V,H}$	Roll resistance moment of the front axle, rear axle [Nm]
$p_{V,H}$	Roll center height of the front axle, rear axle [m]
$s_{FV,H}$	Spring width of front axle, rear axle [m]
$s_{V,H}$	Track width of front axle, rear axle [m]
v	Vehicle speed at center of gravity [m/s]
v_{1-4}	Speed at the wheel centers [m/s].
v_V	Speed at the front axle [m/s]
v_{Vx}	Longitudinal velocity component at the front axle [m/s].
v_{Vy}	Transverse speed component on the front axle [m/s].
v_H	Speed at rear axle [m/s]
v_{Hx}	Longitudinal velocity component at the rear axle [m/s].
v_{Hy}	Transverse speed component at the rear axle [m/s].
x	Longitudinal movement of the center of gravity of the vehicle [m].
$\dot{x}$	Longitudinal speed of the vehicle center of gravity [m/s].
$\ddot{x}$	Longitudinal acceleration of the vehicle center of gravity [m/s^2]
y	Lateral movement of the center of gravity of the vehicle [m].
$\dot{y}$	Lateral speed of the vehicle center of gravity [m/s].
$\ddot{y}$	Lateral acceleration of the vehicle center of gravity [m/s^2]
z	Vertical movement of the center of gravity of the vehicle [m].
z_{1-4}	Vertical movement of a wheel [m]
$\dot{z}$	Vertical speed of the vehicle center of gravity [m/s].
$\dot{z}_{1-4}$	Vertical velocity of a wheel [m/s].
$\ddot{z}$	Vertical acceleration of the vehicle center of gravity [m/s^2]
$\ddot{z}_{1-4}$	Vertical acceleration of a wheel [m/s].
α_{1-4}	Slip angle of a wheel [°]
α_V	Slip Angle of the front axle [°].
α_H	Slip angle of rear axle [°]
β	Side Slip angle [°]
$\dot{\beta}$	Angular velocity of the side slip angle [°/s]
γ_{1-4}	Camber angle of a wheel [°]
$\varepsilon_{V,H}$	Angle of inclined suspension on front axle, rear axle [°].
δ_{1-4}	Steering angle of one wheel [°]
δ_L	Steering wheel angle [°]
δ_V	Front axle steering angle [°]
κ	Roll angle [°]
$\dot{\kappa}$	Roll angle speed [°/s]
$\ddot{\kappa}$	Roll angle acceleration [°/s^2]

μ_{1-4}	coefficient of friction
$\mu_{V,H}$	Friction coefficient of the front axle, rear axle
ρ	Radius of curvature [m]
φ	Pitch angle [°]
$\dot{\varphi}$	Pitch angle velocity [°/s]
$\ddot{\varphi}$	Pitch angle acceleration [°/s²]
ψ	Yaw angle [°]
$\dot{\psi}$	Yaw angular velocity, yaw rate [°/s]
$\ddot{\psi}$	Yaw angle acceleration [°/s²]
ω_{1-4}	Wheel speed [°/s]
$\dot{\omega}_{1-4}$	Wheel acceleration [°/s²]

Chapter 6

A	Frontal area [m²]
$A_{1,2}$	Inlet cross section, outlet cross section [m²]
A_B	Aerodynamic balance
A_E	Aerodynamic efficiency
A_F	Transverse span area of an airfoil [m²]
b	Wing span, underbody width at diffuser [m].
c_W	Drag coefficient for an angle of incidence of 0 °
c_x	Drag coefficient
c_{xF}	drag coefficient of an airfoil
c_{Mx}	Coefficient for aerodynamic moment about the x-axis
c_{My}	Coefficient for aerodynamic moment about the y-axis
c_{Mz}	Coefficient for aerodynamic moment about the z-axis
c_y	Coefficient of lateral air force
c_z	lift or downforce coefficient
c_{zF}	coefficient of lift or downforce of an airfoil
$c_{zV,H}$	Lift or downforce coefficient on front axle, rear axle
C_p	Dimensionless pressure coefficient
F_{Lx}	Drag force [N]
$F_{Lx,1-4}$	Aerodynamic longitudinal force component on the wheel (aerodynamic drag) [N].
F_{Ly}	Lateral aerodynamic force [N]
$F_{Ly,1-4}$	Aerodynamic lateral force component on the wheel (lateral aerodynamic force) [N].
F_{Lz}	Lift force or downforce [N]

$F_{Lz,1-4}$	Aerodynamic vertical force component on the wheel (lift, downforce) [N]
F_{xF}	Drag force on a wing profile [N]
F_{zF}	Lift or downforce on a wing profile [N]
$F_{xV,H}$	Axle longitudinal force on front axle, rear axle [N]
$F_{yV,H}$	Axle lateral force on front axle, rear axle [N]
$F_{zV,H}$	Axle load on front axle, rear axle [N]
h	Distance of a wing profile to the ground [m]
h_1	Ground clearance of the underbody in front of the rear diffuser [m]
h_2	Height of the diffuser end above the roadway [m]
h_d	Diffuser rise height [m]
l	Wheelbase, underbody length [m]
$m_{1,2}$	Inlet mass, outlet mass [kg]
$\dot{m}_{1,2}$	Inlet mass flow, outlet mass flow [kg]
M_{Lx}	Aerodynamic moment about the x-axis [Nm]
M_{Ly}	Aerodynamic moment about the y-axis [Nm]
M_{Lz}	Aerodynamic moment about the z-axis [Nm]
p	Air pressure [bar]
p_0	Ambient pressure [bar]
P_{max}	Maximum power or rated power of the vehicle [kW]
Re	Reynolds number
t	Wing depth [m]
T	Air temperature [°C]
v	Vehicle speed or incident flow velocity [m/s]
$v_{1,2}$	Inlet velocity, outlet velocity [m/s]
v_W	Wind speed [m/s]
α	Angle of attack of a wing profile [°]
δ	Angle between vehicle and wind direction [°]
θ	Rise angle at diffuser [°]
λ_F	Span ratio
$\mu_{V,H}$	Friction coefficient of the front axle, rear axle
ρ_L	Air density at sea level and 15 °C air temperature, 1.225 kg/m^3
τ_L	Sliding angle [°]
υ_L	Kinematic air viscosity [m^2/s]

Chapter 7

a	Vertical distance between instantaneous pole and wheel contact point [m]
A	Frontal area [m^2]
$AKolben$	Area of brake piston (in German: Kolben) [m^2]

b	Horizontal distance between instantaneous pole and wheel contact point [m]
B_V	Braking force on the front axle [N]
c_F	Component-related spring stiffness [N/m]
c_R	Wheel-related spring stiffness or vertical stiffness[N/m]
$c_{R,i}$	Wheel-related spring stiffness or vertical stiffness at wheel i (i = 1, 2, 3, 4) [N/m].
c_T	Torsional stiffness of a torsion spring [Nm/°]
f_i	Natural frequency at wheel i (i = 1, 2, 3, 4) [Hz]
f_{VH}	Weight share of the front axle
f_{LR}	Weight share of the left side of the vehicle
f_{CW}	Weight proportion on the vehicle diagonal
F_{Ai}	Resulting control arm force on the body [N]
F_{Brems}	Total braking force [N]
F_F	Component-related spring force [N]
F_{HA}	Force on the brake cylinder, brake circuit of the rear axle [N]
F_{Jack}	Jacking force [N]
F_{Pedal}	Foot force on brake pedal [N]
$F_{xV,H}$	Axle longitudinal force [N]
F_{yi}	Longitudinal force on wheel i (i = 1, 2, 3, 4) [N]
F_{yi}	Lateral force on wheel i (i = 1, 2, 3, 4) [N]
$F_{yo,u}$	Force in upper, lower wishbone [N]
F_{VA}	Force on the brake force cylinder, brake circuit of the front axle [N]
F_{zi}	Wheel load on wheel i (i = 1, 2, 3, 4) [N]
$Fcylinder$	Force on the brake cylinder [N]
G	Shear modulus [N/m^2]
h	Centre of gravity height [m]
i	Braking force proportion on the front axle
i_{ideal}	Ideal braking force proportion on the front axle
i_F	Spring ratio [N]
I_P	Polar moment of inertia [m^4]
l_A	Distance between damper connection and body-side control arm bearing [m]
l_R	Control arm length [m]
L	Length of a torsion spring [m]
m_i	Proportionate body mass at wheel i (i = 1, 2, 3, 4) [Hz]
$M_{L,Zst}$	Steering torque of the rider at wheel level [Nm]
M_{Br}	Braking torque of the brake disc [Nm]
M_R	Steering system reset torque [Nm]
M_{Servo}	Servo torque of the steering system at wheel level [Nm]
M_{Visco}	Locking torque of a Visco coupling [Nm]
n_K	Mechanical trail [m]
n_R	Pneumatic trail [m]

$p_{Br,VA,HA}$	Brake pressure on front axle, rear axle [m]
q_{ideal}	ideal adhesion utilization ratio [m]
r_B	Effective friction radius of the brake disc [m]
r_{Dr}	Support radius of the differential pin in the thrust collar [m]
r_L	Steering roll radius [m]
r_S	Disturbing force lever arm [m]
s_F	Component-related spring deflection [m]
$w_{VA,HA}$	Balance bar lever for front axle, rear axle brake circuit [m]
z	Wheel travel [m]
z_A	Vertical movement at the articulation point of the damper [m]
δ	Angle of inclination of the suspension strut [°]
ϵ	Torsion angle on a torsion spring [°]
μ_B	Coefficient of friction between brake disc and brake pad
$\mu_{V,H}$	Friction coefficient of the front axle, rear axle
ρ_L	Air density at sea level and 15 °C air temperature, 1.225 kg/m^3

Chapter 8

a_x	Longitudinal acceleration [m/s^2]
a_y	Lateral acceleration [m/s^2]
F_D	Force on differential pinion bolt [N]
$F_{KR3,4}$	Cutting force between differential pinion gear and side gear on left, right wheel [N]
$F_{x3,4}$	Longitudinal force on the left, right rear wheel [N]
$i_{T,i}$	Gear ratio in gear i
i_{FD}	Final drive ratio
J_C	Rotational inertia of the coupling [kg-m^2]
J_M	Rotational inertia of the motor [kg-m^2]
J_R	Rotational inertia of a wheel [kg-m^2]
J_{Ti}	Rotational inertia of the gearbox input shaft [kg-m^2]
J_{To}	Rotational inertia of the gearbox output shaft [kg-m^2]
$M_{3,4}$	Wheel torque at the left, right rear wheel [Nm]
$MBrake$	Braking torque [Nm]
M_{FD}	Axle drive torque [Nm]
M_{Ki}	Clutch torque in clutch i [Nm]
M_M	Engine torque [Nm]
M_{Sperr}	Locking torque [Nm]
M_{Ti}	Transmission input torque [Nm]
M_{To}	Transmission output torque [Nm]

n_M	Motor speed [rpm]
$PBrake$	Braking power [kW]
r	Tire radius, disk radius [m]
r_{AR}	Pitch circle radius of the differential pinion gear [m]
r_{KR}	Pitch circle radius of the side gear [m]
s	Gap width between two disks [m]
s_H	Track width at rear axle [m]
S	Locking value
v	Vehicle speed [m/s]
$v_{3.4}$	Peripheral speed at the left, right rear wheel [m/s]
v_D	Peripheral speed at the differential pinion bolt [m/s]
$v_{K3.4}$	Rigid body speed of the body at the height of the wheel center [m/s]
$v_{KR3.4}$	Peripheral speed of the left, right side gear [m/s]
z	Number of effective pairs of disks, wheel travel [m]
α	Ramp angle [°]
η	Dynamic viscosity of a fluid [Ns/m^2]
μ_{low}	coefficient of friction at the wheel with lowest grip level
ρ_H	Distance between instantaneous center and rear axle [m]
$\dot{\psi}$	Yaw rate [°/s]
ω_{1-4}	Wheel speed at wheel i (i = 1, 2, 3, 4) [°/s]
ω_{AG}	Angular velocity of the differential cage [°/s]
ω_{AR}	Angular velocity of the differential pinion gear [°/s]

Chapter 9

A	Frontal area [m^2]
c_T	Torsional stiffness [Nm/°]
E	Modulus of elasticity N/m^2]
F	Tensile or compressive force [N]
G	Shear modulus [N/m^2]
I_y	Area moment of inertia [m^4]
I_P	Polar moment of inertia [m^4]
l	Component length [m]
M_T	Torsional moment [m]
ϵ	Elongation [m]
σ	Stress [N/m^2]
υ	Poisson's ratio

Chapter 10

a_S	Longitudinal acceleration of the skull [m/s^2]
E_{kin}	Kinetic energy [Nm]
F_{Defo}	Deformation force [N]
HIC	head injury criterion
m	Vehicle mass [kg]
t	Time [s]
v	Vehicle speed [m/s^2]
$\ddot{x}$	Longitudinal deceleration [m/s^2]

Abbreviations

ABS	Anti-Blocking System
ACF	Automobile Club de France
ACO	Automobile Club de l'Ouest
ADR	Accident Data Recorder
ASR	Traction Control System
CFRP	Carbon Fiber Reinforced Plastic
CSI	Comission Sportif Internationale de l'Automobile
CW	Crossweight
DSG	Dual Clutch Transmission
DTM	Deutsche Tourenwagen Masters
FiA	Fédération Internationale de l'Automobile
FRIC	Front Rear Interconnected
FTT	Front Torque Transfer
GT	Gran Turismo
HA	Rear Axle
HIC	Head Injury Criterion
KERS	Kinetic Energy Recovery System
LMP	Le Mans Prototype
MEB	Modular Electric Drive Matrix (german: Modularer-E-Antriebs-Baukasten)
MGU	Motor-Generator Unit (suffix K for Kinetic, H for Heat)
MZ	Roll Center
NASCAR	National Association for Stock Car Auto Racing
PDK	Porsche Double Clutch transmission
SAFER	Steel and Foam Energy Reduction
TCR	Touring Car Racing
UNESCO	United Nations Educational, Scientific and Cultural Organization

VA	Front Axle
WRC	World Endurance Championship
WRC	World Rally Championship
WSC	World Sportscar Championship

Introduction

Powerful and sporty vehicles exert a special fascination on many people, both on the road and on the racetrack, and many childhood dreams revolve around or on four wheels. One dream that can be fulfilled is the Volkswagen Golf VII GTI Performance, which was launched in 2013. It combines sporty driving performance with unrestricted suitability for everyday use. Figure 1.1 compares some of its technical data with those of the Red Bull RB6, the world champion car of the 2010 Formula 1 season. The Golf VII GTI is powered by a 2.0-liter turbocharged inline four-cylinder engine with direct injection that delivers a maximum output of 230 hp. Prescribed by the regulations at the time, the Red Bull RB6 has a direct-injection 2.4-liter naturally aspirated V8 engine that puts out about 725 horsepower. The maximum torque of the TFSI engine is 350 Nm, while the Renault engine of the Red Bull RB6 delivers "only" about 290 Nm. At this point, the Red Bull RB6 would have to admit defeat to the GTI in a quartet of cars. However, the Formula 1 engine reaches a maximum speed of 18,000 rpm. In the Golf VII GTI, the rev limiter is already activated at around 6800 rpm. The engine's power is determined as the product of engine speed and torque, which straightens out the power ratios – even in the quartet of cars. A disadvantage for the Red Bull RB6 in this case, however, would be the consumption of 47–70 l per 100 km, which occurs under racing conditions. The standard consumption of a Golf VII GTI seems to be much better with 6 l per 100 km. In this form, however, the comparison has no significance. Racing engines are more efficient than standard engines, and their consumption is usually considerably lower than that of standard engines in relation to the power output. On the Nordschleife of the Nürburgring, the consumption of a Golf VII GTI rises to well over 20 l per 100 km.

Much of the fascination of racing cars is based on their performance, with the different top speeds of 340 and 250 km/h being more of a side note. What is most impressive are the longitudinal and lateral accelerations achievable by high-performance racing vehicles.

L. Frömmig, *Basic Course in Race Car Technology*, https://doi.org/10.1007/978-3-658-38470-8_1

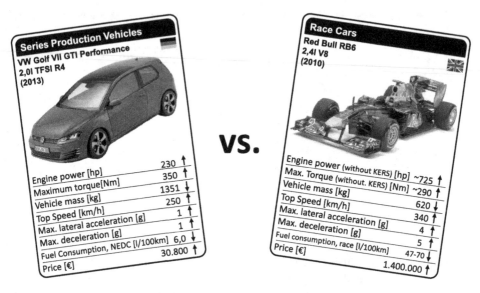

Fig. 1.1 Volkswagen Golf VII GTI Performance (2013) vs. Red Bull RB6 (2010)

Production cars such as a Golf VII GTI achieve maximum lateral accelerations and braking decelerations of around 1 g, which are usually accompanied by a significant squeal from the tyres. In contrast, the maximum lateral acceleration of an F1 car is more than 4 g. In the case of full braking from top speed, about 5 g is reached. This corresponds to four or five times the acceleration due to gravity and is often described as four to five times the driver's body weight in such driving situations. Such acceleration values otherwise only affect pilots of fighter aircrafts.

However, this driving performance is only possible if all the components of the vehicle, essentially the structure, the drivetrain, the chassis with the tires and in particular the aerodynamic components, are developed specifically for motorsport. The development and production of high-performance racing cars is therefore associated with enormous costs. For example, the material costs of an F1 vehicle already amount to more than € 1.4 million. A Golf VII GTI is listed in the price lists starting at € 30,800. This price includes material or purchasing costs, but also the development, personnel and distribution costs as well as other costs and profit margins of the manufacturing company.

Achieving the driving performance possible today has resulted from various development steps, as Fig. 1.2 indicates in the form of the maximum lateral acceleration of production and racing vehicles plotted over time. Technical advances in tyres, lightweight construction and chassis development have roughly doubled the maximum lateral acceleration of production and production sports cars since 1950. An enormous spread between the maximum lateral accelerations of production and racing cars begins in the late 1960s, due to aerodynamics, or the generation of aerodynamic downforce with the help of wings (2) and the underbody (3). Subsequently, the aerodynamic properties of a racing vehicle

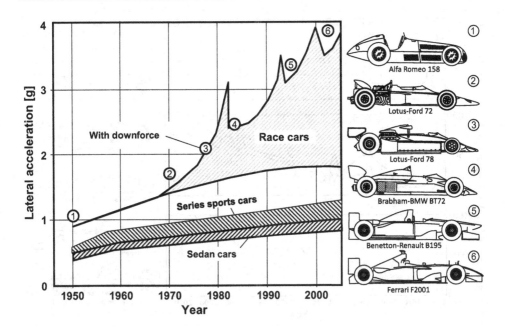

Fig. 1.2 Historical development of maximum lateral acceleration. (Adapted from [1]; courtesy of © Springer Fachmedien Wiesbaden 2018, All Rights Reserved)

have had a lasting influence, particularly on the structural and chassis requirements. The aerodynamics of racing cars are therefore dealt with extensively in this book. The recurring setbacks in maximum lateral acceleration can be explained by changes in the technical regulations. For Formula 1 cars, for example, the elimination of the ground effect by a flat underbody was introduced in 1983 (4), significant changes were also made in 1994 and 1995 after the fatal accidents of Roland Ratzenberger and Ayrton Senna (5), and the number of permitted wing elements was limited in 2001 (6). The development of racing cars is therefore always a competition between the creativity of engineers and the regulating institutions, so that the maximum lateral accelerations of F1 cars today remain at an approximately constant level. The basic nature of technical regulations is also discussed in this book.

The influence of the technical regulations can also be seen in Fig. 1.3, which compiles the characteristic features of the engines – displacement, power and maximum speed – of Grand Prix vehicles between 1906 and 2014. Similar to aerodynamics, technical progress can be read from the continuous increase in power and maximum speed. It can also be seen that after the First World War, with a brief interruption in the 1930s, the limitation of permissible engine capacity became established as a key instrument of technical regulation. The power explosion due to the turbo engine, which started in 1983 and which also clearly exceeded the 1000 hp. limit in qualifying, finally leds to the ban of this engine concept in 1989. In the mid-2000s, maximum engine speeds of around 20,000 rpm were achieved to compensate for the gradual reduction in displacement that began in 1995. Since 2009,

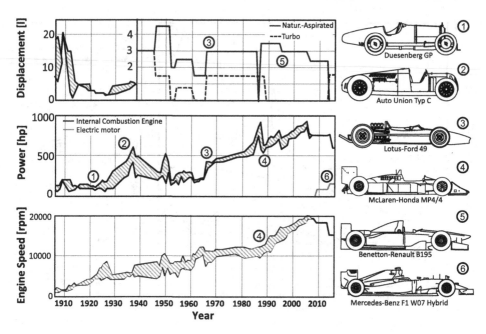

Fig. 1.3 Engine characteristics of Grand Prix and Formula One cars from 1906 to 2014. (Engine data taken from [2–5])

maximum engine speed has also been a component of the technical regulations. The high engine outputs of racing cars are a prerequisite for overcoming the air resistance induced by downforce. In 2009, Formula 1 introduced a powertrain hybridized by an electric motor, known as KERS (Kinetic Energy Recovery System), and since 2014 the use of a turbocharged engine is mandatory. Motorsport is thus following the general trends towards powertrain electrification and engine downsizing. These trends will continue to change motorsport in the coming years and decades due to the ongoing discussions about CO_2 emissions.

Progress in the field of safety has been just as impressive as it is necessary, although it is only since the mid-1990s that the subject of safety has been considered as scientifically as the design of racing cars. This change in thinking began above all after the fatal accidents involving Roland Ratzenberger and three-time Formula 1 World Champion Ayrton Senna at the 1994 San Marino Grand Prix in Imola. Comparable tragic events have also led to serious safety consequences in the US racing series. Many drivers owe their lives to the resulting measures. Figure 1.4 illustrates the downward trend in fatal accidents at racing events in some international racing series. However, events from the recent history of motor sport also show that efforts to optimize safety in motor sport must never slacken. This applies to high-performance racing as well as to junior racing series, and to popular sport, which are less in the media attention. For this reason, a separate chapter is dedicated to the topic of safety.

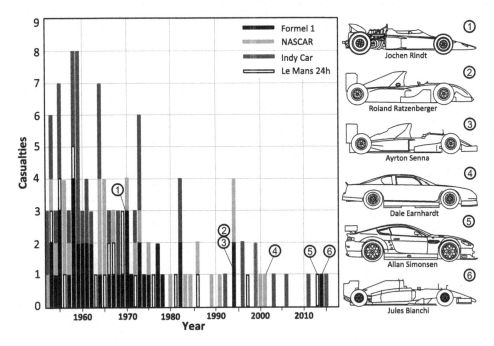

Fig. 1.4 Fatal accidents within selected international racing series from 1952 to 2017

The essential aim of this book is to provide knowledge about the contribution of different components of a racing vehicle to its overall performance. The thematic focus is on the driving dynamics of racing vehicles. Engines and transmissions are for this reason only marginally considered. Beforehand, a historical review summarizes important milestones in motorsport history, which hopefully can transfer some of the emotions associated with motorsport to this book.

References

1. Pfadenhauer, M.: Hochleistungsfahrzeuge. In: Hucho, W.H. (Hrsg.) Aerodynamik des Automobils, S. 452–551. Vieweg, Wiesbaden (2005)
2. Benzing, E.: Dall' aerodinamica alla potenza in Formula 1. Giorgio Nada Editore, Vimodrone (Milano) (2004)
3. Cimarosti, A.: The Complete History of Grand Prix Motor Racing. Aurum Press Limited, London (1997)
4. Lehbrink, H.: 2014 – Turning silver into gold. teNeues Publishing Group, Kempen (2014)
5. http://www.allf1.info/engines. Accessed on 25.06.2017

History and Motivation

2.1 History of Motor Sport

The idea of motor racing is practically as old as the automobile itself, whose history began in 1885 with the maiden voyage of Benz Patent Motor Car Number 1 (Fig. 2.1a) – a three-wheeled vehicle with a gasoline-powered internal combustion engine and electric ignition. On 29 January 1886, Carl Benz applied for a patent for his invention at the Reich Patent Office under number 37435. This patent is now a UNESCO World Heritage Site.

Despite advancing industrialization, at the end of the nineteenth century there was still a lack of a fast, reliable and flexible means of transport for the movement of goods and people. The pioneers of automobile construction, such as Carl Benz, Gottlieb Daimler, Rudolf Diesel, Armand Peugeot, Émile Levassor, René Panhard or Albert de Dion, were still smiled at by the general public and their work was viewed with suspicion. In 1894 the Frenchman Pierre Giffard, editor of "Le Petit Journal" in Paris, announced "a race for vehicles without the use of horses"in order to speed up the development of the automobile. He offered a prize of 5000 francs for the fastest car to cover the distance Paris-Rouen by road. Other conditions were that the vehicles should be easy to handle and reliable, and that they should not cost too much to operate [1].

On 22 July 1894, 21 vehicles finally lined up at the start to cover the 126 km route. Count de Dion was the first to cross the finish line after 6 h. His vehicle had a steam engine and not an internal combustion engine. He was followed by Albert Lemaître with a vehicle designed by Peugeot, whose combustion engine had been supplied by Daimler. The third and fourth-place finishers also covered the full distance with Peugeot vehicles in combination with Daimler engines. Of the 21 participants, 15 reached the finish line, ten of which used internal combustion engines, nine of which had been designed by Daimler and which produced 3 to 4 hp. In accordance with the advertised criteria, the prize money was finally

L. Frömmig, *Basic Course in Race Car Technology*, https://doi.org/10.1007/978-3-658-38470-8_2

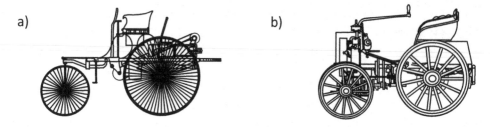

Fig. 2.1 (**a**) Benz Patent Motor Car (1886), (**b**) Panhard-Levassor (1894)

divided between the manufacturers "Panhard et Levassor" (Fig. 2.1b) and "Les Fils des Peugeot Fréres", whose vehicles were powered by an internal combustion engine [2]. Thus, the unstoppable triumph of the automobile began with the first race, and automobile races became an important driver of automotive innovations, especially in the early years.

In the following years, various other city races were held on public roads. These included the 1895 race from Paris to Bordeaux and back (1192 km), which Levassor won on a Panhard-Levassor. It was the first race to use a pneumatic tire, developed by André Michelin. However, due to the poor road conditions, the pneumatic tire was still prone to punctures. In November 1895 the first automobile race in the USA took place, organized by the "Chicago Times Herald", which led over 92 miles from Chicago to Waukegan and back. Similarly, in 1896, the Paris-Marseille-Paris race was held over a distance of 1728 km. Finally, in April 1898, the first automobile race in Germany took place on the route Berlin-Potsdam-Berlin. The aim of these events was to accelerate the development of reliable vehicles suitable for everyday use.

The leading car manufacturers of the early years were Panhard-Levassor, Peugeot, De Dion-Bouton, Renault, Benz, Daimler and later Fiat [3], with most major racing events won by French designs.

2.1.1 1900–1910

The year 1900 marks a milestone in motorsport history. It is the year in which the American journalist Gordon Bennett, who lived in Paris, proclaimed the Gordon Bennett Trophy. The race was designed as a nations' competition in which each participating nation was allowed to enter up to three vehicles.

▶   The Gordon Bennett Trophy was the first race to be held according to a "racing formula" and fixed technical boundary conditions.

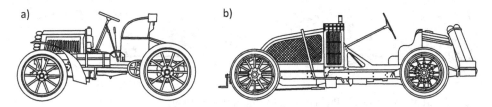

Fig. 2.2 (**a**) 1900 Panhard, (**b**) Renault Grand Prix (1906)

The weight of the vehicles – without service fluids, driver and other load – had to be between 400 and 1000 kg. The crew of the vehicles had to consist of a driver and a co-driver, each with a minimum weight of 60 kg. Deviations from the prescribed minimum weight had to be compensated by ballast [3]. In addition, the participating vehicles and their components had to be entirely manufactured in the respective country of origin. The concept of the country-of-origin color scheme with blue for France, white for Germany, green for Great Britain, yellow for Belgium and, for the time being, red for the United States also goes back to this competition [4]. The race celebrated its premiere on 14 June 1900 on the Paris-Lyon route. However, only five vehicles were entered, which was probably due to the too high demands on the permissible weight. The winner of the race was Fernand Charron on a Panhard & Levassor (Fig. 2.2a) with an average speed of 62.1 km/h. The Gordon Bennett Trophy was held until 1905 and can thus be regarded as the first racing series in the world. In 1904 it was held in Germany, after a Mercedes from Daimler-Motorenwerke had won the previous year. It was the first internationally significant success for a Mercedes.

Numerous city races took place in this era, such as the Paris-Berlin race in 1901, which was still dominated by French designs. The average speed of the winner was 64.14 km/h. One of the last classic city races was the Paris-Madrid race in 1903, but it was cancelled by the French Minister of the Interior after numerous fatal accidents. The event cost the lives of two drivers, including Marcel Renault, two mechanics, two spectators and two soldiers. Very early on, these tragic events showed the dangerous side of motorsport. City races on public roads were subsequently gradually banned and the races moved to – at least provisionally – closed-off tracks.

The French automobile manufacturers increasingly felt that the regulations of the Gordon Bennett Trophy, which only allowed three national constructions at a time, were a severe restriction. They proposed organizing a "Grand Prix de l'Automobile Club de France" instead, in which the same starting conditions applied to all participating manufacturers [4]. In 1906, the Automobile Club de France (ACF) advertised such a "Grand Prix" for the first time. The race was held at the Sarthe Circuit near Le Mans. Today, this combination is inseparably linked with the 24 h of Le Mans. Each participating manufacturer was allowed to enter three cars. For Grand Prix cars, the ACF prescribed a maximum weight of 1000 kg and a maximum fuel consumption of 30 l/100 km [3]. The first Grand Prix was won by Ferenc Szisz driving a 90 hp. Renault (Fig. 2.2b), achieving an

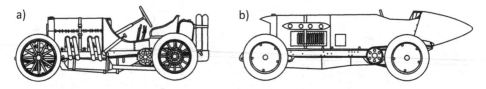

Fig. 2.3 (**a**) Mercedes Grand Prix (1908), (**b**) Blitzen-Benz (1909)

average speed of 101.2 km/h. Also in 1906, the first Targa Florio was held in Italy. The following year, Felice Nazzaro won the French Grand Prix in a Fiat.

In 1908, the German Christian Lautenschlager won the French Grand Prix in a Mercedes manufactured by Daimler-Motoren-Gesellschaft (Fig. 2.3a), whose engine produced 135 hp. from 12,780 cm^3 of displacement. Victor Hémery and René Hanriot, who both drove a Benz, finished second and third respectively. At this time, Benz & Cie. and Daimler-Motoren-Gesellschaft were still competing companies. They did not merge until 1926.

At the same time the racing action took place, the competition for the world speed record began. The Benz & Cie. company constructed a vehicle (Fig. 2.3b) whose four-cylinder engine generated 200 hp. from 21.5 l of displacement. Despite the huge engine, the vehicle weighed only about 1200 kg. This vehicle exceeded a top speed of 200 km/h for the first time in 1909. During the official record drive, a maximum of 205.7 km/h was measured. Officially, this record was not broken until ten years later, even though the vehicle already reached a top speed of 228.1 km/h in the USA in 1911. In the meantime, the vehicle was given the name "Lightning Benz" and later "Blitzen-Benz" by its owner. The car became famous under this name.

2.1.2   1911–1920

In 1912, the Peugeot L76 (Fig. 2.4a) was presented. It was one of the first vehicles to use a new valve system. The 7.6-liter 4-cylinder in-line engine had four valves per cylinder, which were arranged in a V-shape in the combustion chamber and controlled by a desmodromic system via two overhead camshafts. Modern engines are also based on this configuration. At the time, however, this valve train was used exclusively in racing cars. Georges Boillot won the French Grand Prix in both 1912 and 1913 with the L76, which produced 148 hp. with a displacement of 7598 cm^3. Jules Goux also won the Indianapolis 500-mile race with the L76.

The success of the L76 also forced the other manufacturers to adopt the complex valve technology for their engines in order to remain competitive. Grand Prix racing professionalized motorsport in many ways. In the early years of motorsport, race cars were still prototypes for the production cars to come. Motorsport was already an important marketing tool for car manufacturers to advertise the performance of their products.

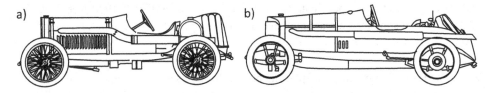

Fig. 2.4 (**a**) Peugeot L76 (1912), (**b**) Mercedes Grand Prix racing car (1914)

Although the vehicles were almost exclusively provided by the vehicle manufacturers in the early years, they were often driven at racing events by wealthy amateurs. However, the growing importance of individual racing events meant that factory drivers and factory mechanics were increasingly employed. At Peugeot, there was already at that time a development team separate from the production business, which was exclusively responsible for the construction and operation of the racing cars. Today such an organizational structure is common among all vehicle manufacturers.

Furthermore, the trend of using technologies specialized for racing was increased. This trend has not been reversed in high-performance racing to this day. The preliminary peak of development was marked by the last pre-war Grand Prix in 1914, which many motorsport experts still consider to be one of the greatest races of all time [4]. By now, the technical standard of the cars included removable wheels that could be fastened and loosened via a wing-shaped central screw using a hammer. Tyre failures, caused by overheating or stones on the unpaved roads, were still part of everyday life. Peugeot established braking of all four wheels via a foot pedal in Grand Prix cars. Previously hand brake levers, which only acted on the rear axle, were common. Meanwhile the engines reached speeds of up to 3500 rpm. Christian Lautenschlager won the 1914 Grand Prix on a Mercedes with 115 hp. (Fig. 2.4b). Mercedes celebrated a triple victory with the second and third placed drivers Wagner and Salzer. But from then on, the guns were talking in Europe.

With the start of the First World War, the first great era of motorsport came to an end. It was the era of city races, which were held over very long distances and on public roads. This was followed by the first provisional circuits. Enormous physical exertion was demanded of the drivers at that time. They were at the mercy of the dust kicked up by the unpaved roads and the wind and weather. Nevertheless, they did important pioneering work for the development of the automobile.

2.1.3 1921–1930

With the Targa Florio and the Indianapolis 500-Mile Race, the first significant post-war races took place as early as 1919. The first post-war Grand Prix was held in 1921 by the ACF in France on the Le Mans circuit. This circuit has been home to the famous 24 h of Le Mans since 1923. The first Italian Grand Prix was also held in the same year. For Grand

Prix races, the technical specifications of the Indianapolis 500 from 1920 were initially adopted, which stipulated a maximum engine capacity of 3.0 l and a minimum weight of 800 kg.

▶ The restriction of engine displacement established itself in the 1920s as an essential instrument of the regulations for limiting engine size and maintaining equality of opportunity. Even in today's regulations, the maximum permissible engine displacement is often a fixed parameter.

The First World War had brought with it drastic further developments in the fields of materials and drive train technology, from which the new generation of racing cars now also benefited. Double camshafts and drive shafts instead of chain gears were now standard. The large displacement four-cylinder in-line engines were replaced by eight-cylinder in-line engines. Front-mounted engines with long bonnets characterized the appearance of Grand Prix vehicles in the following decades. A Duesenberg (Fig. 2.5a) won the French Grand Prix in 1921 with such an engine concept.

▶ The main advantage of these multi-cylinder engines is the reduction of the mass of individual components, such as pistons and connecting rods, so that they and their bearings are subjected to less stress at identical speeds. Conversely, this concept allows higher speeds and thus the achievement of higher engine outputs.

The engine of the Duesenberg could be operated at up to 5000 rpm and reached a maximum output of 115 hp. The Duesenberg also had a hydraulically operated brake system. Nevertheless, the mechanical brake actuation remained the standard until the end of the decade.

In 1922, the maximum engine capacity of Grand Prix cars was reduced to 2.0 l, and the minimum weight had to be 650 kg. Bugatti contested its first Grand Prix in this year. At the French Grand Prix, favored by the new engine regulations, the Fiat Tipo 805 (Fig. 2.5b) used an supercharged engine for the first time. In the same year, the first victory of a supercharged engine was celebrated with this 130 hp. car at the Italian Grand Prix in Monza. Similarly, streamlined bodies began to gain acceptance. An example of this is the Benz RH "Tropfenwagen" (Fig. 2.6a), whose drive concept was a mid-engine design. However, this design did not become established until 30 years later. As a further innovation, this vehicle, in the form of a rear axle with swinging arms, had independent wheel suspension on all four wheels for the first time.

In 1924 and 1925, Grand Prix racing was dominated by the Alfa Romeo P2, which also won the first World Championship title in 1925. The Alfa Romeo P2 had a 2.0-liter eight-cylinder in-line engine with Roots supercharger, the first version of which produced 145 hp. at 5500 rpm [3].

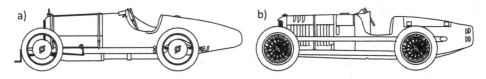

Fig. 2.5 (**a**) Duesenberg Grand Prix (1921), (**b**) Fiat Tipo 805 (1922)

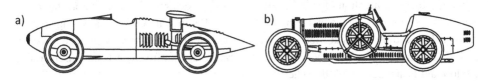

Fig. 2.6 (**a**) Benz RH drop car (1923), (**b**) Bugatti Type 35 (1924)

In 1924, the Bugatti Type 35 (Fig. 2.6b), the most successful vehicle of this brand, celebrated its racing debut. It won over a 1000 races in total, including the Targa Florio five times in succession (1925–1929). The myth of the Bugatti brand, whose design and naming rights have been held by the Volkswagen Group since 1998, is largely based on this vehicle. The Type 35 could also be purchased by private individuals. Over 400 vehicles were produced, making it a commercial success for Bugatti as well.

For the year 1926, the technical regulations for Grand Prix vehicles were adapted once again. The maximum displacement was now 1.5 l with a minimum weight of 600 kg. Bugatti celebrated its first Grand Prix victory with the Type 39, which was based on the Type 35, and also won the world championship title that year. In 1926, a piece of German economic history was written. In the 1920s, almost all car manufacturers were fighting for survival. The Deutsche Bank encouraged a merger between Benz and Daimler. The companies Benz & Co Rheinische Gasmotorenfabrik Mannheim (from 1899: Benz & Cie.) and Daimler-Motoren-Gesellschaft, which had emerged from the work of both automotive pioneers, merged on 28 June 1926 to form Daimler-Benz AG, headquartered in Berlin.

From 1928, a free engine formula came into force. The following year, a minimum weight of 900 kg was required and a fuel consumption limit for Grand Prix races of 14 l/100 km was recommended. This formula remained valid until 1934. Also in 1929, the first Monaco Grand Prix took place, where safety was already a major point of discussion at that time.

The 1920s were also marked by the trend towards building purpose-built racetracks. In 1921, for example, the AVUS was built in Berlin, where the first German Grand Prix was held in 1926, and in 1927 the legendary Nürburgring. In Italy, the Autodromo Nazionale Monza was opened in 1922.

2.1.4 1931–1940

This decade was marked above all by the historic duel between the Grand Prix vehicles of Auto Union and those of Mercedes-Benz, which began with the "750 kg formula" proclaimed in 1932 and valid from 1934. This racing formula stipulated a maximum vehicle weight of 750 kg. During this period, a highly professionalized racing team was being built up in parallel at both the Auto Union and Mercedes-Benz plants. At Auto Union this was under the direction of Willy Walb and his chief designer Ferdinand Porsche. Hans Stuck, Alfred Momberger and Hermann Prinz zu Leiningen were initially engaged as drivers. The leading heads at Mercedes were Alfred Neubauer, who was the race director, and his constructor Dr. Hans Nibel. Rudolf Caracciola, Manfred von Brauchitsch, Luigi Fagioli and Hans Geier took the wheel respectively.

On 17 March 1934, Auto Union first presented its so-called Type A, which was based on the "P-Wagen" designed by Ferdinand Porsche. The Type A had a supercharged 16-cylinder engine in V-construction, which generated 295 hp. from a displacement of 4360 cm^3 at 4500 rpm [5]. A characteristic feature of the Auto Union racing cars was the mid-engine drivetrain arrangement, which was unusual for the time.

Mercedes-Benz first introduced the W25, which was powered by a supercharged eight-cylinder in-line engine with 3360 cm^3 displacement. At 5800 rpm, 345 hp. was achieved.

Both vehicles featured a number of technical innovations that laid the foundation for the dominance of German racing cars in the years from 1934 to 1939. These innovations included specially developed fuels, lightweight frames constructed from hollow sections, independent wheel suspensions and the use of lightweight alloys in engine construction. The relatively stiff frames allowed the use of a relatively soft suspension, resulting in superior traction and handling, and reversing the previous philosophy of soft frames and hard suspensions into its opposite [3]. In particular, the technical superiority in the field of lightweight construction and compressor technology, combined with the prescribed maximum weight, established the superiority of the German racing teams.

The (future) rivals met for the first time at the Eifel Race in 1934. According to the lore of the then Mercedes race director Alfred Neubauer [6], the white-painted W25 weighed 751 kg at the official weigh-in, which would have led to a race exclusion. Manfred von Brauchitsch is said to have responded to this with the words "Why don't you come up with one of your famous tricks. Otherwise we'll be the painted ones . . .", which is said to have given Alfred Neubauer the idea of sanding off the white paint of the cars to reduce the weight of the now aluminum-colored W25 to the required 750 kg. This is referred to as the birth of the Silver Arrows. The first duel was won by Mercedes-Benz with driver Manfred von Brauchitsch. Hans Stuck took second place for Auto Union. From now on, the cars of the two German manufacturers dominated international motorsport. Figure 2.7 shows an overview of the vehicles used in this era. In addition to the Eifel race, drivers von Brauchitsch and Fagioli won two Grand Prix races for Mercedes-Benz with the W25. Hans Stuck took three Grand Prix victories for Auto Union with the Type A.

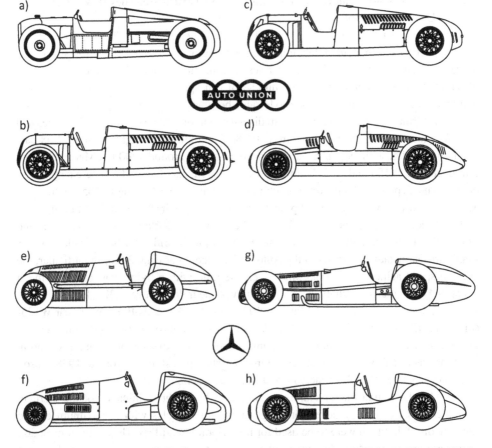

Fig. 2.7 Auto-Union Grand Prix cars: (**a**) Type A (1934), (**b**) Type B (1935), (**c**) Type C (1936–1937), (**d**) Type D (1938–1939). Mercedes-Benz Grand Prix cars: (**e**) W25 (1934–1936), (**f**) W125 (1937), (**g**) W154 (1938–1939), (**h**) W165 (1939)

In 1935, Auto Union introduced the Type B, whose engine now produced 375 hp. with a displacement of 4950 cm^3. The further developed W25 had a maximum of 430 hp. in this year, which resulted from an increase in displacement to 3990 cm^3. This performance advantage of the W25 earned the Mercedes-Benz driver Caracciola the newly introduced European championship title.

The following year Auto Union used the Type C, whose maximum output was 520 hp. The Mercedes-Benz W25 produced 494 hp. in this year. The power figures of the competition were well below 400 hp. The legendary Auto Union driver Bernd Rosemeyer dominated Grand Prix racing that season, winning the title of European champion at the age of 27.

For the 1937 season, Mercedes-Benz presented the newly developed W125, which generated 646 hp. maximum output from 5660 cm^3 displacement. Comparable performance was not achieved again until the mid-1980s. Mercedes driver Caracciola became European champion for a second time.

From 1934 to 1937, a total of 23 Grand Prix were held with the "750 kg formula", of which the two German manufacturers won 19. Auto Union accounted for seven and Mercedes-Benz for 12 Grand Prix victories. During this period, the use of limited-slip differentials, torsion and coil springs (instead of leaf springs), hydraulic shock absorbers and mixed tires on the front and rear axles also became established, among other things.

A new formula prescribed a maximum displacement of 3 l for supercharged engines and 4.5 l for naturally aspirated engines from 1938. The minimum weight of Grand Prix cars had to be 850 kg, with cars in racing trim weighing in at around 1200 kg. Mercedes-Benz introduced the W154 and Auto Union the Type D. Both cars featured a supercharged V12 engine. The Type D had a maximum output of 420 hp. in 1938, rising to 485 hp. in 1939. The output of the W154 was 468 hp. in 1938, but could be increased to 485 hp. in 1939.

In addition to racing, speed records were of enormous importance in this era. At the end of the 1937 racing season, Bernd Rosemeyer set various world records in his class on the Frankfurt-Darmstadt motorway in the Auto Union record car (Fig. 2.8a). In particular, the average speed achieved of 406.3 km/h over one mile with a flying start attracted much attention [7]. Caracciola achieved 432.4 km/h over one mile and even 432.7 km/h over one kilometer on the same track on 28 January 1938 with his Mercedes-Benz record car W125- (Fig. 2.8b) [8]. This is still the world speed record on public roads. Bernd Rosemeyer was killed in an accident while trying to break this record again. Auto Union did not take part in any more record drives after that. The European championship title for the 1938 season went once again to Rudolf Caracciola and Mercedes-Benz.

In 1939, the Italian manufacturers attempted to break the dominance of the German manufacturers by secretly developing regulations for vehicles with a maximum engine capacity of 1.5 l, which were to be used for the Tripoli Grand Prix. However, information about this leaked to Mercedes-Benz, where the W165 type was developed under great secrecy and only registered for the race at the last moment. The W165 produced 254 hp. at 8250 rpm. Hermann Lang finally won this Grand Prix for Mercedes-Benz. It was the only race held under this formula. The dominance of the 3.0-liter vehicles remained unbroken.

On September 3, 1939, Tarzio Nuvolari won the last Grand Prix for Auto Union in Yugoslavia for the time being. Two days earlier, the German Wehrmacht had invaded Poland, and the start of the Second World War silenced the racing engines for a long time.

The critical roles played by Auto Union and Mercedes-Benz under the rule of the National Socialists cannot be dealt with in this book, but they should not remain unmentioned for the sake of completeness. For a discussion of this topic, please refer to the relevant literature.

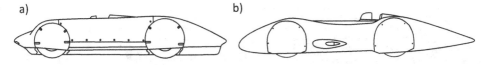

Fig. 2.8 (**a**) Auto-Union record car (1938), (**b**) Mercedes-Benz record car W125 (1938)

2.1.5 1941–1950

The first post-war races were held in Europe as early as September 1945. In 1946, the international umbrella organization of automobile clubs was newly founded as the "Fédération Internationale de l'Automobil" (FIA) and its sports commission "Commission Sportive Internationale" (CSI). A new set of regulations was drawn up for Grand Prix racing, which came into force in 1948. This formula provided for a maximum of 4.5 liters of displacement for naturally aspirated engines and 1.5 liters for supercharged engines. The ratio of 1:3 between supercharged and naturally aspirated engines was intended to break the previous superiority of supercharged engines. Supercharged engines with up to 1.5 l displacement had already been used in the Voiturette class in the 1930s, which was the reason why this engine concept dominated Grand Prix racing in the first years after the end of the war. In parallel, a formula with 2.0 l for naturally aspirated engines for voiturettes or sports cars was being considered. The different regulations were initially differentiated according to Formula A and Formula B. However, the terms Formula 1 and Formula 2 quickly developed from this. At the end of the 1940s, Grand Prix racing practically only spoke of Formula 1 [4].

Motor racing in the first post-war years was dominated by Italian designs from the houses of Alfa Romeo, Maserati and Ferrari. Ferrari took part in a Grand Prix for the first time in 1948 in Italy at the fourth Grand Prix of the year. Since that year, there has not been a season in which Ferrari has not been active in Grand Prix racing. The Ferrari 125 (Fig. 2.9a) of 1948 was equipped with a supercharged 12-cylinder engine in V-construction. The first Grand Prix victory for Ferrari was taken by Alberto Ascari at the 1949 Italian Grand Prix in Monza. That year, the Ferrari 125's engine produced 300 hp. at 7200 rpm. The Ferrari Type 166 also celebrated its first triumph at the 24 h of Le Mans, which was held for the first time since 1939.

However, by far the most successful vehicle during this period was the Alfa Romeo Tipo 158 (Fig. 2.9b), also known as the "Alfetta". As with many vehicles in this period, it was a voiturette class model developed in the pre-war period. The Alfa Romeo 158 and its derivative 159 competed in 26 Grand Prix races, starting with the 1946 Grand Prix des Nationes and ending with the 1951 French Grand Prix, winning them all [4]. The model designation Alfa Romeo 158 is derived from the engine capacity of 1.5 l as well as the number of cylinders eight. As a voiturette, the Alfa 158 was first used in 1938 and at that time produced 195 hp. with the help of a Roots supercharger. The later Grand Prix versions

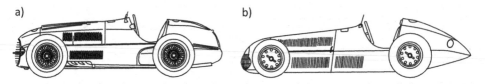

Fig. 2.9 (a) Ferrari F1 125 (1948), (b) Alfa Romeo Type 158 (1950)

produced between 300 and 430 hp. The high performance and extreme reliability of the vehicle were the essential cornerstones for the dominance of Alfa Romeo at that time.

In 1950, Formula 1 awarded the world championship title to the driver for the first time. Alfa Romeo won all the Grand Prix held that year, with the exception of the Indianapolis 500. Consequently, it was an Alfa Romeo driver, Giuseppe "Nino" Farina, who won the first drivers' title.

2.1.6   1951–1960

In 1951, the legendary Juan Manuel Fangio won his first world championship title in an Alfa 159, which at that time produced 430 hp. at 9300 rpm. But in this year Ferrari could inflict three sensitive defeats on the Alfas. The special thing here was that these victories were achieved with the Ferrari 375 (Fig. 2.10a), which had a naturally aspirated engine with 4493 cm^3 displacement and produced 350 hp. at 7000 rpm. These victories and the FIA's decision to hold the 1952 and 1953 World Championships under Formula 2 regulations ended the dominance of turbocharged engines in Formula 1 for a long time. The Formula 2 regulations stipulated the use of non-turbocharged engines with a maximum displacement of 2.0 liters. Alberto Ascari won the drivers' title in both years with the Ferrari 500 (Fig. 2.10b). For 1954, the displacement was limited to 2.5 l for naturally aspirated engines and 0.75 l for supercharged engines, which further increased the displacement advantage for naturally aspirated engines.

In 1953, the FIA introduced the World Sports Car Championship, the inaugural season of which was won by Ferrari. At Le Mans, however, the Jaguar XK 120C triumphed, becoming the first racing car to successfully use a disc brake system.

The first half of the 1950s was again dominated by the star. On 15 June 1951, the Board of Management of Daimler-Benz AG decided to develop a sports car and to return to Grand Prix racing from 1954 when the new regulations came into force [9]. Under the direction of Alfred Neubauer and Rudolf Uhlenhaut, the W194 was initially constructed on the basis of the 300 Mercedes, which bore the legendary designation 300 SL. Its trademark was the gullwing doors, which also made the production model launched in 1954 famous. The engine of the 300 SL produced 171 hp. at 5200 rpm. In January 1952, the return to motorsport was officially announced. On 2 May, three 300 SLs (Fig. 2.11a) finally lined up at the start of the Mille Miglia. After a 12-year abstinence, this was the first time a

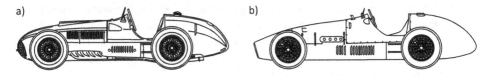

Fig. 2.10 (**a**) Ferrari 375 (1951), (**b**) Ferrari 500 (1952)

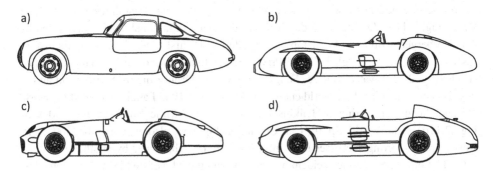

Fig. 2.11 Mercedes-Benz racing cars of the 1950s: (**a**) W194 or 300 SL (1952), (**b**) and (**c**) W196 (1954), (**d**) 300 SLR (1955)

German manufacturer took part in a major international motor sport event. Drivers Kling and Caracciola finished second and fourth for the newly formed racing team.

June of the same year saw the first factory participation by Mercedes-Benz in the 24 h of Le Mans. Three 300 SLs were entered against competition from Ferrari, Jaguar, Aston Martin and Talbot. The maximum power of the cars was throttled back to 166 hp. in favor of reliability, which was reached at 5100 rpm. A roof-mounted and hand-operated air brake was tested in practice, but was not used in the race. In the race it looked for a long time like a sensation. After 22 h the Frenchman Pierre Levegh was leading on a Talbot TS26 GS. Levegh had driven through the 22 h without a break. He was celebrated by the spectators, who were hoping for an all-French triumph. But then Levegh's Talbot broke down with engine failure. Both 300 SLs took the lead. The driver duo of Hermann Lang and Fritz Riess eventually won the 24 h of Le Mans with a record average speed of 155.574 km/h. The 300 SL driven by Theo Helfrich and Helmut Niedermayr completed the double victory. In November 1952 Mercedes-Benz crowned its successful motorsport year with a double victory for the 300 SL at the Carrera Panamerica Mexico.

In January 1954, the new Grand Prix vehicle with the type designation W196 was officially presented. The W196 featured gasoline direct injection, a technology that had until then been used primarily in aircraft engines. The speed-dependent behavior of the carburetor was thus avoided and at the same time acceptable torque was achieved even at low engine speeds. Further advantages were more even cylinder filling and reduced fuel consumption. The valve train was controlled by a desmodromic system, which prevented valve jump at high engine speeds. The inline eight-cylinder engine was tilted 20 ° about its longitudinal axis, which lowered the center of gravity and produced less aerodynamic drag. In the 1954 season, 280 hp. at 8500 rpm was achieved and increased to 290 hp. in 1955.

The Argentinian Juan Manuel Fangio was signed on for the 1954 season. However, as it was foreseeable that the W196 would not be ready for the start of the season, Fangio was allowed to drive a Maserati 250F until then.

The W196 celebrated its racing debut on 4 July 1954 at the French Grand Prix in Reims. It was the fourth Grand Prix of the season. At the start, Fangio and Kling occupied the first row of the grid and in the end took a double victory with the streamlined variant of the W196 (Fig. 2.11b). On the same day in Bern, the German national football team, led by Sepp Herberger, won a world championship title for the first time. German sports history was written. From the English Grand Prix at Silverstone onwards, the variant with open wheels (Fig. 2.11c) was also used. Fangio won at the Nürburgring, in Switzerland and in Italy. He celebrated his third world championship title. In 1955 Fangio defended his world title and the driver pairing of Jenkinson/Moss won the Mille Miglia in the 300 SLR (Fig. 2.11d). The World Sports Car Championship was also won by the 300 SLR. However, the sporting successes from this year were overshadowed by another event.

On 11 June 1955, a tragic accident occurred during the 24 h of Le Mans. The Mercedes-Benz racing team entered three 300 SLRs and the driver pairings Fangio/Moss, Kling/Simon and Fitch/Levegh. The signing of Pierre Levegh was supposed to be a sign of reconciliation. The Mercedes 300 SLR's biggest rival was the Jaguar D-Type, driven by Mike Hawthorn and Ivor Bueb among others. After two hours, Hawthorn formed the front of the field ahead of Fangio. They were only a few seconds apart. On the 35th lap, disaster struck. Mike Hawthorn drove up to Pierre Levegh and the Briton Lance Macklin, who were to be lapped. With his Jaguar D-Type he desperately wanted to pit in front of Fangio. The pit area was not structurally separated from the start and finish straight at the time. Hawthorne overtook Levegh and then had the Austin-Healey of Macklin in front of him. He overtook this, then sheared in from the left close in front of him and braked sharply to pit on the right. To avoid a collision Macklin pulled sharply to the left, crossing the line of Levegh behind. Levegh was unable to take evasive action. The 300 SLR hit the rear of the Austin-Healey and was catapulted into the air. The car flew into the grandstands and exploded on impact. Levegh and over 80 spectators died. This has been the greatest tragedy in motorsport to date. Nevertheless, the race went on! After consultation with corporate headquarters, Alfred Neubauer withdrew the remaining SLR from the race as a show of sympathy. Mike Hawthorn entered the D-Type once more for the final lap and won the race. Mercedes withdrew from international racing at the end of 1955. However, this decision had been taken for economic reasons long before the Le Mans race. Mercedes-Benz did not return to the international motorsport stage until 1989.

Another tragedy occurred during the 1957 Mille Miglia when Alfonso de Portago lost control of his Ferrari 335S and went off the track. De Portago, his co-driver Ed Nelson and ten spectators lost their lives [10]. The staging of the Mille Miglia was subsequently banned – at least in its original form.

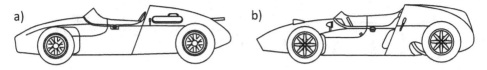

Fig. 2.12 (**a**) Cooper T43 (1957), (**b**) Cooper T51 (1959)

At the end of 1956, a technical revolution in motorsport began. The Cooper racing team presented the T43 Mk II (Fig. 2.12a), an F1 vehicle with a mid-engine. Although the design was already known from the Benz Tropfenwagen and the Auto Union Grand Prix cars of the 1930s, it was not until this car that this design began to gain widespread acceptance. Among the classic front-mid-engined F1 cars, the Cooper T43 Mk II still looked like a foreign body in 1957, something Enzo Ferrari commented on by saying, "I've only ever seen horses pull a car, but not push it with their nose." In 1958 the T43 at the hands of Stirling Moss was the first mid-engined race car to win a Formula 1 race .

▶   The conceptual advantages of the mid-mounted engine are the lower mass inertia and the higher proportion of weight on the driven rear axle. The compact design and the elimination of the drive shaft reduce the overall weight and thus improve handling. There are also aerodynamic advantages, as the drive shaft does not have to be routed to the rear axle below or to the side of the driver, resulting in a reduced frontal area.

Despite inferior engine power, the Cooper T43s were able to win two races that season. Already in 1959, Jack Brabham won the drivers' title on the Cooper T51 (Fig. 2.12b), which was also of mid-engine design. The maximum power of the Climax engine used in the Cooper T51 was 243 hp. at 6800 rpm. The rest of the field still relied on the front mid-engine concept. Brabham also won the drivers' title in 1960 in the Cooper T53. By now, however, Lotus and Porsche were also running the mid-engine concept. It was the last year in which front-engined vehicles took part in Formula One races.

2.1.7   1961–1970

The Formula 1 technical regulations stipulated a maximum displacement of 1.3 to 1.5 l for the 1961 season. For the first time, supercharged engines were not permitted. Furthermore, a minimum weight of 450 kg was set for safety reasons, which also resulted from the skepticism towards the British lightweight constructions. Dual circuit braking systems and roll bars were also prescribed. Enclosed wheels, like those of the streamlined Mercedes W196, were banned for Formula One cars. Double wishbone axles on the front and rear axles and disc brakes became standard. The previous standard solution of a driven DeDion rear axle had its day [3]. Another design trend in the coming years was the construction of a

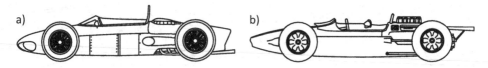

Fig. 2.13 (**a**) Ferrari 156 (1961), (**b**) Lotus 25 (1962)

narrow, cross-sectionally reduced body in order to compensate for the loss of displacement by lowering aerodynamic drag.

The new engine regulations initially favored Ferrari, as they were familiar with this engine dimension from Formula 2. Consequently, Phil Hill won the drivers' title that year with a Ferrari 156 (Fig. 2.13a), whose engine produced 185 hp. at 9300 rpm.

In 1962, the Lotus 25 (Fig. 2.13b) used in Formula 1 initiated a revolution in the design of racing car structures. The Lotus 25 was one of the first racing cars to make consistent use of the superior stiffness-to-weight ratio of thin-walled hollow sections. Its body structure was formed for the first time by two box framed chassis beams instead of a classic tubular frame (see also Sect. 9.3). These two chassis beams ran parallel to each other and connected the front and rear axles. The box frames functioned as longitudinal chassis beams and consisted of aluminum sheets riveted together to form several thin-walled hollow sections. The driver and cockpit, as well as the engine-transmission unit, were located between these two chassis beams. Inside the longitudinal chassis beams were the fuel tanks. In the transverse direction, the structure was reinforced by transverse bulkheads. Until then, this concept had only been used in aircraft construction.

The advantages of this construction method are a reduced overall weight with higher torsional stiffness and better safety. The torsional stiffness of the Lotus 25 was about two to three times as high as the torsional stiffness of a vehicle in classic steel tubular frame construction. The Lotus 25 can thus be regarded as a pioneer for the monocoque construction method used today. The high stiffness-to-weight ratio allowed for a very narrow design of the driver's cell. In 1963 the Briton Jim Clark won his first world championship title with the Lotus 25, which had a maximum output of 200 hp. at 9800 rpm with its Climax V8 engine. Jim Clark's world championship title marked the beginning of Team Lotus' rise to become one of the most legendary and innovative racing teams of all time.

Comparable designs were also used by the competition and in other racing classes in the following years and were permanently further developed. The next evolutionary step in racing car design was to integrate the engine-transmission unit into the structure as a load-bearing element. This means that the loads of the chassis are fed directly to the gearbox, which eliminates the need for recessed beams or subframes and thus reduces the overall weight. In such a design, the body structure consists of an enclosed driver or safety cell at the front and a load-bearing engine-transmission unit at the rear, which is flanged directly to the safety cell.

▶   The division of the structure into a safety cell and a load-bearing engine-transmission unit corresponds to the current design of modern high-performance racing vehicles.

The first attempts to use this principle were made in 1964 with the Ferrari 158 and in 1966 with the Lotus 43 and the BRM P83, both of which used the BRM H16 engine. However, this design only became established with the development of the Ford-Cosworth DFV engine.

A new set of engine regulations came into force for Formula 1 in 1966, prescribing a maximum displacement of 1.5 l for turbocharged engines and 3.0 l for naturally aspirated engines. These key data of the regulations remained unchanged until 1986 and opened the door for new dimensions of engine performance. In the first year of the new engine regulations, the V12 engine in the Ferrari 312 produced 360 hp. at 9800 rpm. However, the new regulations also led to the withdrawing from Formula 1 of the British engine manufacturer Climax-Coventry, which also affected the English team Lotus.

Lotus team boss Colin Chapman succeeded in forging a cooperation between Ford, Cosworth and Lotus, which provided for a development of a new racing engine financed by Ford and carried out at Cosworth. This power unit was allowed to be used exclusively by Lotus in the 1967 season. This alliance resulted in the Lotus 49 (Fig. 2.14a) and the Ford-Cosworth DFV (Double Four Valve), whose basic concept was based on a load-bearing engine-transmission unit. The Ford-Cosworth DFV was a naturally aspirated V8 engine with a bank angle of 90 °, four valves per cylinder and an aluminum engine block, producing 408 hp. The DFV-powered Lotus-Ford 49 made its racing debut on 4 June 1967 at the Dutch Grand Prix at Zandvoort. It was the first of four season wins Jim Clark and Ford achieved that year. In 1968, Graham Hill won the world championship with the Lotus 49-Ford. By 1985, the Ford-Cosworth DFV and its development stages, which were henceforth also offered as a customer engine, had been used to achieve 155 victories in 267 Grand Prix starts. In total, nine different drivers won 12 world championships with this power unit. In 1969 and 1973, every single race was won with a DFV engine [11]. In addition to Lotus, McLaren and Williams, among others, also used the DFV. The Ford-Cosworth DFV is still the most successful racing power unit of all time and shaped an era of Formula 1.

In 1966, a groundbreaking innovation in aerodynamics began in the American CAN-AM series. The Chaparral 2E (Fig. 2.14b) designed by Jim Hall featured a wing profile mounted above the rear axle, which was attached directly to the wheel carrier via two high-leg struts. It was the first sustained attempt to optimize the performance of a racing car by generating downforce. The angle of attack of the wing could be changed via a foot pedal. The same principle was adopted for the Chaparral 2F. Many observers initially saw the design as simply a means of increasing braking deceleration and stability. The increase in grip level due to downforce was not immediately recognized [12], which was certainly also due to the fact that the two vehicles could only celebrate isolated successes.

In 1968, at the Belgian Grand Prix, Ferrari was the first team to use a wing in Formula 1. Similar to the Chaparral 2E and 2F, the wing could be adjusted via hydraulics. However,

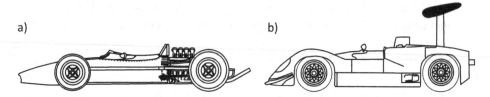

Fig. 2.14 (**a**) Lotus-Ford 49 (1967), (**b**) Chaparral 2E (1966)

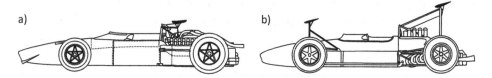

Fig. 2.15 (**a**) Ferrari 312 (1968), (**b**) Brabham BT26 (1968)

in the Ferrari 312 (Fig. 2.15a), the wing was located near the center of gravity and thus more centrally located. Overnight Brabham came up with a similar solution. Team Lotus equipped the 49B with a high wing profile, which was attached to the wheel carrier by filigree struts. Such designs quickly spread throughout the field. The aerodynamic revolution was now unstoppable. In extreme cases, such as the Brabham BT26 (Fig. 2.15b), both rear and front wings were attached to high-mounted struts.

Due to an accumulation of serious accidents caused by the failure of aerodynamic components, the regulating associations enacted some basic rules that are still valid today. These include the requirement that aerodynamic components must be attached to the sprung masses and the prohibition of movable aerodynamic components. Likewise, regulations were issued for the permissible width and height of wing profiles.

A defining novelty of 1968 was the placement of advertising space on Formula One vehicles, which the CSI permitted for the first time. As a result, the traditional country-of-origin color scheme of the racing cars became less important; the formerly dark green Lotus cars, for example, advertised the Gold Leaf cigarette brand in a red and white color scheme that year.

In 1970, Porsche took the first of 19 overall victories in the 24 h of Le Mans with the legendary 917 K (Fig. 2.16b). The Porsche 917 K was driven by the drivers Hans Hermann and Richard Attwood. The development of the Porsche 917 was carried out with the clear objective of finally achieving overall victory in this race. The project was under the direction of Ferdinand Piëch. The winning car featured a V12 engine with a displacement of 4.5 l, which produced 520 hp. at 8000 rpm and allowed a top speed of 341 km/h. In addition to Le Mans, the Porsche 917 won the 1970 endurance races at Daytona, Brands Hatch, Monza, Watkins Glen and the Österreich-Ring, which earned Porsche the World Sports Car Championship at the end of the season.

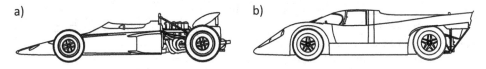

Fig. 2.16 (a) Lotus-Ford 72 (1970), (b) Porsche 917 K (1970)

In Formula 1, the Austrian Jochen Rindt led the world championship superiorly in 1970. With his Lotus-Ford 72 (Fig. 2.16a) he won five of the first nine Grand Prix. The Lotus 72 was the harbinger of the design of modern Formula 1 cars and other monoposti.[1] The key features of the Lotus 72 were the radiators positioned to the side behind the driver and its extreme wedge shape, which provided reduced drag. The Lotus 72 was powered by a Ford-Cosworth DFV, which now produced 426 bhp at 10,000 rpm. On 5 September, practice for the Italian Grand Prix took place at the Autodromo di Monza. To achieve a higher maximum speed, Rindt had the rear wing removed from his car. In the Parabolica, Rindt lost control of his Lotus-Ford 72, presumably due to the breakage of a front sideshaft,[2] at 322 km/h. The car swerved to the left, wedged its nose in a barrier, and decelerated abruptly before skidding back into the gravel. Rindt was killed instantly. He died at just 28 years old. In the remaining races, no other driver managed to catch up with Rindt's points lead. Rindt was the first and only world champion not to live to see his title win. The world championship trophy was awarded to him posthumously and accepted by his widow Nina Rindt. In 1970 also Piers Courage died as well as the Australian Bruce McLaren who had founded the racing team of the same name in 1966.

2.1.8 1971–1980

With a further developed 917 K (Fig. 2.17a) in the famous "Martini" livery, Porsche was able to repeat its success at Le Mans in 1971. The winning car, driven by Helmut Marko and Gijs van Lennep, achieved an average speed of 222 km/h. This record was not broken until 2010.[3] The Porsche 917, which was also successful in numerous versions in the American CAN-AM series, was voted racing car of the century by a jury of experts in 1997. In 1973, the 12-cylinder turbocharged engine of the Porsche 917/30 KL (Fig. 2.17b) produced 1100 hp. at 8000 rpm from a displacement of 5374 cm³ as standard; it was secretly said that for a short time even 1500 hp. could be delivered. To this day, the Porsche

---

[1] Monoposto (Italian for *single-seater*) = racing car with a single seat located in the middle of the car. As the term is used for formula cars, monopostos do not have a driver's cabin and have free-standing wheels (cf. Sect. 3.2).

[2] The Lotus 72 had an inboard brake system.

[3] The track layout at Le Mans was changed several times after 1971. Among other things, the originally 6 km long Hunaudières straight was interrupted by two chicanes in 1990.

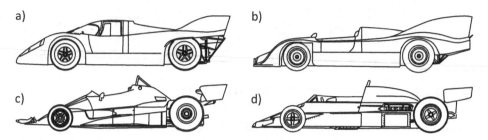

Fig. 2.17 (**a**) Porsche 917 K (1971), (**b**) Porsche 917/30 KL (1973), (**c**) Ferrari 312 T2 (1976), (**d**) McLaren-Ford M23 (1976)

917/30 KL remains the most powerful racing sports car of all time. In 1971, non-grooved tyres (slicks) were used for the first time in Formula 1 [13] and subsequently became the standard tyres for all racing cars on dry roads.

In 1975, the Austrian Niki Lauda became Formula 1 World Champion for the first time in his Ferrari 312 T (Fig. 2.17c). It was the first drivers' title for Ferrari since 1964, when the Briton John Surtees triumphed. However, Niki Lauda made racing history mainly through the events of 1976, the year in which his greatest rival was the Briton James Hunt, driving a McLaren-Ford M23 (Fig. 2.17d). It was the duel of two completely different characters. On the one hand the ambitious, cool and sometimes arrogant Niki Lauda and on the other hand the charismatic playboy James Hunt, who was also not averse to the world of women and parties.

On August 1, the German Grand Prix took place on the Nürburgring's Nordschleife. Jackie Stewart nicknamed the circuit the "Green Hell" due to its location in the middle of the Eifel region and its dangerous nature. It was the tenth round of the season. The first half of the season was dominated by Lauda, who won five of the nine races. The weather conditions on race day were changeable. All but one of the drivers started the race on wet tyres. However, the track dried quickly, so that the drivers switched to slicks after the first lap. After a rather moderately fast pit stop, Lauda was only in the rear midfield on the second lap and had to catch up with the leaders. In the Bergwerk section of the track, one of the most famous racing accidents in motorsport history finally occurred. Amateur footage taken by a French spectator showed Lauda's Ferrari suddenly crashing to the right into a rock face, skidding along the track and bursting into flames. The impact caused Lauda to briefly loose consciousness. Due to inadequate padding in his helmet, it detached from Lauda's head. Behind the wreckage of Lauda, the following cars came to a stop. The drivers Brett Lunger, Guy Edwards and Harald Ertl tried to free Lauda from his Ferrari. However, it was former Ferrari driver Arturo Merzario who succeeded in opening Lauda's seat belt system. Lauda was almost defenseless against the flames for more than half a minute before John Watson extinguished them with a fire extinguisher. Brett Lunger helped Niki Lauda, who had regained consciousness in the meantime, out of the wreckage. In addition to severe burns to his face, Lauda suffered life-threatening chemical burns to his lungs from inhaling toxic fumes. In hospital Lauda fell into a coma, his condition was critical, and rumors of his demise began to spread. Lauda was even given the last rites by a priest in hospital.

Contrary to expectations, Lauda recovered relatively quickly from his serious injuries. He missed only two races and returned to the cockpit of a Formula One car after just 42 days, albeit in extreme pain, earning the respect and admiration of the motorsport world. James Hunt had taken advantage of the brief absence and the weak phase of Lauda, who was still severely impaired, to narrow the gap in the drivers' standings by winning four races. The final race of the season in Japan saw a showdown between the two rivals. The weather conditions at Fuji were extreme, with incessant rain. The race was postponed for a long time, but then started before darkness would have made driving impossible. Lauda parked his car after the second lap for safety reasons. The second former world champion still active, Emerson Fittipaldi, did the same, as did Carlos Pace and Larry Perkins. Hunt led at times, but then fell back after a pit stop. He finished third at the end of a furious chase to win the world title by one point. It was one of the most spectacular duels in Formula 1 to this day. Lauda became world champion again in 1977, driving a Ferrari. After internal squabbles, he switched to the Brabham team in 1978.

The year 1977 marked Renault's return to Grand Prix racing, from which it had withdrawn in 1908. The Renault RS01 (Fig. 2.18a) used in Formula 1 was the first Formula 1 car with exhaust gas turbocharging. Its V6 turbocharged engine generated a maximum power of 500 hp. at 11,000 rpm from the prescribed displacement of 1.5 liters. However, regular damage to the turbochargers, pistons and piston rings meant that the Renault RS01 did not finish once in 1977. Turbo technology was not yet taken seriously by the competition due to its lack of reliability. Another problem with the turbo engine was the pronounced turbo lag, which was responsible for poor drivability of these cars in tight corners and in duels. Together with Renault, the tyre manufacturer Michelin also returned to Grand Prix racing and equipped a Formula 1 car with radial tires for the first time with the Renault RS01.

The years 1977/78 were again lastingly influenced by Team Lotus. For the fourth race of the 1977 season Lotus used the Type 78. This had a revolutionary aerodynamic concept (see Sect. 6.6.1). The underside of the sidepods was shaped like a wing profile, so that the underbody was significantly used for downforce generation. The Lotus-Ford 78 became the forefather of all ground effect cars. In 1978, Mario Andretti finally won the drivers' title in the Lotus-Ford 78 and its more advanced successor, the Lotus-Ford 79 (Fig. 2.18b). It was the last title for the tradition-rich Team Lotus. The Lotus-Ford 79 showed the concept for modern Formula 1 cars. It had a narrow safety cell and a fuel tank located at the center of gravity. The radiators were mounted on the side of the car and the wheel suspension was outside the airflow.

The superiority of the two Lotus-Ford 78 and 79 made the competition aware of the importance of underbody aerodynamics, so that for the 1979 season only ground effect cars were on the grid. Pandora's box had been opened. From this time at the latest, aerodynamics were given the same importance as engine and chassis. The performance of the cars increased rapidly, as did the stresses on the driver and the material, which led to serious

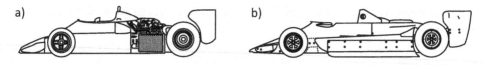

Fig. 2.18 (**a**) Renault RS01 (1977), (**b**) Lotus-Ford 79 (1978)

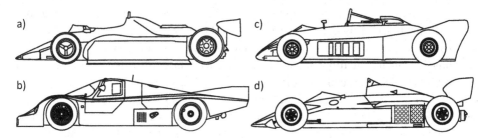

Fig. 2.19 (**a**) McLaren-Ford MP4/1 (1981), (**b**) Lotus-Renault 88 (1981), (**c**) Porsche 956 (1982), (**d**) Brabham-BMW BT52 (1983)

accidents time and again. In 1979 Jody Scheckter won the drivers' title in Formula 1, which was the last drivers' title for Ferrari until a certain Michael Schumacher, who was only ten years old at that time, was highly successful again with this traditional racing team. Renault scored the first Grand Prix victory of a car with a turbo engine that year with the RS01.

2.1.9   1981–1990

The year 1981 also began with a groundbreaking innovation in race car construction. McLaren and Lotus introduced the McLaren-Ford MP4/1 and the Lotus-Renault 88 (Fig. 2.19a, b), respectively, which for the first time used carbon fiber reinforced plastics in combination with aluminum honeycomb structures for the construction of the monocoque (see Sect. 9.3.7). The use of these materials reduced the overall weight of the vehicles and increased their stiffness. The higher stiffness of composites allowed the monocoques to be made progressively narrower. Another advantage of this design was the significant increase in safety due to an extremely stable safety cell.

At Le Mans, the Group C regulations came into force from 1982. In accordance with these, Porsche first designed the Porsche 956 (Fig. 2.19c) and later the Porsche 962. The Porsche 956 was Porsche's first ground-effect vehicle. At its very first appearance at Le Mans, Porsche took the overall victory. In total, the Porsche 956 and 962 took six Le Mans victories in a row until 1987. One notable innovation was Porsche's PDK dual-clutch gearbox, the same principle first brought into series production by Volkswagen as a direct-shift gearbox (DSG) from 2003. In 1994, a Porsche 962 road-legalized by Dauer Sportwagen GmbH won the 24 h of Le Mans one more time.

The increasing downforce and the high lateral accelerations, which now amounted to up to 4 g, brought the pilots of the Formula 1 cars to their physical limits. In order to control the ground clearance, the suspension was sometimes so hard that the damping and suspension was practically only via the tires. Drivers increasingly complained of headaches and back pain, and calls for limits on downforce grew louder.

▶ In 1983, Formula 1 finally mandated a flat underbody between the axles of the car. This marked the end of the era of ground effect cars in Formula 1. Today, all technical regulations contain provisions that severely restrict the design of the underbody geometry.

The drivers' title of the 1983 F1 season was won by Nelson Piquet in a Brabham-BMW BT52 (Fig. 2.19d). It was the first drivers' title in Formula 1 history to be won with a turbo engine. The 1.5-l in-line four-cylinder monoturbo produced up to 750 hp. in the race at 3.4 bar boost pressure [14]. The power advantage over the naturally aspirated engines was up to 200 hp. The breakthrough of the turbo engine had already become apparent in the early 1980s, after the number of pole positions and race victories achieved by turbo engines increased continuously. In 1981, Ferrari converted its drive concept to turbocharged engines and was able to win the constructors' title in 1982. For the 1983 season, two more manufacturers entered Formula One with turbo engines, Porsche (on behalf of TAG) and Honda. The power units of these two manufacturers won all drivers' and constructors' titles until 1988. Renault replaced the mechanical valve springs with a pneumatic valve system in its 1.5-l turbo engines in Formula One in 1986. The maximum engine speed was 12,500 rpm. Pneumatic systems were an important prerequisite for further increases in engine speed and thus power output. At speeds beyond 10,000 rpm, mechanical springs react too sluggishly to still ensure precise opening and closing times. This disadvantage can be circumvented by pneumatic valve systems, but this requires an external actuator to generate pressure. Pneumatic valve systems are still standard in today's F1 engines. The Benetton-BMW BT86, as the most powerful F1 car of all time, briefly generated over 1300 hp. in qualifying with 5.5 bar boost pressure [14].

The use of carbon-fiber-reinforced materials and the development of the turbo engine led to an enormous increase in costs in addition to an extreme explosion in performance. To counteract these developments, fuel stops were banned in 1984 and the maximum tank capacity and maximum boost pressure were gradually reduced. In 1987, the permissible displacement for naturally aspirated engines was increased to 3.5 liters, and from 1989, turbo engines were banned in Formula 1. However, the intensive development work to minimize turbo lag, which initially caused delayed response times of 500 to 1000 ms, made a significant contribution to optimizing turbo technology in production engines.

For the 1988 Formula 1 season McLaren engaged the exceptional driving talent Ayrton Senna, who drove the McLaren-Honda MP4/4 (Fig. 2.20a) alongside the two-time world champion Alain Prost. The maximum boost pressure had been limited to 2.5 bar that year.

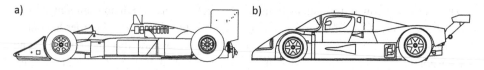

Fig. 2.20 (a) McLaren-Honda MP4/4 (1988), (b) Sauber-Mercedes C9 (1989)

The Honda V6 unit produced 630 hp. at 12,300 rpm. With the McLaren-Honda MP4/4 15 of 16 races were won in the last year of the turbo engine (until its return in 2014). Ayrton Senna accounted for eight and Alain Prost for seven victories. At the end of the season, Ayrton Senna won his first drivers' world championship as a rookie in the team. Alain Prost and Ayrton Senna became bitter rivals in the following. From 1989 onwards, only naturally aspirated engines with a maximum displacement of 3.5 liters were permitted in Formula 1.

Also in 1988, Mercedes-Benz officially announced its return to international motorsport after an absence of more than 32 years. Since the end of 1984, the Swiss Sauber racing team had been supported in its private use of the C8 Group C sports car prototype by the provision of a racing power unit. For 1987, with increasing support from Mercedes, Sauber developed the Sauber C9 (Fig. 2.20b), which won the Supercup at the Nürburgring that year. For the 1988 season, the now factory-supported Sauber-Mercedes C9s were painted in the livery of AEG, a subsidiary of Daimler AG. Five rounds of the World Sports Car Championship (WSC) were won. In 1989, the Sauber-Mercedes C9s were back in the traditional silver livery for the first time. The Silver Arrows were back, winning both the 24 h of Le Mans and the World Sports Car Championship that year. The Sauber-Mercedes C9 crossed the finish line first seven times in eight race appearances. It was powered by a V8 biturbo engine with 4973 cm^3 displacement. At Le Mans, the maximum output of this unit was 925 hp. at 2.4 bar boost pressure, which helped the Sauber-Mercedes C9 reach a top speed of over 400 km/h [9]. The win of the sports car world championship could be repeated in 1990 with the successor C11. This was driven, among others, by a young racing driver named Michael Schumacher.

Ferrari introduced semi-automatic or automated manual transmissions for its Formula One cars in 1989. The shifts were triggered by paddles on the steering wheel, but the actual shifting was done by a hydraulic actuator that was controlled electronically. The driver no longer had to take his hands off the steering wheel to shift gears, and the gear changes were much faster.

The drivers' title in the 1989 Formula 1 World Championship was decided again in the penultimate race between McLaren drivers Alain Prost and Ayrton Senna. At the Japanese Grand Prix, Alain Prost was leading by 16 points ahead of Ayrton Senna, who had to win the two remaining races to defend his title. At the start Senna was on pole position, but Prost was able to overtake him at the start. Until the 46th lap Prost was in the lead. Before the Casio chicane, the second slowest spot on the entire circuit, Senna moved to the right of Prost to pass him on the inside. Prost moved to the inside to fend off the overtaking attempt. Neither driver gave way, and a collision occurred. Both cars came to a halt in the run-off area. Knowing that a retirement of both McLaren's would make him world champion, Prost got out of his car. Senna gestured wildly to the marshals to push him through the emergency exit in order to restart the engine. The maneuver succeeded and Senna returned to the race. However, he had to pit first due to a damaged front wing. After this stop, he dropped five seconds behind Alessandro Nannini on Benetton. On lap 51, Senna overtook the leading Nannini at the Casio chicane and crossed the finish line first. Prost had rushed to the race control in the meantime. The podium ceremony was conspicuously delayed. Finally Senna was disqualified. By using the emergency exit, Senna had skipped the chicane and thus, according to existing regulations, had taken an unauthorized shortcut. The correct behavior – according to the FIA – would have been to turn around in the emergency exit in order to return to the track first against the direction of travel. Prost ended up winning his third world title and switched to Ferrari for the 1990 season due to the squabbles with Ayrton Senna. At a later press conference, McLaren team principal Ron Dennis showed television footage of examples of other drivers also using the emergency exit in emergency situations and returning to the track from there. However, none of these drivers were subsequently disqualified.

In 1990, the showdown between Senna and Prost took place again at the Japanese Grand Prix in Suzuka. This time, however, the omens were reversed. A retirement by Prost would see Senna secure his second world championship. In qualifying, Senna put his McLaren on pole position, with Prost following right behind in a Ferrari. For reasons unexplained, the race committee moved the pole position from the racing line to the dirty side of the track. Senna again felt himself to be a victim of political events going on in the background. At the start Prost was able to use the traction advantage of the racing line to get in front of Senna. The right-hand bend after the start-finish section was approached by Prost from the outside and Senna from the inside. Before the turn, Senna sat next to Prost, who was turning in to the right. There was another collision and both cars slid into the gravel in a cloud of dust. The race continued. Ayrton Senna was world champion. It was the climax of the bitter rivalry between Prost and Senna, which had to be called enmity in the meantime. However, the duel of the two also increased the media interest in Formula 1 in the long run. The ratings and thus the commercial value of Formula 1 increased.

2.1.10 1991–2000

Michael Schumacher contested his first Formula 1 race at the 1991 Belgian Grand Prix in Spa-Francorchamps. He was standing in for Bertrand Gachot, who was serving a prison sentence at the time after spraying a taxi driver with irritant gas in a dispute. To the surprise of all the experts, Schumacher qualified his Jordan in seventh position. At the start he overtook two competitors, but then dropped out while lying in fifth position with clutch damage. But even with this short appearance Schumacher was able to impress the experts. After his first race, he switched to the established Benetton racing team and became a

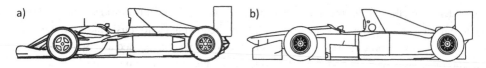

Fig. 2.21 (**a**) Williams-Renault FW14B (1992), (**b**) Benetton-Ford B194 (1994)

teammate of three-time world champion Nelson Piquet, whom he always left behind in qualifying with one exception.

The following year, Schumacher celebrated his first Formula 1 victory, again at the Belgian Grand Prix. He finished the season in third place in the world championship. In the 1993 season, Schumacher also scored a victory and finished fourth overall.

However, the Formula 1 World Championships of 1992 and 1993 were mainly characterized by the technical dominance of the Williams team, which led to the winning of the drivers' title by Nigel Mansell and Alain Prost respectively. The Williams-Renault FW14B (Fig. 2.21a) and Williams-Renault FW15C used in these two years had an active suspension. However, their use was not a novelty in Formula One, as such a system had already been used in 1983 in the Lotus-Renault 93 T, and the first victory with active suspension was achieved in 1987 with the Lotus-Renault 97 T. However, it was with the Williams-Renault FW14B and FW15C that this technology was first developed to sustainable racing maturity. However, traction control and the Renault V10 engine, which provided between 730 and 760 hp., also made an important contribution to the superiority of the Williams-Renault cars. After 1993, active suspensions or any form of driving dynamics control systems were banned in Formula 1.

The 1994 motorsport year was dominated by the tragic events that took place during the San Marino Grand Prix at Imola. In the run-up to the season, three-time Formula 1 world champion Ayrton Senna had switched from McLaren to Williams. However, the major competitive advantage of the 1992 and 1993 Williams cars, the active suspension, had, as mentioned, been banned for the 1994 season. Michael Schumacher won the first two races in his Benetton-Ford B194 (Fig. 2.21b), while Ayrton Senna remained without points in each case. The Brazilian's tension was correspondingly great at the third race in Imola.

On Friday 29 April, the first sensational incident occurred during free practice. The young Brazilian driver Rubens Barrichello lost control of his Jordan-Hart in the Varianta Bassa. At 240 km/h the car shot straight over a curb and was catapulted into the air. With a huge bang, the Jordan slammed into the tyre piles. Barichello flipped over and eventually landed upside down in the gravel. Television footage showed how brutally the Brazilian's head was tossed back and forth. Miraculously, Barrichello survived the accident without any serious injuries, but after this accident, Formula 1's luck ran out.

During the second qualifying practice the front wing of the Simtek-Ford of the Austrian Roland Ratzenberger suddenly broke away. The car became an uncontrollable projectile and raced straight ahead in the Villeneuve bend towards the boundary wall. The impact took place with more than 300 km/h. The vehicle then slid parallel to the track boundary for hundreds of meters before the Simtek-Ford came to a stop after another spin. Ratzenberger's head was hanging lifeless in the vehicle. The rescue team was immediately on the scene of the accident. But the resuscitation attempts visible on the television pictures were in vain. Roland Ratzenberger succumbed to his severe head injuries. Formula 1 was in shock. Nevertheless, the race was started on Sunday.

Already at the start, the eerie series of accidents continued. Lehto's Benetton-Ford came to a halt. From behind, Pedro Lamy slammed into the stationary vehicle. Nine spectators were slightly injured by the flying wreckage. The safety car was deployed. J.J. Lehto and Pedro Lamy were not injured. The safety car was practically a standard production saloon at the time, which could not reach the speed required for formula cars to condition the tyres and brakes. Senna asked the safety car to speed up several times. At the end of the fourth lap, it went off the track, clearing the race for the fifth lap. Senna was in the lead ahead of Schumacher. On lap 6, the cameras showed the Williams-Renault suddenly slamming straight into the boundary wall in the left-hand Tamburello and being thrown back into the run-off area. Also in this case the rescue forces around Sid Watkins were immediately on the spot, but also for Senna any help came too late. He was flown by helicopter to the Maggiore Hospital in Milan, but the news of Senna's death was announced there at 6.40 p. m.

Michael Schumacher in his Benetton-Ford emerged from the dramatic 1994 season B194 as the first German Formula 1 world champion and triggered a real motorsport boom in Germany. The following year he won his second world title in the Benetton-Renault B195 and switched to Scuderia Ferrari in 1996.

▶ Global coverage of the tragic events at Imola led to an ongoing discussion about motorsport safety due to public pressure, and launched a campaign led by Max Mosley to make sustained and successful improvements to motorsport safety. In the 20 years that followed, no driver suffered a fatal accident in Formula One. One of the immediate measures adopted for the 1995 season was to reduce the maximum engine capacity to 3.0 liters.

In 1998 and 1999, Mercedes-Benz continued its successful past as McLaren's engine partner. Mika Häkkinen won the Formula 1 drivers' title in both years. In 1998, the McLaren-Mercedes team also won the constructors' championship. The McLaren-Mercedes MP4/13 (Fig. 2.22a) of the 1998 season, designed by Adrian Newey, had 760 hp. maximum power, which the Mercedes-Ilmor V10 engine reached at 16,500 rpm. In 1999, the power output of the McLaren-Mercedes MP4/14 was increased to 765 hp. [15].

Mercedes-Benz's return to Formula One was paved by its cooperation with the Sauber racing team. In 1993, Sauber entered Formula 1, financed by Mercedes-Benz. The Sauber C13 was initially powered by an Ilmor V10 engine. At the end of 1993, Mercedes joined Ilmor as a shareholder and became the official engine partner of the Sauber racing team for

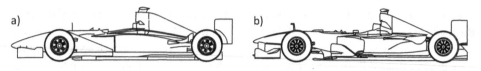

Fig. 2.22 (a) McLaren-Mercedes MP4/13 (1998), (b) Ferrari F2000 (2000)

the 1994 season. However, due to unsatisfactory results, the cooperation with Sauber was discontinued at the end of 1994, and a strategic partnership was entered into with the traditional English racing team McLaren from 1995. In 1995 and 1996 the McLaren-Mercedes were still painted in the red and white livery of the main sponsor. It was not until 1997 that the traditional silver livery was used, adding a new chapter to the history of the Silver Arrows.

Michael Schumacher brought long-awaited success back to Maranello in 2000. He won the first drivers' title for Ferrari since 1979. However, this success was preceded by a structural and personnel upheaval initiated by Luca di Montezemolo, who was appointed to the Ferrari board in 1991. Ferrari had lost touch with the leading British racing teams in the age of aerodynamics, turbo engines, active chassis systems and electronics. Harvey Postlethwaite and John Barnard were brought in to head engine development, and for the 1996 season they changed the Ferrari's power unit from a V12 to a V10 design. In 1993 Jean Todt was appointed head of the Formula One team. As well as signing Michael Schumacher for the 1996 season, Todt also managed to bring in senior engineers from the Benetton team, Ross Brawn and Rory Byrne, for 1997. These changes eventually led to the development of the Ferrari F-2000 (Fig. 2.22b), which led Schumacher to the drivers' title and the Scuderia to the constructors' title. The Ferrari's V10 engine produced 810 hp. at 17,600 rpm and won a total of ten races that year, nine of which went to Michael Schumacher. By 2004, Schumacher had won four more world championships with Ferrari. His seven world championship titles made him the most successful Formula 1 driver of all time.

Audi clinched its first overall victory in the 24 h of Le Mans with the R8 in 2000. The vehicle was powered by a V8 biturbo engine that generated 610 hp. of power from 3596 cm^3 displacement. It was the beginning of a long success story in endurance racing.

2.1.11 2001–2010

The Audi R8 (Fig. 2.23a) also won the 24 of Le Mans in 2001 and 2002. Since 2001, the V8 biturbo engine had gasoline direct injection, which now bore the name suffix FSI (Fuel Stratified Injection) like the production engines from Volkswagen and Audi. In 2003, this engine was also used by the Group's sister company Bentley, which won the 24 h of Le Mans that year with the Bentley EXP Speed 8 (Fig. 2.23b). In 2004 and 2005, this race was

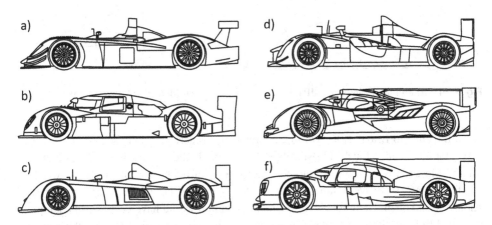

Fig. 2.23 Volkswagen Group Le Mans winners: (**a**) Audi R8 (2000), (**b**) Bentley EXP Speed 8 (2003), (**c**) Audi R10 (2006), (**d**) Audi R15 plus (2010), (**e**) Audi R18 (2011), (**f**) Porsche 919 (2015)

again won by an Audi R8, but now privately entered by the Japanese Goh team and the American Champion Racing team. In total, the Audi R8 won 63 of 80 races.

In 2006, Audi set another milestone in the history of motorsport. With the Audi R10 (Fig. 2.23c), a vehicle with a diesel engine won the 24 h of Le Mans for the first time. The V12 diesel engine with biturbo had a displacement of 5499 cm^3 and produced 675 hp. The 24 h of Le Mans was also won by an Audi R10 in 2007 and 2008. After Peugeot inflicted defeat on Audi in 2009, the race was won again in 2010, this time by an Audi R15 plus (Fig. 2.23d).

A curious scene occurred at the US Grand Prix in Indianapolis during the 2005 F1 World Championship. There were only six cars on the grid. In practice, Ralf Schumacher's Toyota had smashed heavily into the track barrier after a tyre puncture when driving through Turn 13. Due to the heavily banked turn, the tyres were exposed to particularly high loads in Turn 13. Michelin could not guarantee the stability of the tire for the race. It was not possible to agree on a mitigation of the curve. At the end of the formation lap, all cars with Michelin tires turned into the pit lane for safety reasons. Only the cars with Bridgestone tires took up their grid positions. Ferrari took an unchallenged double victory. It remained the only Ferrari success this season.

The maximum cubic capacity of Formula 1 cars was reduced to 2.4 liters for the 2006 season, and the engine had to have a V8 layout.

In the 2008 F1 season, as in the previous year, the world champion was decided at the final race in Brazil. The remaining title contenders were Felipe Massa in the Ferrari and Lewis Hamilton in the McLaren-Mercedes. In 2007, Hamilton had lost the world title in dramatic fashion to Kimi Raikkonen. Hamilton would have had to score only four points in the last two races, but he failed to do so. In the penultimate race in Shanghai, while leading, he lost control of his car on the approach to the pit lane and retired. In the last race in Brazil, Hamilton finished seventh after a botched start and technical problems. In 2008, Hamilton

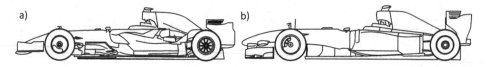

Fig. 2.24 (a) McLaren-Mercedes MP4/23 (2008), (b) McLaren-Mercedes MP4/24 (2009)

again travelled to Brazil as the world championship leader. To finally secure the title this time, Hamilton had to finish at least fifth if Massa won. In the starting line-up Hamilton was fourth, Massa was on pole. It turned out to be a difficult race in changing weather conditions. Massa drove a superior race at the front. On lap 63 the order was Massa, Alonso, Raikkonen, Hamilton, Vettel, Glock. Hamilton would have been world champion if he had finished like that. Then rain set in. The five leading cars came into the pits and changed to intermediates. Glock stayed out and moved up to fourth. Hamilton was now fifth. The rain increased, Hamilton was finally overtaken by Vettel and could not follow Vettel. With sixth position, the title was all but lost. Massa crossed the finish line with a big lead. There was jubilation in the stands and in the Ferrari pit. But in the last corner before the start and finish, Hamilton was able to overtake Glock, who was driving on slicks, and Hamilton crossed the finish line in fifth place and as the new world champion. The Ferrari cheers died down. Massa's would-be world championship lasted just 38.9 seconds. It remains to this day one of the most dramatic decisions of all time. The world champion car was a McLaren-Mercedes MP4/23 (Fig. 2.24a), whose V8 engine provided about 780 hp.

▶   General automotive development is increasingly being shaped by an ongoing discussion about CO_2 emissions. Two significant trends in this context are downsizing, i.e. the reduction of engine displacement and cylinder numbers through turbocharged engines, and the electrification and thus hybridization of the powertrain.

Even high-performance motorsport could not escape this discussion. Formula 1 was the first racing series to introduce a hybridized powertrain, better known as KERS (Kinetic Energy Recovery System), for the 2009 season. The KERS essentially consists of a battery and an electric motor. During braking, the electric motor is used as a generator and feeds energy into the battery. This stored energy is used to power the electric motor during acceleration. The first hybridized Formula 1 car to win a Grand Prix was Lewis Hamilton's McLaren-Mercedes MP4/24 (Fig. 2.24b).

After the end of the 2009 season, Daimler AG (so renamed since 2007), together with its major shareholder Aabar Investments, took over the former Brawn GP racing team and founded an F1 works team under the name "Mercedes Grand Prix" for the first time since 1955. Seven-time world champion Michael Schumacher returned as a Mercedes works driver for the 2010 season, before finally retiring from motorsport after the 2012 season.

2.1.12 2011 To Date

In 2011, a black streak of tragic accidents in high performance racing began. On October 16, 2011, Briton Dan Wheldon died after a mass crash that occurred during an IndyCar race at Las Vegas Motor Speedway. Wheldon's vehicle was catapulted into the air by the rear of a car ahead and struck the catch fence. He succumbed to his severe head injuries at the hospital, which was announced 2 h after the accident. In lieu of a restart, the drivers completed five laps of honor in memory of the 2005 IndyCar champion. Dan Wheldon left behind a wife and two children.

In 2012, André Lotterer, Benoit Treluyer and Marcel Fässler won the 24 h of Le Mans in an Audi R18 (Fig. 2.23e) e-tron quattro. It was the first victory clinched with a hybrid drive. The vehicle had a V6 diesel engine with a displacement of 3.7 l that was turbocharged by a monoturbo and produced a maximum of 490 hp. Two electric motors worked on the front axle, producing a combined 218 hp. The car's all-wheel drive powertrain thus produced a total output of 708 hp. Audi won the 24 h of Le Mans a total of 13 times until 2014. In 2014, the Group's sister company Porsche returned to Le Mans and the World Sports Car Championship with the 919 Hybrid (Fig. 2.23f). In 2015 and 2016, Porsche won the sister duel. The 2016 Porsche 919 Hybrid produced nearly 900 hp., of which about 500 hp. came from the V4 turbocharged gasoline engine and about 400 hp. from the motor-generator unit. In 2017, before retiring from endurance racing, Porsche took its third consecutive overall victory. Thus, between 2000 and 2017, only once did a car win the 24 h of Le Mans that did not belong to the Volkswagen Group.

Volkswagen entered the World Rally Championship (WRC) in 2013. As a newcomer, it managed to win the title straight away, which it defended in 2014, 2015 and 2016. These are the Volkswagen brand's most significant motorsport successes to date. The VW Polo R WRC (Fig. 2.25a) went down in motorsport history as one of the most successful rally cars of all time. The car took 43 victories in 52 World Rally Championship events, four drivers' titles with Sébastian Ogier, four co-drivers' titles and four constructors' titles. The all-wheel-drive Polo WRC produced a good 300 hp. from its 1.6-l TSI engine. Volkswagen ended its involvement in the WRC at the end of 2016, and Audi also announced its withdrawal from endurance racing.

On 22 June 2013, the Dane Allan Simonsen suffered a fatal accident on the third lap of the 24 h of Le Mans. Accelerating out of the Tertre Rouge, Simonsen lost control of his Aston Martin Vantage on the partially wet track and hit the side of the crash barriers. Shortly after the impact, Simonsen was still responsive. He died on the way to the medical center. It was the first fatal accident at Le Mans since 1997. Later investigations at the point of impact revealed that there was a tree directly behind the crash barrier. This prevented the crash barrier from deforming and thus from reducing the force of the impact. This accident rekindled discussions about the safety of the "Circuit des 24 Heures".

In 2014, Formula 1 followed the general trends in automotive engineering and the turbo engine returned to Formula 1. As in 1986, it has been the only approved drive option ever since. Today, the maximum displacement of a turbocharged F1 engine is set at 1.6 liters

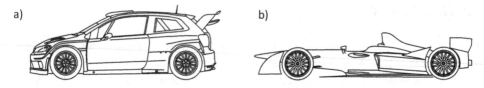

Fig. 2.25 (**a**) Volkswagen Polo R WRC (2013), (**b**) Formula E (2014)

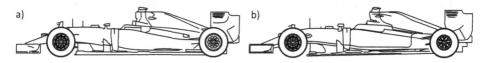

Fig. 2.26 (**a**) Mercedes-Benz F1 W05 Hybrid (2014), (**b**) Mercedes-Benz F1 W07 Hybrid (2016)

and the maximum engine speed is limited to 15,000 rpm. In both 2014 and 2015, Lewis Hamilton clinched the F1 world championship for Mercedes. The Mercedes F1 W05 Hybrid (Fig. 2.26a) from the 2014 season had a total power output of about 761 hp. [16], which was 600 hp. from the turbocharged engine and 161 hp. from the electric motor. The changes in the regulations for Formula One cars and for Le Mans prototypes significantly increased the importance of fuel-efficient technologies. In this context, the FIA also founded Formula E, which is the first racing formula for purely electrically powered vehicles (Fig. 2.25b). The first Formula E race was held in Beijing on 13 September 2014.

On 5 October 2014, Frenchman Jules Bianchi lost control of his Marussia on a wet track during the Japanese Grand Prix. His car hit a recovery vehicle at high speed, which was still being used to recover Adrian Sutil's Force India that had crashed earlier. The Frenchman suffered severe head injuries to which he eventually succumbed on 17 July 2015. Jules Bianchi was the first fatality at an official Formula One event since 1994, and the motorsport world was once again made aware that the battle for safety can never considered be won.

This fact came aware to the motorsport world again in August 2015 when former F1 driver Justin Wilson was killed in an accident during the IndyCar race at Pocono Raceway. He was hit directly in the head by wreckage that came loose during a collision involving the car in the lead. A similar accident had taken place in 2009 during a Formula 2 race at Brands Hatch in the UK. There, Henry Surtees, son of former Formula 1 World Champion John Surtees, was hit on the helmet by the torn-off wheel of a crashed car. Henry Surtees also suffered fatal head injuries as a result. These accidents of the past years led to a renewed debate about the safety of open-wheel racing vehicles, which continues to this day.

At Le Mans, the "Garage 56" is reserved for projects with special technical innovations. In 2016, this was awarded to the "SRT41 by OAK Racing" team. The special feature here was the driver trio around the severely disabled Frenchman Frederic Sausset. He had to have both legs and both hands amputated after a bacterial infection. The Nissan Morgan,

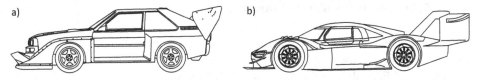

Fig. 2.27 (**a**) Audi Sport Quattro S1 E2 (1987), (**b**) Volkswagen I. D. R Pikes Peak (2018)

entered in the LMP2 category, had been modified so that it could be driven by both Sausset and his non-disabled team-mates. The vehicle finished 38th overall, but more important was the demonstration of how modern technology can give people with disabilities access to mobility and thus greater independence.

In 2016, Nico Rosberg became the third German driver after Michael Schumacher and Sebastian Vettel to win the Formula 1 World Championship in a Mercedes-Benz F1 W07 Hybrid (Fig. 2.26b). However, he was the first German driver to achieve this in a German vehicle. It was also the second time that a driver followed in his father's footsteps. Keke Rosberg had won the drivers' title in a Williams-Ford FW07 in 1982. Graham Hill and his son Damon Hill succeeded in 1962, 1968 and 1996. In a complete surprise, Nico Rosberg announced his immediate retirement from Formula 1 only five days after his title win in order to retire into private life.

Audi entered Formula E as a factory team for the 2017/2018 season. BMW followed suit for the 2018/2019 season. Porsche and Mercedes will be involved in Formula E from the 2019/2020 season, which is a further expression of the increasing importance of electric or electrified drive concepts. The same applies to VW's participation in the 2018 "Pikes Peak International Hill Climb" (also known as "Race To The Clouds"), which was won by the Group's sister company Audi a total of six times in the 1980s and for the last time in 1987 by Walter Röhrl on an Audi Sport Quattro S1 E2 (Fig. 2.27a). The Volkswagen I. D. R Pikes Peak (Fig. 2.27b) has an all-electric four-wheel drive powertrain that produces a total output of 680 hp. The designation I. D. is also used here for the production models based on the Modular Electric Drive Matrix (MEB), which will be launched on the market from 2020. On 24 June 2018, Romain Dumas makes history with the Volkswagen I. D. R Pikes Peak when he conquered Pikes Peak in just 7:57.148 min, breaking the absolute record from 2013 that had previously been held by a vehicle with an internal combustion engine.

It remains to be seen how the current megatrends in automotive engineering, electromobility as well as automated driving will change the technology in motorsport. What is certain, however, is that motorsport will continue to write exciting and emotional stories that continue to fascinate millions of people around the globe. High-performance motorsport thus remains an important marketing platform for companies that want to demonstrate their high-tech expertise, innovative strength, competitiveness and/or sporting spirit or benefit from the global media interest in motorsport.

2.2   Motivation for Motor Sports

Involvement in high-performance racing requires budgets worth millions for the development and operation of the racing vehicles used. Figure 2.28 shows the estimated budget available to the Formula 1 racing team maintained by Mercedes-Benz AG in the 2014 season. Mercedes-Benz divides the operation of the racing business between two companies. The first is Mercedes-Benz Grand Prix Ltd., which operates under the name "MERCEDES AMG PETRONAS Formula One Team" and is based in Brackley, England. The second company is Mercedes AMG High Performance Powertrains, which is also based in England. Together, the two companies had a total budget of around €429 million in 2014. This amount is comparable to the budgets of other top racing teams, such as Red Bull or Ferrari. The budget includes all costs that are necessary for the development of the vehicle and the powertrain as well as for racing operations. Among other things, this includes the personnel costs for around 750 employees, including both drivers.

The budget consists of various components. A large part of the budget, around €173 million, is raised through participation in the marketing rights and sponsorship money. HPP Ltd. generates its budget of €179 million independently by selling the powertrain to other racing teams (McLaren, Williams, Force India), even making a small profit of about € 7.7 million. The amount provided by Daimler AG to cover costs thus amounts to an estimated € 69.3 million, a relatively small proportion of the total costs. Nevertheless, even this enormous amount must be compensated by Daimler AG's core business, namely the sale of automobiles and automotive services. In this context, it is worth mentioning that, in parallel to Formula 1, Mercedes-Benz also became involved in the DTM as well as in GT racing in 2014.

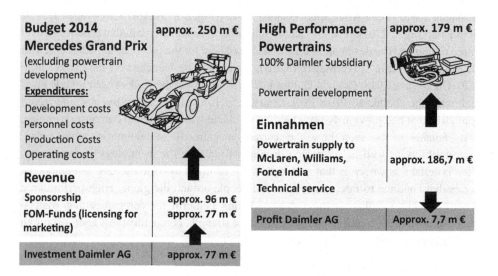

Fig. 2.28 Mercedes GP and High Performance Powertrains budget in 2014. (After [17])

As Sect. 2.2 shows, motor sport is a competition that takes place in a highly technological but also emotional environment, where triumph and tragedy are sometimes very close together. Today, this mixture binds millions of viewers in front of the screen on every race weekend or attracts them in their tens of thousands directly to the race track. High-performance motorsport receives a global media presence through live broadcasts, summaries, articles in print media as well as online coverage, which ultimately represent the equivalent value for the multi-million investments by profit-oriented companies. Formula 1 and other high-class racing series are thus an important instrument for image cultivation and branding of a company.

A very good example of this is the term "Silver Arrows", which today almost everyone immediately associates with the Mercedes-Benz brand. Audi's successes in rallying in the 1980s also contributed to the term "quattro" now being synonymous with this brand's all-wheel-drive vehicles. Companies not directly associated with motorsport or the automotive industry also seek to benefit from this environment through sponsorship. A successful involvement in motorsport can thus be quite profitable economically, even if this value can often not be measured by reliable economic figures. This is symbolized by the US saying:

▶   "Win on Sunday, sell on Monday!"

To a limited extent, and thus no longer to the same extent as in past decades, engineering testing and method development also play a certain role for companies involved in motorsport. Nowadays, the focus is probably on lightweight materials and simulation programs. The close exchange between series production and motorsport is particularly promoted at Porsche AG, for example. The development departments of the production and racing vehicles are both located at the Weissach site and maintain a lively exchange in various projects. In this context, the promotion of young engineers may also be of interest.

A multi-billion euro industry has grown up around motorsport, whose business model is the profitable provision of services and manufacture of components and accessories for motorsport. The numerous racing series in which amateurs can participate are also part of this environment. Many of these amateurs are driven by pure enthusiasm for motorsport.

▶   **Important**
    In summary, the motivation for direct or indirect engagement arises mainly from the following factors:

    • Image cultivation and branding
    • Testing and method development
    • Promotion of young talent
    • Business case and economic factor
    • Enthusiasm

The reasons for taking up motorsport are complex, but the central starting point is the fascination that the automobile and motorsport exert on millions of people. The following chapters of this book deal with the technical aspects that contribute significantly to this fascination.

References

1. Rosemann, E., Demand, C.: The Big Race – The Story of Motor Racing. Nest, Frankfurt am Main (1955)
2. Giffard, P.: Voitûres sans Cheveaux. Le Petit J. **24**, 1 (1894)
3. Cimarosti, A.: The Complete History of Motor Racing. Aurum Press, London (1997)
4. Ludvigsen, K.: Classic Grand Prix Cars – The Front-Engined Era. Haynes Publishing, Sparkford (2006)
5. Edler, K.H., Roediger, W.: Die Deutschen Rennfahrzeuge – Technische Entwicklung der letzten 20 Jahre. VEB Fachbuch, Leipzig (1990)
6. Neubauer, A., Rowe, H.T.: Männer. Frauen & Motoren – Die Erinnerungen des Mercedes-Rennleiters Alfred Neubauer, Motorbuch, Stuttgart (2011)
7. Knittel, S.: Auto Union-Grand-Prix-Wagen. Schrader & Partner GmbH, München (1980)
8. Pritchard, A.: Silberpfeile – Die Duelle der Grand-Prix-Teams von Mercedes-Benz und Auto Union von 1934–1939. Motorbuch, Stuttgart (2009)
9. Ludvigsen, K.: Mercedes-Benz Renn- und Sportwagen. Bleicher, Gerlingen (1993)
10. Pritchard, A.: Mille Miglia – the World's Greatest Road Race. Haynes Publishing, Sparkford (2007)
11. Noakes, A.: The Ford Cosworth DFV – the Inside Story of F1's Greatest Engine. Haynes Publishing, Sparkford (2007)
12. Bamsey, I.: The Anatomy & Development of the Sports Prototype Racing Car. Motorsports International, Osceola (1991)
13. Hughes, M.: Speed Addicts – Grand Prix Racing. Dakini Books, London (2005)
14. Bamsey, I.: The 1000 BHP Grand Prix Cars. Haynes Publishing, Sparkford (1988)
15. Benzing, E.: Dall' aerodinamica alla potenza in Formula 1. Giorgio Nada Editore, Vimodrone (Milano) (2004)
16. Lehbrink, H.: Turning silver into gold. teNeues publishing group, Kempen (2014)
17. http://www.motorsport-total.com/f1/news/2014/11/die-wahren-kosten-der-formel-1-14110519. html. Accessed on 27 January 2017.

Organization and Regulation

3

3.1   Associations and Vehicle Categories

The professional staging of racing series or motor sport events requires a structured interaction between motor sport or automobile associations, marketing companies, organizers and participating racing teams. Figure 3.1 shows the main roles and tasks involved in staging a racing series. Figure 3.2 provides a selection of some of the major automobile and motorsport associations and their associated racing series.

The best-known automobile association is the Fédération Internationale de l' Automobile (FIA for short). The FIA, based in Paris, is the global governing body of motor racing. It was founded on 20 June 1904 as the Association Internationale des Automobile Clubs Reconnus (AIACR) and renamed the FIA in 1946. The original aim of the federation was to regulate and guarantee safety in motor sport in the long term. Numerous national automobile associations are members of the FIA. In Germany, these are the Allgemeine Deutsche Automobil Club (ADAC), the Automobilclub von Deutschland (AvD) and the Deutsche Motor Sport Bund (DMSB). Among other things, the FIA is responsible for the sporting and technical regulations of the Formula 1 World Championship, the World Endurance Championship (WEC) and various other world championships in a wide range of categories. In addition, it has issued numerous safety standards for motorsport that are valid worldwide.

In Germany, the DMSB and ADAC in particular are involved in motorsport, sometimes in close cooperation. The DMSB draws up the technical regulations for the Class 1 racing cars that are used in the German Touring Car Masters (DTM). The DTM is the highest-class racing series under German responsibility and is mainly held on international race tracks in Europe. The ADAC hosts, among others, the national GT Masters as well as the

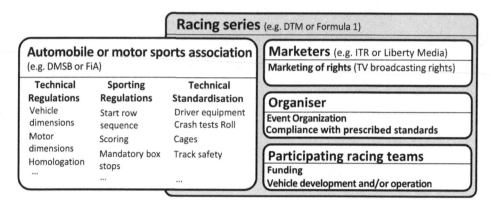

Fig. 3.1   Distribution of roles in motorsport

FiA	Fédération Internationale de l' Automobile	Formula 1 (F1), World Endurance Championship (WEC), World Rallye Championship (WRC), World Touring Car Championship (WTTC), Formula E, Formula 3 (F3), GT3, TCR
ADAC	General German automobile club	GT Masters, Rally Germany, Formula 4, TCR Germany, Rally Masters, Kart Masters
DMSB	German Motor Sport Federation	German Touring Car Masters (DTM)
NASCAR	National Association for Stock Car Auto Racing	Monster Enery Cup Series
INDYCAR		Verizon Indycar Series
ACO	L' Automobile Club de l' Ouest	24 Heures du Mans
JAF	Japan Automobile Federation	Super GT
CAMS	Confederation of Australian Motor Sport	V8 Touring Car Series

Fig. 3.2   Motorsport associations and racing series

TCR Germany and promotes young motorsport talents through the German Formula 4 Championship and various kart championships.

In North America, the National Association for Stock Car Auto Racing (NASCAR for short) is the most important organization in motorsport. It is primarily responsible for the Monster Energy Cup, which was previously held as the Sprint Cup (2008–2016), Nextel Cup (2004–2007), Winston Cup (1972–2003) and Grand National (1949–1971). The championship has been named after its primary sponsor since 1972. Monster Energy Cup races are among the largest sporting events in the United States.

INDYCAR organizes the Verizon IndyCar Series, which is also very popular in the USA. The International Motorsport Association (IMSA for short) is generally less well known. Its most important event is the WeatherTech SportsCar Championship, which also includes the famous 24 h of Daytona. However, it also organizes the Porsche GT3 Cup USA, the Ferrari Challenge North America or the Super Trofeo North America (Lamborghini).

Worth mentioning in this context is the French Automobile Club de l' Ouest (ACO), which stages the most important single race in motorsport with the 24 h of Le Mans (24 Heures du Mans). For a long time, the ACO itself defined the technical regulations for the racing cars. Currently, however, the FIA draws up the technical regulations for the participating vehicles because the 24 h of Le Mans are integrated into the World Endurance Championship.

In addition to these associations, there are numerous other national associations and organizations that host top-class racing series. These include, for example, the Super GT Series in Japan or the V8 Supercars in Australia. In Great Britain, the British Touring Car Championship has a long tradition.

▶ **Important**
The main task of an automobile or motor sport association is to provide the technical and sporting framework for the organization of a racing series, with the main objectives being

- ensure the safety of spectators, pit crew members, marshals and drivers,
- to create equal opportunities for exciting competition,
- to contain development costs,
- to grant freedom of development and
- to establish development security.

It is obvious that there are conflicting objectives between the lower four objectives and, to some extent, the wishes of large manufacturers or private teams.

Freedom of development allows large manufacturers in particular to stand out from the competition with their own technical solutions. A positive marketing effect can be effectively achieved if the manufacturer competes for victories in a racing series. Three types of racing series can be distinguished on the basis of the degree of development freedom:

- Open racing series without significant restrictions on the degree of development freedom, e.g. Formula 1 and WEC
- Spec series (or "one make" series) without any freedom of development, e.g. IndyCar, Formula 3
- Racing series with limited development freedom, e.g. DTM and Formula E

With relatively unrestricted technical regulations, the necessary know-how advantage can often be achieved through the use of corresponding financial resources. As a result, equality of opportunity suffers, as smaller racing teams do not have these financial resources to develop a vehicle capable of winning. The lack of success leads to a drop in sponsorship income and the threat of insolvency. Such a racing series is often dependent on the commitment of large manufacturers. Such tendencies could be observed in recent years, especially in Formula 1 and the WEC. The containment of development costs is therefore an important aspect of the technical regulations today. Cost-reducing measures include the restriction of test drives, the prohibition of certain materials or the compulsion to use standardized components. In a spec series, practically all components are standardized, so that all teams have identical material at their disposal. The philosophy that in such series only the driver and not the material makes the difference is very popular, especially in the American region. In such racing series, access to high-class motorsport is often possible with a relatively low budget. This approach is the opposite of that of Formula 1 and the WEC. In between, there are racing series such as DTM and Formula E, in which the vehicles are built to a high degree with standard components, but in which proprietary developments are nevertheless permitted in defined areas.

Consistency in the regulations guarantees that the high investments required for the development of a racing car have a lasting effect. An example of this are the engine regulations of Formula 1, which, particularly in the last two decades, have always pre-scribed the same engine concept over several years. The aspects mentioned above are significantly influenced by the technical regulations, the basic structure of which is considered in Sect. 3.2. In addition, there are numerous standardizations which, like technical norms, prescribe the nature of certain components, such as roll cages and fuel tanks.

The complete set of rules for a racing series consists of technical and sporting regulations. The sporting regulations govern the aspects of the sporting competition, which include, among other things, the determination of the starting order, the system for awarding points (25–18–15-12–10-8-6-4-2-1 in Formula 1 and DTM, in DTM points [3–2-1] are still awarded for the three fastest in qualifying), speed limits in the pits, specifications for avoiding collisions or the behavior in safety car phases. The sporting regulations also include, for example, the penalties for non-compliance with the rules.

3.2   Technical Regulations

The technical regulations specify the essential characteristics of a racing vehicle. Depending on the nature and degree of freedom of development of the regulations, three basic categories of racing vehicles can be distinguished (Fig. 3.3):

- Close-to-production race cars
- Prototypes
- Monoposti

Vehicles that are close to series production are characterized by the fact that their structure and other components originate to a significant extent from series production or are at least

Close-to-production race cars	Prototypes	Monoposti
Racing vehicles, the structure and components of which originate to a significant extent from series production or are at least based on it.	Racing vehicles, the structure and components of which have been specially developed for motor sport and which have closed wheel arches.	Open and single-seater racing vehicles whose structure and components have been specially developed for motor sport and which have free-standing wheels.

Fig. 3.3 Classification of racing vehicles

based on it. The so-called homologation[1] of these vehicles and the permissible modifications to the homologation vehicle are strictly regulated. Examples of production-based racing vehicles are the FIA GT3 category or the TCR category. Various national racing series and one-make cups are held according to these regulations. Close-to-production racing cars are also particularly suitable for mass sports, as shown, for example, by their use in the VLN series.

Prototypes are racing vehicles whose structure and components have been developed exclusively for motorsport. As an external feature, they have closed wheel arches. The prototype category includes, among others, the LMP1 and LMP2 cars used in the WEC as

---

[1] Homologation (<u>French</u> *homologation* "approval, admission; type testing"; from Ancient Greek ὁμολογεῖν *homologein* "to agree, to conform"; here in the sense of "to comply with the regulations") = registration and approval of racing vehicles or also <u>racing circuits</u> by national and international motorsport authorities. Homologation is the formal prerequisite in many motorsport categories to be able to participate in competitions.

well as the Class 1 racing cars used in the DTM. The stock cars used in the Monster Energy Cup also belong to this category.

Monoposti are also racing cars whose structure and components have been developed exclusively for motorsport. A monoposto is a single-seater and has an open driver's seat and free-standing wheels. The best known vehicles in this category are Formula 1 cars. Indycars and Formula E cars also fall into this category.

Depending on the intended use, further distinctions are possible, e.g. between circuit, endurance, off-road or record vehicles. Since this book concentrates on circuit vehicles, no further categorization is made.

A complete treatment of the technical regulations is not possible within the scope of this book. In the following, some basic features and basic concepts of various technical regulations are discussed.

▶   The technical and sporting regulations of a race series can usually be downloaded free of charge from the homepage of the responsible associations.

3.2.1   FIA Technical Regulations for Formula One Cars (2018)

Formula 1 is considered the premier class of motorsport, which is largely due to the impressive technical specifications and performance of these vehicles (Fig. 3.4). The technical regulations of Formula 1 for the 2018 season comprise only 105 pages. Figure 3.5 shows its structure. Basically, most technical regulations are structured according to this or a comparable scheme.

Article 1 ("Definitions") first defines the terminology. One of the most important definitions is, for example, the description of the bodywork. The bodywork includes all parts that belong to the sprung masses and are in contact with the external airflow. Excluded from this are cameras, camera housings and parts that ensure the mechanical function of the engine and transmission. Airbox, radiator and exhaust system are considered as part of the bodywork. Article 1 is of particular importance for the regulation of aerodynamics. Weight is defined as the weight of the vehicle and the driver together with his equipment.

Article 2 contains important general principles. One of these is that the stewards[2] are allowed to exclude dangerous constructions from the competition. An example of the application of this paragraph can be found in 2001, when Arrows installed a high-mounted auxiliary wing on the nose of its vehicle, which was in the driver's field of vision. This paragraph prevents the extreme exploitation of loopholes in the regulations.

---

[2] Steward = race observer or technical delegate, who controls controversial situations as well as the technical conformity of a racing vehicle and, if necessary, pronounces penalties.

Fig. 3.4 Key technical data of a Formula 1 car

FiA Formula 1 Technical Regulations (2018)	
1 Definitions	11 Brake System
2 General Principles	12 Wheels and Tyres
3 Bodywork and Dimensions	13 Cockpit
4 Weight	14 Safety Equipment
5 Power Unit	15 Car Construction
6 Fuel System	16 Impact Testing
7 Oil and Coolant System and Charge Air Cooling	17 Roll Structure Testing
8 Electrical Systems	18 Static Load Testing
9 Transmission System	19 Fuel
10 Suspension and Steering Systems	20 Engine Oil
	21 Television Cameras and Timing Transponders
	22 Final Text

Fig. 3.5 Structure of the Formula 1 technical regulations (according to [1])

Article 3 specifies the nature of the bodywork and the basic vehicle dimensions ("Bodywork and Dimensions"). Among other things, a maximum width of 2000 mm is specified, whereby the width of the body between the axle center lines may only be 1600 mm. The maximum height above a reference plane must not exceed 950 mm. Behind the rear axle centerline the height may only be 800 mm, limiting the maximum height of the rear wing. The maximum front overhang is 1200 mm. The dimensions of the front wing are defined by the area marked in Fig. 3.6. This gives the front wing a sweep which gives the cars a more aggressive appearance. This has no aerodynamic function here. The maximum length of a Formula 1 car is not regulated. Regulations governing the specification of the front and rear wings and the underbody, which are also part of article 2, have a significant influence on the aerodynamic performance of the vehicle.

Article 4 sets the minimum weight of the vehicle at 733 kg. This relatively high minimum weight takes into account the high additional weight due to the components of

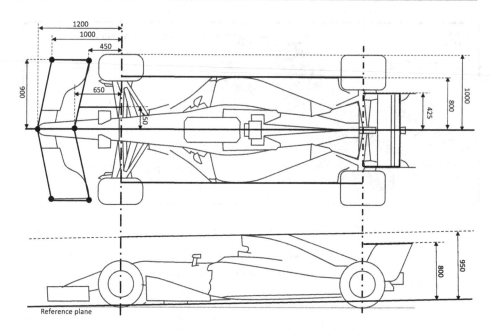

Fig. 3.6 Dimensions of an F1 vehicle

the hybrid system (batteries and electric motors). It is intended to prevent extreme competition in the field of lightweight construction or an advantage due to unhealthy underweight of the pilot. The weight on the front and rear axles must not fall below 333 and 393 kg respectively, which means that the weight distribution is fixed to within a small tolerance.

Article 5 specifies the design of the internal combustion engine ("power unit"), whereby almost all characteristic features of the engine are prescribed. There are therefore hardly any visible conceptual differences. The internal combustion engine must be a four-stroke Otto engine with six cylinders and a maximum displacement of 1.6 liters. The cylinder bank angle must be 90 °. Two intake and exhaust valves must be used per cylinder. The maximum engine speed is set at 15,000 rpm. A single-stage monoturbo must be used. The maximum petrol flow rate is 100 kg/h. The minimum weight of the drive unit including the energy recovery systems is 145 kg.

▶   In the 1990s, V8, V10 and V12 engines were used in Formula 1, so that there was still
     a real concept competition on the engine side. Although other concepts were permit-
     ted, however, only V10 engines have been used by all teams since 1998. The number
     of cylinders and engine design has been clearly prescribed in Formula 1 since 2000.
     Between 2000 and 2005, a V10 engine with a maximum displacement of 3.0 liters
     had to be used. However, the cylinder bank angle was still freely selectable. There
     were approaches between 72 ° and 111 °. The advantage of a high bank angle is the
     low engine center of gravity. However, depending on the geometric constraints of the

underbody, a high cylinder bank angle prevents engine components from entering the diffuser channels. Between 2006 and 2013, a V8 engine with a 90 ° cylinder bank angle and a maximum displacement of 2.4 l was to be used. Other characteristic features of the engine were also regulated, so that since 2006 at the latest, all F1 cars have had an identical engine concept.

Article 6 regulates the fuel system. The use of a safety fuel tank in accordance with FIA standards is mandatory. Among other things, it is also required that the tank is housed within the safety cell.

Article 7 specifies the requirements for the lubrication and cooling system as well as the charge air cooling of the turbo engine ("Oil and Coolant System and Charge Air Cooling"). The area in which fuel pipes and reservoirs may be located is restricted.

Article 9 ("Transmission System") requires two wheels to be driven and prohibits the use of traction controls. Eight forward gears and one reverse gear are prescribed. The minimum width and weight of the gears and the use of steel as a material are also specified. Drive torque distribution between the two rear wheels may be provided by a limited slip differential. Other systems for distributing the drive torque between the two rear wheels are not permitted. Torque distribution systems are generally prohibited on the front axle.

Article 10 ("Suspension and Systems") essentially prohibits the use of active suspension systems, such as rear axle steering or camber adjustment. The energy source of the power steering must not be electrical. The use and condition of the safety ropes for the wheel and wheel carrier are also prescribed.

Article 11 regulates the configuration of the braking system ("Brake System"), which is of great importance in connection with the hybridized powertrain. Two brake circuits are prescribed, which are operated via an identical pedal. Active brake boosting and an anti-lock braking system are not permitted. At the rear axle circuit, the brake pressure set by the driver via the pedal may be measured and set to the rear axle via an actuator. This serves to coordinate the braking effect of the electric motors in generator mode and the riction brake. This is a brake-by-wire system, but its use is severely restricted (see Sect. 7.10.3).

Article 12 regulates the condition of tires and rims ("Wheels and Tires"). The use of exactly four wheels is mandatory. Any modification to the tires provided by the tire supplier is prohibited.

Article 13 prescribes important characteristics for the design of the driver's position ("cockpit"). These features include, among others, the opening cross-section and the position of the pedals behind the front wheel center line.

Article 14 contains requirements for the safety and restraint systems on the vehicle ("Safety Systems"). Among other things, it regulates the characteristics of the safety belts and the on-board fire extinguishing system.

Article 15 lays down some general rules on the construction of the vehicle ("car construction"). Among other things, it regulates which materials are permissible for the construction of the vehicle in order to prevent the use of excessively expensive materials.

Articles 16 to 18 describe the crash and load tests that the vehicle must pass in order to be approved for racing.

Article 19 regulates the chemical composition of the fuel ("Fuel").

Article 20 ("Engine Oil") has been added for the 2018 season, as it was suspected in the previous season that some teams injected part of their oil volume into the combustion chamber to increase performance. This use of oil is prohibited by this article.

Article 21 ("Television Cameras and Timing Transponders") sets out requirements for the installation of cameras and transponders.

Article 22 ("Final Text") states that the English version shall prevail for the interpretation of the Regulations.

3.2.2   FIA Technical Regulations for LMP1 Hybrid Cars (2017)

The key data of an LMP1 hybrid vehicle are shown in Fig. 3.7. The technical regulations for the LMP1 vehicles participating in the WEC and the 24 h of Le Mans (Fig. 3.8) are structured similarly to the technical regulations for Formula 1 vehicles, although the vehicles differ significantly in their nature. Here, too, excerpts of some characteristic elements of the regulations are reproduced and commented on.

Article 1 ("Definitions") comprises the general definitions of terms. A distinction is made between "Le Mans Prototypes 1" *without* and *with* energy recovery systems. Le Mans Prototypes without energy recovery systems may only be used by privateers. The vehicles must have closed cockpits. Also of great importance is the definition of mechanical components, which includes all components required for the powertrain, suspension, steering and brakes. The vehicle body includes all sprung components that are in contact with the external airflow. It excludes components responsible for the operation of the engine, driveline and chassis. The gross vehicle weight is determined without the driver and fuel.

Article 2 ("Regulations") provides important guidelines for the interpretation of the technical regulations. The guiding principle is that anything that is not expressly permitted is prohibited. Such a statement can usually be found in the regulations of near-series racing vehicles in particular. Active systems are prohibited if they are not expressly permitted. This mainly includes automatic transmissions and chassis control systems of any kind. The vehicle must be under the control of the driver at all times. No materials may be used that have a higher specific modulus of elasticity than 400 GPa/(g/cm^3). Minimum wall thicknesses are prescribed for certain materials. This is to discourage the high cost of competing in the field of lightweight construction. Safety-related changes in the regulations can be made at any time and without advance notice.

Article 3 ("Bodywork and Dimensions") regulates the vehicle dimensions and the structure of the vehicle body and thus, above all, the aerodynamic characteristics of the vehicle. A maximum vehicle length of 4650 mm, a minimum width of 1800 mm and a maximum width of 1900 mm as well as a maximum height of 1050 mm above the reference

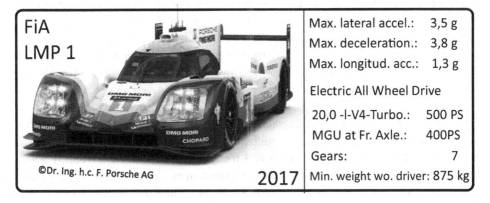

Fig. 3.7 Key data of an LMP1 hydride vehicle

FiA Prototype LMP1 Technical Regulations (2017)	
1 Definitions	11 Transmission System
2 Regulations	12 Suspension
3 Bodywork and Dimensions	13 Steering
	14 Brake System
4 Weight	15 Wheels & Tyres
5 Power Unit	16 Cockpit
6 Fuel System	17 Safety Equipment
7 Oil System	18 Safety Structures
8 Hydraulic System	19 Fuel
9 Coolant System	20 Final Text Disputes
10 Conventional Electric Equipment	

Fig. 3.8 Structure of the technical regulations for Le Mans prototypes (according to [2])

plane are specified. The maximum front overhang is limited to 1000 mm and the rear overhang to 750 mm. In plan view, side view, front view and rear view, the mechanical components defined in Article 1 shall not be visible. This Article ensures the characteristic appearance of these vehicles with closed wheel arches. Some exceptions to this rule are also laid down. The rear diffuser must consist only of one inclined flat surface. The maximum height of any point on this surface must not exceed a distance of 150 mm from the reference plane. No component of the sprung masses may be below the reference plane. No component other than the approved front and rear wings may have a wing profile. The rear wing may consist of a maximum of two wing sections. A vertical fin is mandatory to increase aerodynamic stability. Once the vehicle is in motion, aerodynamic components must not be movable. Also in connection with prototypes, this article significantly determines the aerodynamic performance potential of the vehicles.

Article 4 ("Weight") sets the minimum weight (including 3 kg heavy cameras) for LMP1 hybrid cars at 878 kg and for LMP1 cars at 858 kg. The vehicle must be designed in such a way that an additional weight of 20 kg can be installed. This additional weight is a component of the "Balance of Performance".

Article 5 ("Power Unit") describes the permissible characteristics of the engine and the energy recovery systems. The engine concept is, in contrast to Formula 1, freely selectable except for some predefined boundary conditions, which depend, among other things, on the recuperation class. The engine displacement is freely selectable for LMP1 hybrid vehicles. For LMP1 vehicles without an energy recovery system, it may be a maximum of 5.5 liters. It must be a four-stroke internal combustion engine or a diesel engine with reciprocating pistons. A maximum of two intake and exhaust valves per cylinder may be used. Electro-magnetic valve actuation is prohibited. Variable valve lift, valve opening and valve closing times are not permitted. Geometric specifications are not made. However, the lap-related volume flow and energy consumption are regulated, which, according to Fig. 3.9, depend on the recuperation class in which the vehicle is classified. In addition to the combustion engine, the use of two further motor-generator units (MGU: Motor Generator Unit) or electric motors is permitted. In the Porsche 919 Hybrid, a combination of a mechanical MGU on the front axle and an MGU coupled to the turbocharger was used. Toyota used one mechanical MGU each on the front and rear axles in the TS050. The LMP1 hybrids thus have the most complex powertrains ever used in motorsport.

Article 6 ("Fuel System") regulates the nature of the fuel system. The maximum permissible tank volume is limited. The use of a safety tank according to FIA standards is also mandatory for these vehicles. A firewall must separate the cockpit and the fuel tank. There must be no fuel pipes in the cockpit.

Article 7 ("Oil System") sets out requirements for the design of the oil system, primarily for safety in the event of an accident.

Article 8 ("Hydraulic System") limits the oil pressure of the hydraulic system to 300 bar and prescribes the use of self-sealing pipes.

Article 9 ("Coolant System") deals with the cooling system. The pressure in the coolant system must not exceed 4.75 bar. Only coolant pipes belonging to the energy recovery system are allowed in the cockpit.

Article 10 ("Conventional Electrical Equipment") regulates the conventional on-board electrical system. An essential part of this article are, for example, the specifications for the condition of the lighting system.

Article 11 ("Transmission System") contains technical specifications for the drive train. Dual clutch, automatic and CVT transmissions are prohibited. A maximum of seven forward gears may be used. Only one gear may be engaged at any given time. Gears must be made of steel. The use of passive limited slip differentials is permitted on the front and rear axles.

a)

	LMP1	LMP1 hybrid			
Released Energy [MJ/Lap]	0	<2	<4	<6	<8
Max Cubic capacity [l]	5,0	-	-	-	-
Car Mass [kg]	855	878	878	878	878
Released Power [kW]	0	<300	<300	<300	<300
Petrol Energy [MJ/Lap]	157,2	136,3	131,7	127,2	124,9
Max Petrol Flow [kg/h]	101,4	87,9	85,0	82,0	80,6
Maximum Petrol Capacity [l]	75,0	62,3	62,3	62,3	62,3
Diesel Energy [MJ/Lap]	147,1	130,4	126,2	121,9	117,4
Max Diesel Flow [kg/h]	86,4	76,3	73,6	71,1	68,5
Maximum Diesel Capacity [l]	-	50,1	50,1	50,1	50,1

b)

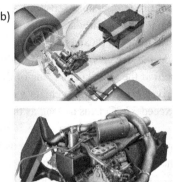

Fig. 3.9 (**a**) Drive concept-dependent regulations for LMP1 vehicles, (**b**) Motor-generator units of a Porsche 919 (2014). (Courtesy of © Dr. Ing. h.c. F. Porsche AG 2018. All Rights Reserved)

Article 12 ("Suspension") regulates the suspension of the chassis. Any construction that can be used to adjust the ground clearance is prohibited. The adjustment of springs, dampers and stabilizers may not be changed from the driver's seat. The chassis components must be made of a homogeneous metallic material.

Article 13 ("Steering") covers the steering system. A mechanical connection between the steering wheel and the front wheels is mandatory. Steer-by-wire systems are not permitted. The only permissible function of the power steering system is the reduction of steering forces. In contrast to Formula 1, electromechanical steering systems are also permitted. A quick release must be used on the steering wheel.

Article 14 ("Brake System") places some restrictions on the braking system, which can otherwise be designed relatively freely. The use of at least two brake circuits operated by the same pedal is mandatory. The use of a mechanical system to adjust the brake force distribution is permitted. Only one brake caliper with a maximum of six brake pistons and one brake disc each may be used per wheel. Anti-lock braking systems and brake boosters are prohibited. In conjunction with energy recovery systems, the use of brake-by-wire systems is permitted. In this case, however, the total braking force must always be clearly assigned to the pedal actuation. Additional auxiliary functions are prohibited. In the event of failure of the brake-by-wire system, the braking effect must still be achieved which the driver would achieve with a hydraulic brake application and a pure pedal actuation. Such a braking system is presented in paragraph 7.10.3.

Article 15 ("Wheels & Tires") states that the number of wheels is fixed at four. Identical tires must be used on the left and right side of the vehicle. Only one specification, which must be homologated, may be used on the front and rear axles during the entire season. The permissible tire dimensions and the minimum weights of the wheels are also prescribed. The use of wheel tethers according to FIA standards is mandatory. The use of tire pressure and tire temperature sensors is strongly recommended. This serves to detect possible tire damage at an early stage.

Article 16 ("Cockpit") contains the specifications for the condition of the driver's seat and the driver's cab. Among other things, it is prescribed that the driver's feet must be behind the wheel center line of the front axle. The driver's position must be such that the driver can leave the vehicle within 7 seconds. The temperature in the cockpit must not exceed 32 °Celsius as long as the outside temperature does not exceed 25 °Celsius. Above an outside temperature of 25 °Celsius, the temperature in the cockpit must not be higher than the outside temperature plus 7 °Celsius. Further specifications concern, for example, the driver's field of vision.

Article 17 ("Safety Equipment") describes the safety equipment of the driver and the vehicle.

Article 18 ("Safety Structures") sets out important requirements for the structure of the vehicle in order to minimize the risk of injury to the driver in the event of an accident.

Article 19 ("Fuel") stipulates that the race organizer shall provide only one type of petrol and one type of diesel, which must be commercially available. Petrol is blended with 20% biofuel and diesel with 10% biofuel.

Article 20 ("Final Text – Disputes") states that the French text is decisive for the implementation and interpretation of the technical regulations.

The regulations for LMP1 hybrid cars have only been adjusted minimally for 2018. Following the withdrawal of Audi and Porsche at the end of 2016 and 2017, only Toyota remained in the highest class of LMP1 hybrid cars for 2018. A comprehensive reform of the regulations has been announced for 2020, as the complex hybrid systems and the associated development costs mean that entry into the highest class of sports car prototypes now appears very unattractive.

3.2.3 Technical Regulations of Other Racing Series

Formula 1 cars and Le Mans prototypes today have the highest degree of freedom in technical development, which is associated with corresponding costs. One measure to defuse the conflict of objectives between cost control and exciting and fair competition is the standardization of selected components or modules, such as is applied to the Class 1 cars of the DTM. The 2016 technical regulations contain a list of a total of 90 standardized components, including the safety cell, gearbox, power-assisted steering, pedals or rear wing profile. This puts the DTM somewhere between an open-wheel and a spec series. It is worth mentioning that the DTM cars are pure-bred prototypes, although they have a silhouette reminiscent of the respective production car. Apart from that, the technical regulations are structured similarly to the previously described regulations of Formula 1 and the WEC.

The FIA has drawn up numerous articles and technical regulations for the use of production-based racing vehicles. The most important of these include:

- Article 251: Classifications and Definitions.
- Article 252: General Prescriptions for Production Cars [Group N], Touring Cars [Group A] and GT Production Cars [Group R-GT].
- Article 253: 'Safety Equipment (Groups N, A, R-GT)'.
- Article 254: Specific Regulations for Production Cars [Group N].
- Article 255: Specific Regulations for Touring Cars (Group A).
- Article 257A: Technical Regulations for Cup Grand Touring Cars [Group GT3].
- Article 263: "Specific Regulations for Modified Production Cars on Circuits [Super-2000]".

Another example of an internationally significant series-based racing car category is the technical regulations of the TCR (Touring Car Racing), which refer at various points to the articles issued by the FIA and incorporate them into the regulations.

Homologation is usually required for the approval of a production-based racing vehicle. Homologation comes from the Greek and means conformity. In motorsport, the term describes that the characteristics of the racing vehicle are matched against a production vehicle (the homologation vehicle). For homologation in groups A and N, a homologation vehicle must have been produced in at least 2500 identical units in 12 consecutive months. The number of seats must be at least four. The general regulations for this category of racing vehicle state that any modification with respect to the homologation vehicle is prohibited, unless it is expressly authorized by the regulations. The technical regulations, which are otherwise structured analogously to the regulations already discussed, therefore summarize the permissible modifications.

Further concrete contents of a technical regulation will also be dealt with in connection with the consideration of the technical components of a racing vehicle beginning in the following.

References

1. FIA: Formula One Technical Regulations. Federation internationale de l'automobile, Place de la Concorde (2018)
2. FIA: Technical Regulations for LMP1 Prototype Hybrid. Federation internationale de l'automobile, Place de la Concorde (2017)

Racing Tires

The tires are the only link between the vehicle and the road. The contact surface between the tire and the road, the tirecontact patch, is where all the forces required for propulsion, braking and steering are transmitted (Fig. 4.1). The properties of the tires therefore shape a vehicle's dynamic driving behavior in a decisive way. A serious examination of vehicle dynamics requires the following principle to be observed:

▶ If you want to understand the dynamic behavior of a vehicle and the tuning of its suspension, you first have to understand the tire.

For this reason, this chapter is devoted in detail to the characteristic properties of the tire.

4.1 Friction Circle

The simplest form of describing the basic behavior of a tire is the friction circle shown in Fig. 4.2, which goes back to Wunibald Kamm . The friction circle shows that the maximum force $F_{xy, max}$ (which may be also called the "grip level" of the tire) that can be transmitted from the tire to the road surface lies within a circle. The maximum force that can be transmitted is the product of the wheel load F_z, i.e. the force with which the tire is pressed vertically onto the road surface, and the coefficient of friction μ between the road surface and the tire. Longitudinal forces decelerate or accelerate the vehicle. Lateral forces cause a turning or yawing motion of the vehicle and support the resulting lateral acceleration (see Chap. 5). Mathematically, the friction circle states that the vectorial sum of longitudinal and lateral forces cannot exceed the product of wheel load and coefficient of friction. The following applies:

Fig. 4.1 Transfer of tire forces to the road surface. (Courtesy of © Daimler AG 2018. All Rights Reserved)

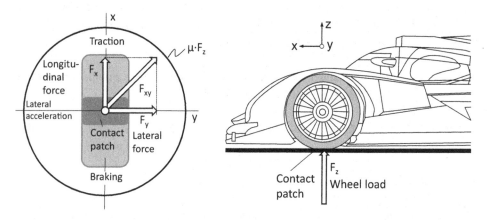

Fig. 4.2 Friction circle

$$\mu \cdot F_z = F_{xy,\,max} \leq \sqrt{F_x^{\,2} + F_y^{\,2}}. \qquad (4.1)$$

However, this relationship only applies as long as the tire does not spin or lock. The wheel load of the vehicle is determined by its mass geometry, the chassis set-up, its aerodynamic properties (see Chap. 6) and the driving condition. The coefficient of friction μ depends on the characteristics of the tire and the condition of the road surface. These relationships are explained in the following paragraph. Typical coefficients of friction between the tires and the road surface are:

- $\mu \approx 1$ for profiled passenger car tires on dry road surfaces
- $\mu \approx 1.5\text{-}1.7$ for racing slicks on dry road surfaces
- $\mu \approx 2.2$ for 1980s qualifying tire (slick) on dry road surface
- $\mu > 5$ for drag racing rear tires on dry roads

In the following chapters, the friction circle will be used again and again to explain the basic interactions between tires, chassis and aerodynamics of a racing vehicle.

4.2 Mechanisms of Force Transmission

The force F_{xy} transmitted from the tire to the road is a frictional force. The tire as an elastomer uses three mechanisms to transmit force (Fig. 4.3a). The frictional force transmitted from the tire to the roadway is composed of adhesion, deformation, and wear resistance forces [1]. The following relationship therefore applies to the force transferred to the road surface:

$$\mu \cdot F_z = F_{xy} = F_{Adh\ddot{a}sion} + F_{Deformation} + F_{Diffusion\backslash ss}. \tag{4.2}$$

The dominant components in force transmission are the adhesion and deformation forces, which is why the following considerations focus on these two mechanisms. *Adhesion forces* arise from molecular bonding forces between the road surface and the tire. Chemically, this is the same process as adhesive tape. The tire literally "sticks" to the road surface. Adhesion forces are proportional to the contact area and also depend on material properties, surface temperature and surface pressure. In this context, the contact area is not synonymous with the contact patch with which the tire stands on the road surface as seen from above, but, as Fig. 4.3b indicates, the microscopic properties of the road surface must also be taken into account. The elastic tire is pressed into the micro-roughness of the road surface by the wheel load due to its compliance. In this process, its tread penetrates deeper into the pavement the higher the wheel load, which is illustrated in Fig. 4.3c. Deep penetration of the tread into the road surface is favored by a soft tire compound. Deeper tire penetration increases the contact area between the tire rubber and the road surface, which consequently increases the adhesion forces and thus the grip level of the tire. Under dry road conditions, adhesion forces contribute about two-thirds of the total transmissible force. However, adhesion forces decrease dramatically when the road surface is wetted with water or covered with dust or similar.

On completely smooth surfaces, frictional forces are transmitted practically exclusively by adhesion. On rough surfaces such as a roadway, the second important component is the *deformation force*. The generation and effect of deformation forces is shown in Fig. 4.4. The deformation of a stationary tire on the road leads to a symmetrical pressure distribution (Fig. 4.4a) in the contact area between tire and road. No resultant force is transmitted in the plane of the road and the vehicle remains stationary. When the wheel is turning, the pressure on the incoming side of the contact surface is higher than on the outgoing side.

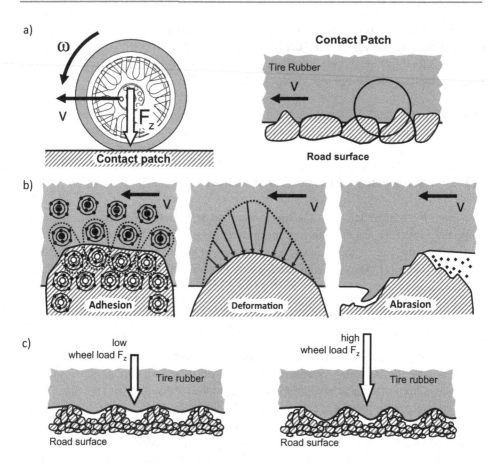

Fig. 4.3 (a) Tire contact patch and microscopic road surface, (b) Mechanisms of force transmission at the tire (based on [1]), (c) Influence of wheel load on the contact area

This pressure difference results in a longitudinal and/or lateral force with which the tire is supported relative to the road surface. The higher the maximum longitudinal and lateral forces can become, the greater the longitudinal accelerations, decelerations and lateral accelerations that can be achieved by a vehicle.

Also in the mechanism of deformation, a higher penetration depth leads to a larger contact area on which the differential pressures act and generate corresponding longitudinal and lateral forces. In this context, the viscoelasticity of the tire plays an important role. Viscoelasticity means that the tire does not return to its original shape immediately after a deformation, such as occurs when penetrating and driving over road surface roughness, but with a time delay. The higher the viscoelasticity of a tire, the longer it takes to return to its original shape. The case of increased viscoelasticity is shown in Fig. 4.4c. Due to the delayed recovery, the rubber cannot transmit pressure on the outgoing side. The pressure difference to the incoming edge increases, and the resulting force and thus the performance of the tire increases. For this reason, racing tires have a significantly higher viscoelasticity

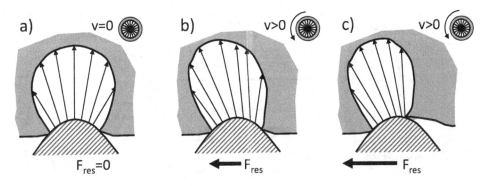

Fig. 4.4 Force transmission by deformation and viscoelasticity (based on [2]). (**a**) stationary wheel, (**b**) rolling wheel with low viscoelasticity and (**c**) rolling wheel with increased viscoelasticity

than tires for production vehicles. On dry roads, deformation forces contribute about one third of the total force of the tire.

As mentioned above, on wet roads hardly any adhesion forces can be transmitted and thus practically only deformation forces are generated. For this reason, wet tires have the softest compounds of all racing tires, which however makes them susceptible to overheating in drying conditions. Under wet track conditions, a deviation from the dry racing line is often observed, which can be explained by Fig. 4.5 and the roughnesses on the dry racing and wet racing line shown there. Due to the frequent use of the dry racing line, the roughnesses of the asphalt are polished and rubber residues are increasingly deposited in the gaps. For this reason, the penetration of the tread into the road surface cannot be as deep as required for optimum use of the deformation potential. On the wet racing line, the roughness of the road surface is usually more pronounced, so that it offers a higher grip level. As with standard tires, contact with the road surface is ensured in wet tires by the tread. The tread displaces the water in front of the tire to the sides, and early onset of aquaplaning is prevented. The wet tires of an F1 car displace about 100 l of water per second at a speed of 300 km/h and a 3 mm high water film [3].

In summary, it can be stated from the previous considerations that the contact area between tire and road is a decisive factor for the performance of the tire. This also shows why a slick tire can transmit more force than a treaded tire with an otherwise identical rubber compound (under dry conditions). For optimum performance, the use of the widest possible tires is advantageous for the same reason. The principle that increased contact area leads to higher performance was exploited by the FIA in 1998-2008 when it reduced the contact area by prescribing grooved tires, thus lowering the grip level of the tires by about 18% (Fig. 4.6). However, the contact area is also significantly influenced by the overall tire design, which is discussed in Sect. 4.7.

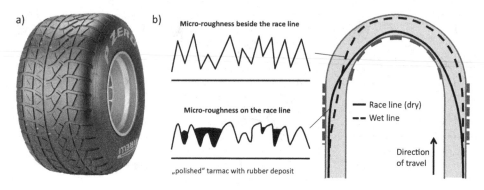

Fig. 4.5 (**a**) Wet tire. (Courtesy of © Pirelli & C. S.p.A. 2018. All Rights Reserved), (**b**) Comparison of il dry and wet racing line

Fig. 4.6 Contact patch of grooved and slick tires on a BMW Sauber F1.08 and BMW Sauber F1.09. (Photos courtesy of © BMW AG 2018. All Rights Reserved)

4.3   Lateral and Longitudinal Forces

The generation of longitudinal and lateral forces by the mechanisms just explained requires a relative or sliding speed between the tire and the road surface. This relationship can be illustrated by considering a flat track tire test rig (Fig. 4.7). The flat track simulates the road surface and the vehicle speed v. By means of an actuator, the wheel is pressed onto the flat

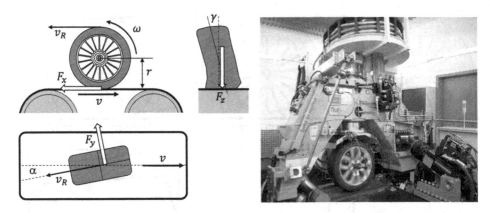

Fig. 4.7 Kinematic variables on a tire test rig. (Photo on the right courtesy of © fka GmbH 2018. All Rights Reserved)

track with a defined wheel load F_z. The actuator can simultaneously apply drive and braking torques as well as a defined angle of rotation α between the wheel center plane and the direction of travel of the treadmill. The angle of rotation α is referred to as the slip angle. Sensors are integrated into the actuator which record the forces and torques in all three spatial directions as well as the wheel speed ω or the longitudinl speed v_R. With the help of these measured variables, the lateral and longitudinal force behavior can be characterized as a function of slip, slip angle and wheel load.

An essential characteristic of the tire is the dependence of the generated longitudinal force on the traction or brake slip. Figure 4.8 illustrates the formation of the longitudinal force-slip diagram for a slip angle of zero degrees. The wheel is free of lateral force. The driven wheel and the braked wheel are considered. For the driven wheel, the sliding speed of the tire is caused by the wheel rolling faster, or having a higher longitudinal speed, than its actual speed over ground. The amount of this sliding speed is given as the traction slip, for which the following definition applies:

$$S_{an} = \frac{\omega \cdot r - v}{\omega \cdot r} = \frac{v_R - v}{v_R} = 1 - \frac{v}{\omega \cdot r} = 1 - \frac{v}{v_R}. \tag{4.3}$$

Based on this definition, a slip value of $S_{an} = 1$ means that the wheel is spinning. Similarly, when braking, the tire rolls slower than its ground speed. The resulting brake slip is defined as:

$$S_{br} = \frac{v - \omega \cdot r}{v} = \frac{v - v_R}{v} = 1 - \frac{\omega \cdot r}{v} = 1 - \frac{v_R}{v}. \tag{4.4}$$

A slip value of $S_{br} = 1$ therefore corresponds to a locking wheel. When measuring the tire, various drive and braking torques are applied via the actuator for a constant wheel load, and the resulting longitudinal forces and the slip occurring are measured. The measurement

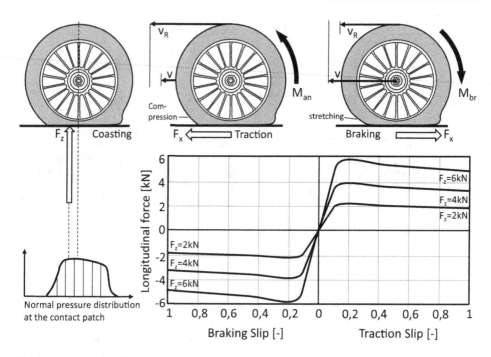

Fig. 4.8 Origin of the longitudianl force-slip diagram

points are entered in the longitudinal force-slip diagram. This measurement procedure is repeated for different wheel loads, resulting in the diagram in Fig. 4.8.

In addition, Fig. 4.8 shows deformation of the contact patch and the normal pressure distribution of a free-rolling wheel. It can be seen that the center of gravity of this pressure distribution lies in front of the wheel center. For this reason, the wheel load on the road is also supported in front of the wheel center. This results in a moment which counteracts the rolling motion of the tire. This phenomenon is called rolling resistance. A rolling resistance force can be calculated from this resistance moment. This rolling resistance force is expressed in the form:

$$F_R = f_R \cdot F_z. \tag{4.5}$$

Here f_R is the rolling resistance coefficient of the tire. An efficient and fuel-efficient tire should have a coefficient of resistance f_R that is as low as possible. In the context of this book, however, rolling resistance is only of secondary importance, which is why we will not go into this topic in depth.

Figure 4.9 illustrates the formation of the lateral force-slip diagram, which is formed in a similar way to the longitudinal force-slip diagram. The wheel load is kept constant and the wheel is deflected by defined slip angles with respect to the flat belt. During this process, the contact patch adheres to the flat belt and the sidewall deforms due to the skew of the rim. The deformation between the contact patch and the rim can be thought of as the winding of a spring, with the resulting spring force corresponding to the lateral force F_y. This measured lateral force is entered in the lateral force slip diagram as a function of the slip angle for various wheel loads. The lateral force does not act at the wheel center, but is offset to the rear by the tire caster n_R. The lateral force and caster form the tire aligning torque M_z around the wheel center, which is also entered in a corresponding diagram. The tire aligning torque attempts to turn the tire back to its original position and plays an important role in connection with the steering system. From the lateral force slip diagram, various properties of the tire can be taken, which are of central importance for the design and tuning of the chassis. A very decisive property is the so-called wheel load sensitivity, the importance of which for driving dynamics is explained in detail in Sect. 5.4. Equation (4.1) gives the following definition for driving conditions with pure lateral force transmission ($F_x = 0$):

$$\mu_y = \frac{F_{y,\max}}{F_z}. \tag{4.6}$$

Wheel load sensitivity is generally understood as the dependence of the coefficient of adhesion μ on the wheel load F_z. Figure 4.10 shows the curve of the coefficient of adhesion μ_y as a function of the wheel load F_z, which is derived from the lateral force slip diagram. The lateral force slip diagram shown is based on tire data from the Ferrari F2000 [4]. It can be seen that the coefficient of adhesion μ_y decreases with increasing wheel load. As long as the coefficient of adhesion decreases slower than the wheel load increases, the diameter of the friction circle and thus the maximum transmissible force become larger. However, the coefficient of adhesion shows a progression. This means that the increase in lateral force with increasing wheel load is less than the loss due to the decrease in wheel load by the same amount. The effect of wheel load sensitivity also occurs in connection with longitudinal forces, but is generally not as pronounced there.

In addition to the maximum adhesion potential μ, the tire has other properties which are important for the driving behavior of production and racing vehicles. Some of these properties are shown in Fig. 4.11 by comparing four different tires. Shown are the lateral force slip diagram of the front and rear tires of a racing vehicle, a tire for production passenger cars, and a hypothetical tire. One distinguishing feature is the cornering stiffness. The cornering stiffness c_α is the initial slope of the lateral force as a function of the slip angle α. The higher the cornering stiffness, the more responsive the vehicle is to steering movements, but this may also make the vehicle more challenging to drive. A high stiffness also means that the losses generated in the tire (see Sect. 4.5) are lower. For these reasons, the highest possible cornering stiffness is sought for racing vehicles. The cornering stiffness of racing tires is significantly higher than that of production vehicles.

Furthermore, the course of the lateral force in the area of the maximum adhesion μ_y is significant for the drivability of a tire. The maximum adhesion of the racing tire is significantly higher than that of the passenger car tire. According to the previous

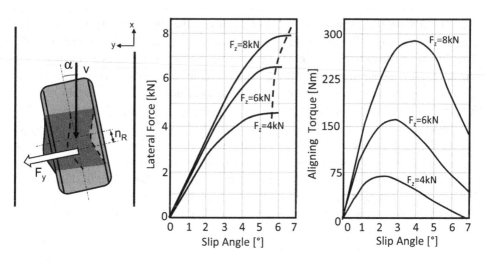

Fig. 4.9 Origin of the lateral force slip angle diagram

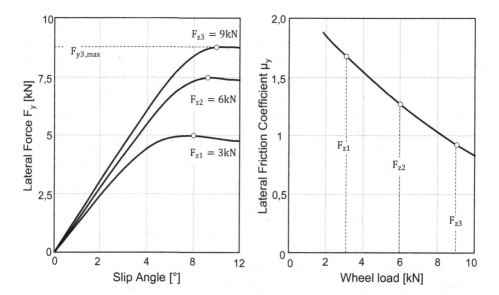

Fig. 4.10 Wheel load sensitivity

explanations, this can be explained by the use of larger tire dimensions, slick tires and a softer tread compound. The rear tire has a higher maximum adhesion than the front tires, which is due to the mixed tires commonly used for rear-axle driven racing vehicles. The maximum adhesion of the passenger car tire is only reached at higher slip angles. The driver is informed when the maximum adhesion is reached by a degressive increase in lateral force. The lateral force level remains nearly constant for a long time after reaching

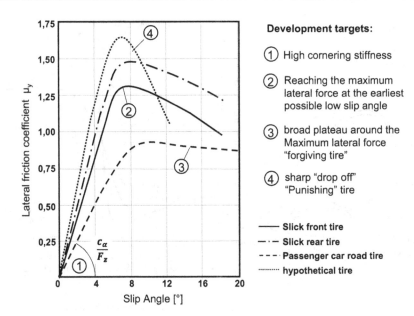

Fig. 4.11 Lateral force behavior of different tires and development targets

the peak of adhesion, which means that the tire is relatively forgiving of errors, since excessive steering angle reactions or too aggressive application of the accelerator or brake pedal do not lead to an abrupt loss of lateral force. The hypothetical tire, on the other hand, exhibits a very pronounced increase or decrease in lateral force both before and after peak adhesion. It is very challenging for the driver to balance his vehicle at this maximum. Driving errors quickly lead to understeer when steering beyond the "drop-off" at the front axle, while a "drop-off" at the rear axle leads to oversteer. In junior racing series and in series production cars, the aim is to have tire behavior that is more forgiving of mistakes, while in high-performance racing series tires with a pronounced "drop-off" are often specifically designed to add an additional element of excitement for the spectators.

Lonitudinal force-slip and lateral force-slip diagrams refer to driving conditions in which a vehicle is either only driven or braked or only steered. In order to permanently exploit the physical limits of driving on a race track, however, the vehicle is moved over long distances in combined driving conditions. In these driving conditions, longitudinal forces or slip and lateral forces or slip angles occur simultaneously. An example of this is acceleration from the apex of a curve. The tire behavior in these situations and its influence on the friction circle is shown in Fig. 4.12.

Based on the clongitudinal force-slip diagram, the mutual influence of the two force components is shown. With a constant wheel load and three constant slip angles ($2°$, $6°$ and $10°$), the longitudinal and lateral forces are measured as a function of slip. The maximum lateral force occurs at a slip angle of $10°$ and a slip of $S = 0$. This point a forms the intersection of the friction circle with the y-axis. If a driving or braking force is transmitted at the same time, the transmissible lateral force must decrease. Similarly, the

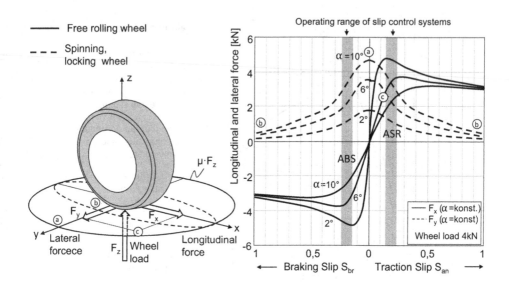

Fig. 4.12 Combined longitudinal and lateral force behavior

maximum transmissible longitudinal force decreases when the slip angle is present. The maximum transmittable total force moves along the arc of the friction circle, which is the case at point c, for example. If the driving or braking torques requested by the driver exceed the available adhesion potential, the driven wheel spins or the braked wheel locks. In this condition (S_{an} or S_{br} = 0, point b) the tire loses its lateral control capability. Only very small lateral forces can be transmitted, while the transmission of braking and driving forces is only slightly affected. In this case, the friction circle takes the form of an ellipse. In modern production cars and some racing cars, the traction control system (ASR) or the anti-lock braking system (ABS) prevent the wheels from spinning and locking. To do this, the slip is regulated to a range in which there is always sufficient potential for lateral control. In most high-performance racing series, however, the use of slip control systems is prohibited or only possible to a very limited extent, so that it is left to the skill of the driver to keep the tire within an optimum slip window.

The mutual dependencies of longitudinal and lateral forces as well as the wheel load can also be represented as a friction ellipse, as done in Fig. 4.13. The friction ellipse essentially corresponds to the representation of the friction circle. The shape of an ellipse results from the fact that for a given wheel load, the maximum longitudinal force is usually slightly greater than the maximum lateral force. The shape of the friction ellipse, or the force transfer behavior of the tire, is influenced by a number of variables, some of which have already been discussed in Sect. 4.2. These variables include the tire dimension and the tread compound. Other influencing variables are the tire construction and the rim geometry, which affect the deformation behavior and thus the contact patch. The friction ellipse also depends on the temperature and wear condition of the tire. Both variables also interact with

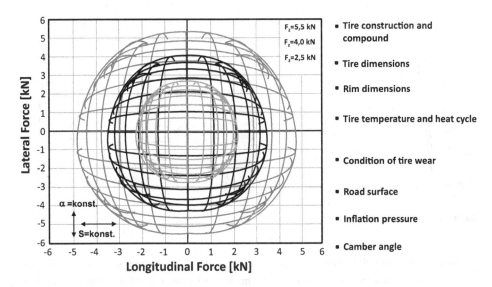

Fig. 4.13 Friction ellipse of the tire and influencing variables

the condition of the road surface. The size of the contact area between the tire and the road surface is also determined by the tire pressure and the wheel camber. There are various interactions between these variables, which are explained below.

4.4   Influence of Wheel Camber

The wheel camber is a parameter of the chassis geometry and is specified in the form of the camber angle γ. Figure 4.14 illustrates the definition of the camber angle. The camber angle is the angle between the longitudinal rim plane and the longitudinal vehicle plane vertical to the road. If the top of the tire rotates toward the center of the vehicle, the camber is negative. The camber angle measured on a stationary vehicle is called the static camber angle. When measuring a single tire on a test rig, the rim geometry must be taken into account when defining the sign, as asymmetric rims lead to asymmetric camber behavior of the tire.

An existing camber angle changes the deformation behavior of the rolling wheel. Figure 4.14 shows the comparison of the contact patch and the shear stress distribution of a cambered and a non-cambered wheel during straight-ahead travel. The contact patch becomes wider in the direction of the overturned upper side. On this side, the deformations in the contact patch caused by running-in and the resulting shear stresses are therefore retained for longer than on the opposite side. This asymmetrical shear stress distribution leads to the so-called camber side force, which also acts in the direction of the overturned upper side.

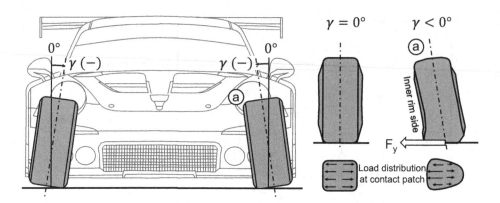

Fig. 4.14 Definition of wheel camber

Figure 4.15 shows the influence of the camber angle on the force transmission behavior of the tire. In the lateral force-camber diagram it can be seen that for a camber angle of $\alpha = 0\,^{\circ}$ there is a camber lateral force directed according to the sign of the camber. For small camber angles, this camber side force is initially proportional to the camber angle. For larger camber angles, however, the lateral camber force then increases only slightly. In particular, it must be taken into account that positive camber angles lead to a reduction in the maximum transmissible lateral force, while negative camber angles cause an increase in the maximum lateral force up to a certain limit. This effect can also be illustrated with the help of the friction circle. It can also be seen from the friction circle that camber angles – both positive and negative – lead to a reduction in the maximum transmissible longitudinal force. For the chassis design, this means that positive camber angles – especially at the wheels on the outside of the curve – should be avoided if possible. It should be noted that the lateral camber force of a wheel is dependent on the wheel load F_z. Excessively high camber angles can therefore lead to unsteady driving behavior on uneven road surfaces or due to wheel load fluctuations caused by the road surface.

In addition to the power transmission behavior, the camber angle also changes the wear and temperature behavior of the tire. The asymmetric load in the contact patch generally leads to higher wear of the tire, as it wears faster at the locally higher loaded points. Similarly, as Fig. 4.16 shows, the heat input occurs mainly at the inner edge of the tire. At the same time, the tire only releases heat to the road over a reduced width. As a result, the inner edge of the tire has a significantly higher temperature than the rest of the tread. Extreme static camber angles can cause local overheating at the inner edge of the tire, which then leads to blistering of the tire surface. Such a phenomenon was observed, for example, in 2011 on Sebastian Vettel's Red Bull RB7, which at the time had a static camber angle of about $-4.5\,^{\circ}$. This camber value significantly exceeded the values recommended by the tire manufacturer, so that the FIA was later forced to limit the maximum camber angles. Typical values for setting the static camber value are given in Sect. 7.9.

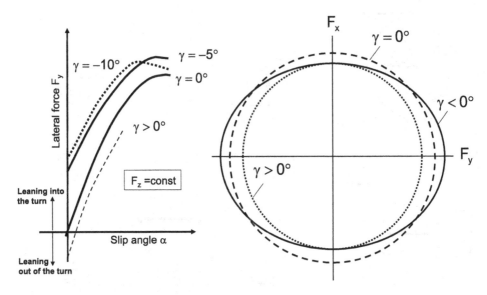

Fig. 4.15 Influence of the wheel camber on the force transmission behavior at a curve outer wheel (cf. [5])

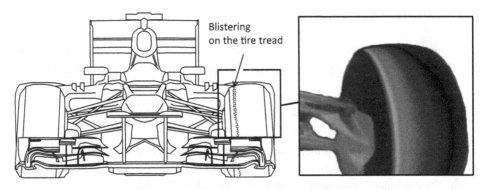

Fig. 4.16 Temperature distribution due to camber and bubble formation on the Red Bull RB7 (2011)

4.5   Power Losses and Heating

The sliding speeds between the tire and the road surface that occur when transmitting longitudinal and lateral forces lead to power losses that the tire absorbs in the form of heat. In the case of a wheel driven in a straight line, shown in Fig. 4.17, the conditions can be clearly explained. The drive power arriving at the two wheels of the rear axle is:

$$P_{an} = F_{xH} \cdot v_R = M_{an} \cdot \omega. \tag{4.7}$$

However, the power converted into propulsion by the vehicle is not calculated from the longitudinal speed of the wheel, but from the actual vehicle speed over ground. Therefore applies:

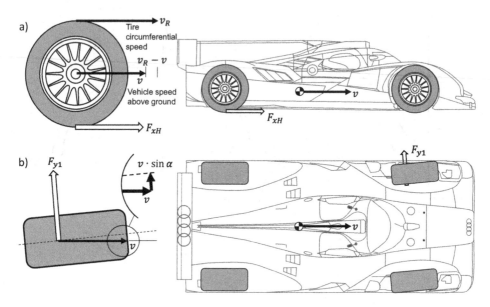

Fig. 4.17 Frictional losses (**a**) due to slip and (**b**) due to slip angles

$$P_{Fzg} = F_{xH} \cdot v. \tag{4.8}$$

The power loss converted to heat is equal to the difference between the incoming power and the power converted to propulsion. The power loss due to drive slip is obtained as:

$$P_{V,S_{an}} = P_{an} - P_{Fzg} = F_{xH} \cdot (v_R - v) = F_{xH} \cdot v \cdot \frac{S_{an}}{1 - S_{an}}. \tag{4.9}$$

The slip-induced power loss is therefore proportional to the generated circumferential force F_{xH} and the sliding speed between the tire and the road surface occurring in the contact patch. The dependence on slip is obtained with the aid of Eq. (4.3).

Analogous to the longitudinal force, the slipping-related power loss is proportional to the lateral force and the lateral sliding speed. According to Fig. 4.17b, this results for the left front wheel:

$$P_{V,\alpha} = F_{yi} \cdot v \cdot \sin(\alpha). \tag{4.10}$$

In order to get an idea of the power losses occurring at the tire, the example maneuvers shown in Fig. 4.18 are considered (cf. [3]). In case (a), the longitudinal acceleration on a straight line is considered, whereby the vehicle reaches a longitudinal acceleration of 1.2 g at a speed of 100 km/h. The rear tires generate a traction slip of 3% in this driving condition. At this point, air resistance can be neglected compared to inertial forces. The drive force required for acceleration is then given by:

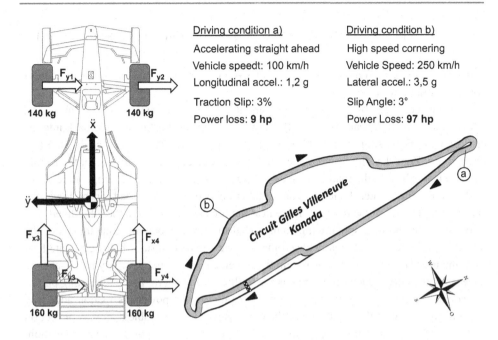

Driving condition a)

Accelerating straight ahead

Vehicle speedt: 100 km/h

Longitudinal accel.: 1,2 g

Traction Slip: 3%

Power loss: **9 hp**

Driving condition b)

High speed cornering

Vehicle Speed: 250 km/h

Lateral accel.: 3,5 g

Slip Angle: 3°

Power Loss: **97 hp**

Fig. 4.18 Examples for the calculation of power losses in tires. (cf. [3])

$$F_{x3} + F_{x4} = m \cdot \ddot{x}. \qquad (4.11)$$

Thus, according to Eq. (4.9), one obtains for the power loss:

$$P_{V,San} = m \cdot \ddot{x} \cdot v \cdot \frac{S_{an}}{1 - S_{an}} = 600 \ \text{kg} \cdot 1.2 \cdot 9.81 \frac{m}{s^2} \cdot \frac{100m}{3.6s} \cdot \frac{0.03}{1 - 0.03}$$

$$= 6.534 \ \text{kW}. \qquad (4.12)$$

During this acceleration process, a power loss of 3.267 kW or 4.5 hp is generated per tire, which the tire absorbs in the form of heat. In comparison, a (formerly) commercially available light bulb has an output of 80 W. Case (b) represents a high-speed curve that is traversed at 240 km/h. During this maneuver, the lateral acceleration is 3.5 g. The tires run at a slip angle of 3 ° to generate the lateral forces. The sum of the lateral forces on the individual wheels is calculated:

$$F_y = m \cdot \ddot{y}. \qquad (4.13)$$

According to Eq. (4.10), the power loss is given by:

$$P_{V,a} = m \cdot \ddot{y} \cdot v \cdot \sin(\alpha) = 600 \text{ kg} \cdot 3.5 \cdot 9.81 \frac{m}{s2} \cdot \frac{240m}{3.6s} \cdot \sin(3^\circ) = 71.878 \text{ kW.} \quad (4.14)$$

The power loss at all four wheels is over 71 kW or over 97 hp and is also referred to as cornering drag power.

The tire therefore absorbs enormous amounts of power through slip and slip angles, which it converts into heat. As Fig. 4.19 shows, some of this heat is released back into the environment. For a simplified view of the thermal processes, the tire can be divided into three zones. Zone 1 is the front contact area of the contact patch. In this zone, the rolling wheel releases heat to the road surface as it enters the contact area. The temperature drops in this zone. The greater the temperature gradient between tire and road, or the lower the surface temperature of the road, the more heat the tire can give off. On the other hand, this also means that higher asphalt temperatures lead to higher operating temperatures of the tire. Also, a large contact area between the tire and the road surface helps dissipate heat from the tire. This is the reason why rain tires overheat very quickly on dry roads, as they have a significantly reduced contact area due to their tread pattern. In the remaining part of the contact patch (zone 2), the temperature rises due to the power dissipation added by friction. After leaving the contact patch, heat is dissipated to the ambient air via convection and the surface temperature drops. Low outside temperatures support the heat dissipation. The temperatures that develop in the tire have a decisive influence on its power transmission and wear behavior.

Figure 4.20 shows on the left the lateral force slip angle diagram of a slick tire for three different tire temperatures and on the right schematically the dependence of the adhesion potential as well as the tire wear on the tire temperature. According to this, there is a certain temperature at which the tire reaches its maximum grip level. Tire temperatures that are too low or too high lead to a loss of grip. However, the wear of the tire increases continuously as the temperature rises. At a certain temperature level, the tire then starts to wear excessively. Reasons for too high temperatures can be too large slip angles, e.g. caused by too high toe-in or toe-out values or by a too aggressive driving style, too high camber values with too high local loads, too low tire inflation pressures or a too soft rubber compound (e.g. rain tires on dry roads).

The range near the maximum grip level and the range of acceptable tire wear determine the optimum temperature window of a tire. The optimum temperature level of the tire depends on its construction and, if applicable, the race strategy. Under certain circumstances it may be advisable to keep the temperature level in the tire lower in order to avoid a pit stop. The optimum operating temperature of racing tires is between 80 and 120 °C. This specification is usually an average operating temperature. A meaningful analysis of the temperature behavior requires the recording of at least three measuring points distributed over the width of the tire. These measuring points should record the temperature in the center of the tire and on the outer and inner shoulders as shown in Fig. 4.21. With these three measured values it is possible to draw conclusions about the tire inflation pressure and the chassis settings.

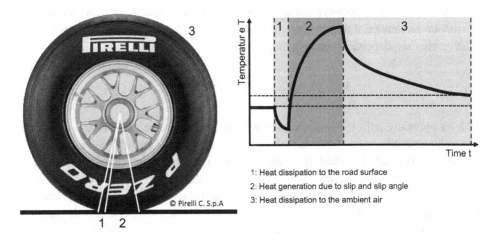

1: Heat dissipation to the road surface
2: Heat generation due to slip and slip angle
3: Heat dissipation to the ambient air

Fig. 4.19 Heat absorption and heat dissipation at the tire

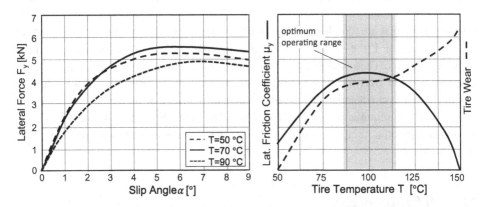

Fig. 4.20 Schematic tire temperature behavior

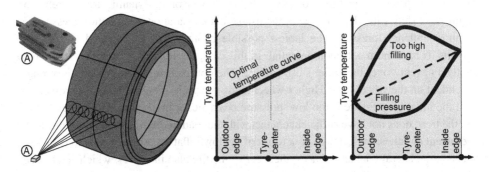

Fig. 4.21 Measuring the temperature distribution

According to [6], two characteristic parameters can be defined by which the inflation pressure or the camber of the tire can be checked. The characteristic value K_p describes the inflation pressure-dependent temperature curve on the tread of a tire. It is defined as:

$$K_p = T_{Mitte} - \frac{T_{au\backslash ssen} + T_{innen}}{2} \approx 0\,^\circ C. \tag{4.15}$$

With an optimally adjusted inflation pressure, this characteristic value is approximately zero. A high positive value indicates that the inflation pressure is too high, as the tire is only stressed in the center of its tread. If the inflation pressure is too low, mainly the outer areas of the tread are stressed, resulting in a high negative value for K_p. The influence of the air pressure on the usable tread is explained in Sect. 4.6. By recording the tread temperature, the set air pressure can be corrected. The second characteristic value K_y allows the analysis of the set camber angle. The following applies to it:

$$K_\gamma = T_{innen} - T_{au\backslash ssen} < 10\,^\circ C \backslash and > -10\,^\circ C. \tag{4.16}$$

Positive values that are too high indicate that the static camber on the corresponding wheel is set too negative. If the value is strongly negative, the camber should be adjusted in the negative direction. These values are based on the assumption that an approximately linear temperature curve over the tread is ideal. This need not necessarily be the case, which is why the optimum characteristic values of common tires have to be determined individually, which, however, requires complex real-time measurements. Such measurements can then be used, among other things, to analyze the axle kinematics or the setting of the limited slip differential.

4.6   Inflation Pressure

The inflation pressure of a tire also has a serious influence on its dynamic driving behavior. Figure 4.22 shows the dependence of the usable contact area on the inflation pressure. At optimum inflation pressure, the largest possible contact area is achieved and the tire exhibits its maximum adhesion potential. In this case, the inflation pressure is distributed relatively evenly over the entire slush surface. The optimum inflation pressure is strongly dependent on the wheel load. Higher wheel loads generally require higher air pressures.

If the inflation pressure is too low, pressure peaks occur in the area of the sidewalls, so that the tread does not have sufficient contact with the road in the area of the tire center. The coefficient of adhesion decreases as a result of the low inflation pressure. Due to the higher loads on the lateral contact surfaces, the wear of the tire also increases, which can be seen after a longer period of operation through increased wear of the lateral treads. The reduced tread also contributes to the heating of the tire, as less heat can be dissipated to the road.

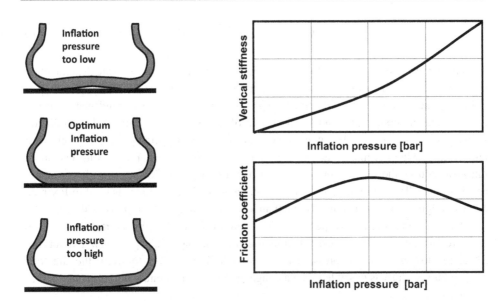

Fig. 4.22 Influence of inflation pressure on the contact patch area and vertical stiffness

If the inflation pressure is too high, the tire takes on a rather bulbous shape so that the lateral treads lose contact with the road surface and the coefficient of adhesion is reduced. However, the inflation pressure not only affects the tire's coefficient of adhesion, but also other properties, such as its vertical and lateral stiffness in particular. The vertical stiffness of the tire increases with an increase in inflation pressure. The increased vertical stiffness reduces deformation of the sidewall, which also reduces heating caused by flexing. The higher inflation pressure also increases the cornering stiffness. The tire inflation pressure is therefore an important parameter when tuning a vehicle.

The inflation pressure of a tire is dependent on the air temperature prevailing inside the tire. As heat is added to the tire by slip, skidding and flexing, the tire heats up during use until an approximately constant level is reached on average. The dependence of the pressure on the temperature results from the thermal equation of state of ideal gases:

$$p \cdot V = n \cdot R_m \cdot T. \tag{4.17}$$

In Eq. (4.17), p is the inflation pressure and V the volume of the tire, n the quantity of substance contained, R_m the universal gas constant and T the temperature of the filling medium inside the tire. If we assume for the sake of simplicity that the volume and the universal gas constant are independent of pressure and temperature, we obtain the increase in the filling pressure as a function of the change in temperature:

$$p = p_0 + \Delta p = p_0 + \frac{n \cdot R_m}{V} \cdot \Delta T. \tag{4.18}$$

The change in inflation pressure for a given temperature increase is therefore dependent on the universal gas constant R_m, i.e. on the properties of the inflation medium and the inflation volume, which is a property of the tire design. Figure 4.23 shows the dependence of the tire pressure on the temperature and the influence of the inflation medium and the inflation volume. When the tire is filled with ambient air, the pressure increases progressively with temperature, which is mainly due to the water content in the air. Filling with dry air or nitrogen reduces the temperature-dependent pressure increase. Racing tires, if permitted, are therefore filled with nitrogen instead of ambient air. Balloon tires, as used in Formula 1, have a significantly higher air volume than low-profile tires. For this reason, they are less sensitive to changes in tire temperature. To ensure that the tire is not subject to excessive pressure fluctuations during operation or racing or qualifying and, in particular, that it is not set at too low an inflation pressure at the start, it is advisable to warm the tire to a temperature as close as possible to the average operating temperature during its use before it is used.

This is achieved, for example, by the mandatory use of electric blankets, which additionally ensure that the tread is at a temperature closer to the optimum. In addition, the drivers bring the tire up to the required operating temperature during the warm-up lap by selectively applying diagonal running. One way to reduce the effects of temperature-induced pressure fluctuations is to use so-called bleeder valves, such as those shown in Fig. 4.24. Bleeder valves are pressure relief valves that are incorporated into the rim in addition to the normal inflation valve. The valve seat consists of a ball which is pressed against the valve seat by a spring. The preload of the spring determines the maximum pressure that can prevail inside the tire. If the pressure rises too high, the valve opens and the excess air escapes from the tire. In this way, it is possible to prevent the tire from having to operate at too low a pressure before it warms up. The scenario of an overpressure developing is completely avoided. It should be noted, however, that a cooled tire, if excess air has been previously blown off, will be traveling at significantly too low an inflation pressure. Illustrative examples of the analysis of tire temperature and air pressure behavior can be found in [7].

4.7   Tire Construction

In the construction of tires, a distinction is essentially made between cross-ply and radial tires. Today, modern passenger car and racing tires are almost exclusively manufactured in radial construction. A comparison of the two types of construction should make their specific advantages clear. Figure 4.25 therefore shows the characteristic features of radial and cross-ply tires. In cross-ply tires, the carcass is formed from cords which are drawn diagonally across the rim and cross at the tread centerline at an angle of approximately

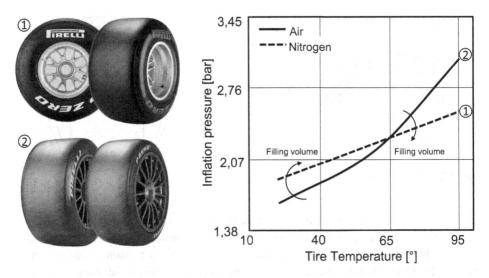

Fig. 4.23 Influence of temperature on tire inflation pressure for different inflation media and tire constructions. (Photos courtesy of © Pirelli & C. S.p.A. 2018. All Rights Reserved)

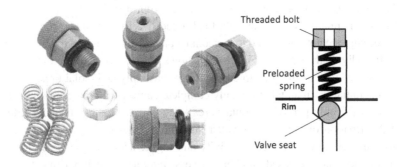

Fig. 4.24 Bleeder valves to prevent overpressure

30-40 °. The carcass is formed by several layers of cords running in opposite directions, creating a diamond pattern. The crossed plies create a rounded shaped carcass. This rounding must be compensated for by the tread, which limits the maximum width of bias-ply tires. The deformations occurring under load or under slip and sideslip lead to relative movements between the individual cord plies. The resulting friction dampens unevenness of the road surface and thus increases driving comfort, but this comes at the price of a significant heat input into the tire. Deformations of the sidewall lead to deformations of the contact patch. The friction between the cords and the deformation of the tread increase the heat-related wear of bias-ply tires.

In radial tires, the cords are drawn from bead to bead with a maximum of two plies at an angle of about 90 ° – i.e. transversely or radially. The required dimensional stability of the

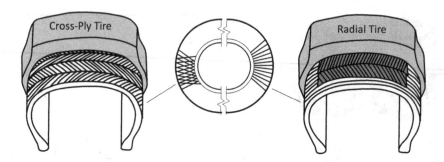

Fig. 4.25 Cross-ply and radial tires

tire is achieved by several belt plies, which lie between the carcass and the tread. The belt plies consist of textile fibers or fine steel wire braids. In general, attempts are made to keep the cord length between the rim and tread relatively short or to increase the effective bending length by stiffening the sidewall. Among other things, this avoids excessively high slip angles. Figure 4.26 shows the structure of a Pirelli P-Zero Formula 1 radial tire of the 2013 season (see also [8]).

The central advantage of radial tires is that the radial arrangement of the cords and the use of belt plies results in a relatively soft sidewall with high dimensional stability of the tread. Due to the relatively soft sidewall, load-induced deformations, as they occur under slip and side slip, are not transferred to the contact patchr, and no relative movements occur between the individual cord plies. Figure 4.27 shows the extreme deformations of an F1 tire under side slip as well as the principle behavior of cross-ply and radial tires.

While the radial tire maintains virtually complete contact with the ground, the sidewall of the bias-ply tire pulls part of the contact area away from the road surface. Avoiding load-induced deformation of the tire sidewall results in reduced rolling resistance, increased cornering stiffness, lower camber sensitivity and higher adhesion potential. The reduced rolling resistance and increased camber stiffness minimize heat input into the tire. The lower heat input allows the use of softer tread compounds, which also results in higher adhesion potential. Disadvantages of the radial tire are its lower comfort, due to the lack of frictional damping, and a sharper drop-off after reaching the adhesion limit. This has to be taken into account by an adapted chassis tuning. In summary, however, the radial tire has a significant performance advantage over the cross-ply tire. Radial tires were introduced by Michelin in 1977. In 1978 and 1979 the first Grand Prix victory and the first drivers' world championship were won with this type of tire. The lateral stiffness or the bending stiffness under the influence of lateral forces is decisively responsible for the driving dynamic properties of a tire (see also Fig. 4.11).

The radial tire also has a damping function. However, its design allows the force-transmitting properties to be separated from the damping function. For this reason, the tread and rubber matrix of the sidewall have different compositions. The composition of the rubber compound for the sidewall focuses on damping and stiffness. When tuning the

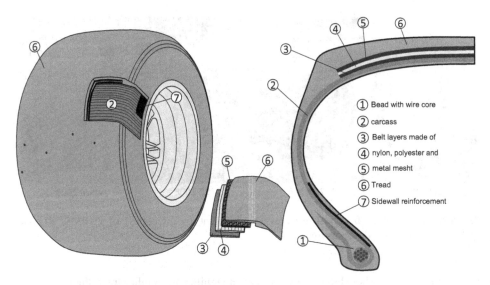

Fig. 4.26 Pirelli P-Zero F1 radial tires of the 2013 season

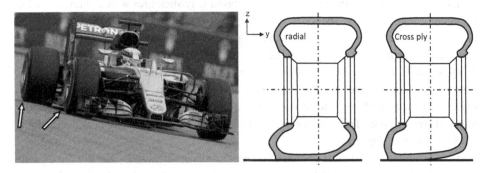

Fig. 4.27 Tire deformation under lateral force. (Photo courtesy of © Daimler AG 2018. All Rights Reserved)

damping properties, heat input must also be taken into account. In the past, incorrect design has often led to tire failure in the high-speed range. The tread compound must provide maximum grip with sufficient durability. In some racing series, such as NASCAR Sprint Cup, or motorcycle racing tires, the tires also have zones of different rubber compounds on the tread. For example, the sidewalls that are subjected to camber loads can be provided with a harder compound.

Another design feature of the tire is its tread thickness. Rubber is a heat insulator. This means that as the thickness of the tread increases, less heat can be dissipated from inside the tire to the environment. This effect is undesirable for continuous and sustained use. For short stresses, such as a qualifying run or a hill climb where there is no warm-up period, it can be useful to keep heat inside the tire by increasing the tread thickness. For this reason,

| until 1970 | 1971-1997 | 1998-2008 | since 2009 |

Fig. 4.28 Treads of F1 front and rear tires from different eras. (Courtesy of © BMW AG and © Daimler AG 2018. All Rights Reserved)

racing tires for very short hill climbs can have a significantly higher tread thickness than tires for endurance races.

This property of tread thickness is also the reason why slick tires were not introduced in Formula 1 until 1971 by the Firestone company. Until then, grooves were needed to dissipate heat via the locally thinner tread (Fig. 4.28). It was not until the development of new manufacturing processes and more heat resistant tread compounds that the use of treadless slick tires became possible [9]. Due to the high thermal stress, racing tires generally have a considerably thinner tread thickness than passenger car tires. The tread thickness of a typical street-legal tire is about 10 mm, not least because of the legally required tread depth. This compares to a tread thickness of only about 3.2 mm for a NASCAR Sprint Cup tire.

The performance of a racing tire is also significantly influenced by the tire dimensions and in particular the tire width. Elastomers have the property that their coefficient of friction μ decreases with increasing surface pressure (wheel load per surface), which is also reflected in the effect of wheel load sensitivity. High tire performance is therefore achieved by maximizing the contact patch. Design-wise, this can be achieved by using the widest possible tire, which is another argument for using radial tires due to their design. A wider tire also has the advantage of providing a larger surface area for both heat absorption and dissipation to the road and surrounding area. Today, however, the dimensions of the racing tire are usually precisely prescribed by the technical regulations. In order to avoid high costs due to a development competition, in numerous racing series the tires are provided by only one supplier. Special qualifying tires are also generally no longer permitted today. So it is often no longer possible to gain a competitive advantage by using different tire textures. This makes it all the more important for the engineers of a racing team to have a sound understanding of how the tire works under varying boundary conditions.

4.8   Wear and Damage Patterns

The wear behavior of a tire or its damage can often be analyzed on the basis of its outer tread pattern. These tread patterns allow conclusions to be drawn about the inflation pressure set, the chassis setting or the driver's driving style. Figure 4.29 shows tread patterns caused by incorrectly set inflation pressure or an incorrectly set or damaged suspension. According to Fig. 4.22, an excessively high inflation pressure causes the tire to roll essentially over the central part of its tread, with the result that the tire wears disproportionately in this area. Similarly, an too low inflation pressure will cause the tire to wear severely in the area of its sidewalls. A negative camber value that is too high will result in increased wear on the inner sidewall of the tire. Too high a toe-in value will wear the tire disproportionately on the outer half of the tread, while too high a toe-out value will cause an identical tread pattern on the inner half of the tread. Irregular wear-related washouts distributed over the circumference indicate a dynamic imbalance or incorrect spring/damper setting. Such tread patterns can also be observed on passenger car tires.

Another tread pattern of racing tires is created by scrubbing (not shown). Scrubbing refers to the "starting up" of the tires by heating them up once or twice. The temperature change influences the chemical characteristics of the tire. This can make the tire harder or remove the top layer of rubber. "Scrubbing" race tires in practice is often beneficial to rid the tire of the release agents from production. In the race, the tire then works faster at a higher grip level. Scrubbing by hand is prohibited in Formula 1.

Figure 4.30 shows some of the wear phenomena that occur on the tread as a result of the load that occurs during racing. Figure 4.30a, b show the phenomenon of graining. Graining occurs when the tire slides, it is then loaded transversely to the direction of travel. A kind of "orange peel" forms on the tread, consisting mainly of groove abrasion in the form of small "grains". In the case of very heavy graining, the abrasion collects in small "beads" or "grains" on the tread. The tire's grip is reduced as the contact area with the asphalt is reduced. The degree of graining generally depends greatly on the track condition. Grooved tires in particular have an increased tendency to graining, as the rubber moves back and forth intensively at the edges.

The edge of the tire in Fig. 4.30b shows the effects of blistering. Blistering is the term used to describe blistering that occurs when too much heat develops between the belt and the tread. This can cause parts of the tire to melt and separate locally from the carcass, resulting in blisters on the outside. Blistering is usually visible to the naked eye, as the areas affected by blistering stand out in dark black against the otherwise charcoal-colored tread. If the tire is blistered over a large area, its performance will be reduced, which usually requires a tire replacement as soon as possible.

"Pickup"can also be observed very frequently on racing tires (Fig. 4.30c). Here, the tire wear lying on the track sticks to the still hot racing tires. Pickup is often deliberately collected to provide a small weight cushion for the final weighing of the vehicle.

The formation and consequences of a brake flat are shown in Fig. 4.31. Brake flats are caused by the wheels locking. On the stationary wheel, one or more layers of rubber are

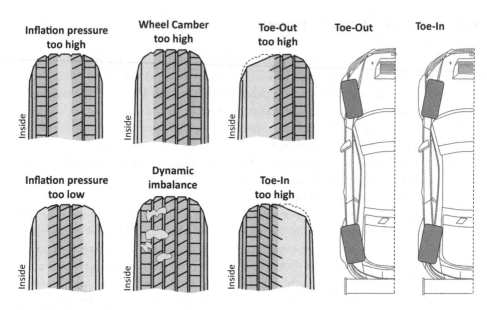

Fig. 4.29 Schematic tread patterns of tires

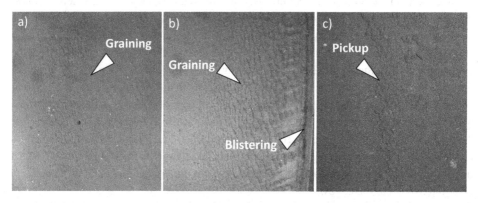

Fig. 4.30 Wear characteristics on racing tires: (**a**) Light graining (**b**) Intensive graining and blistering (**c**) Pickup

worn away or burnt off with the development of smoke because the tire has to absorb very high power for a short time. The energy that normally causes the brake discs of a racing vehicle to glow is completely absorbed by the tire. In extreme situations, even the belt can come to the surface.

Tires with a brake flat usually force the driver to make a pit stop, as the flat can cause secondary damage. Brake flats cause the tire to tend to stop on its already flattened tread when the brakes are applied again. This increases the damage. Furthermore, the imbalance of a brake plate causes the suspension to be subjected to severe vibration stress, which can

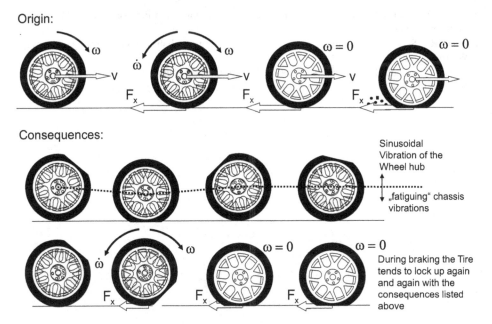

Fig. 4.31 Consequential damage after the occurrence of flat-spotting under braking

lead to failure of the suspension components under certain circumstances. Such an incident was seen on Kimi Räikkönen's McLaren in 2005 during the European Grand Prix. Raikkonen, while in the lead, drove for several laps shortly before the end of the race with a brake plate, fearing that he would lose several places due to a pit stop. On the final lap at the end of the start-finish straight, the right-side suspension finally broke. Chaser Alonso eventually won the race.

References

1. Haney, P.: The Racing & High Performance Tire. Society of Automotive Engineers, Warrendale (2003)
2. Moore, D.F.: The Friction of Pneumatic Tyres. Elsevier, Amsterdam (1975)
3. Wright, P.: Formula 1 Technology. Society of Automotive Engineers, Warrendale (2001)
4. Wright, P.: Ferrari Formula 1 – under the Skin of the Championship-Winning F1-2000. Society of Automotive Engineers, Warrendale (2003)
5. Milliken, W.F., Miliken, D.L.: Race Car Vehicle Dynamics. Society of Automotive Engineers, Warrendale (1995)

6. Weber, W.: Fahrdynamik in Perfektion – Der Weg zum optimalen Fahrwerk-Setup. Motorbuch, Stuttgart (2011)
7. Pütz, R., Serné, T.: Rennwagentechnik – Praxislehrgang Fahrdynamik. Springer Vieweg, Wiesbaden (2017)
8. Piola, G.: Formula 1 Technical Analysis 2013/2014. Giorgio Nada Editore, Vimodrone (2014)
9. Hughes, M.: Speed Addicts – Grand Prix Racing. Dakini Books Ltd., London (2005)

Driving Dynamics Basics

5

The subject area of vehicle dynamics describes, among other things, the relationship between driver actions, vehicle characteristics and the vehicle's state of motion. The motion variables result from the forces acting on the vehicle, which are imposed on the vehicle by the driver through the operation of the steering wheel and pedals via the tires.

5.1   Movement Variables on the Vehicle

Figure 5.1 summarizes the central movement variables of a vehicle and its chassis. The individual movement variables of the vehicle body in relation to the vehicle's center of gravity are:

- x, $\dot{x}$ or v_x, $\ddot{x}$ or a_x: longitudinal motion, velocity, acceleration
- y, $\dot{y}$ or v_y, $\ddot{y}$ or a_y: transverse motion, velocity, acceleration
- z, $\dot{z}$, $\ddot{z}$: Vertical or heave movement, speed, acceleration,
- κ, $\dot{\kappa}$, $\ddot{\kappa}$: roll angle, roll velocity, roll acceleration.
- φ, $\dot{\varphi}$, $\ddot{\varphi}$: Pitch angle, pitch velocity, pitch acceleration.
- ψ, $\dot{\psi}$, $\ddot{\psi}$: yaw angle, yaw rate, yaw velocity, yaw acceleration, respectively.

A vehicle-fixed xyz coordinate system is used to describe the motion variables. The x-axis lies on the longitudinal axis of the vehicle and describes the longitudinal movement of the vehicle, to which the longitudinal speed v_x or $\dot{x}$ and the longitudinal acceleration a_x or $\ddot{x}$ are essentially assigned. The rotation of the vehicle body about the x-axis is called the rolling motion, which is described in terms of the roll angle κ and by its derivatives. Along the y-axis, the transverse motion of the vehicle takes place. Here, the lateral acceleration a_y or $\ddot{y}$ is the central variable for describing the transverse dynamic driving condition. The velocity

L. Frömmig, *Basic Course in Race Car Technology*,
https://doi.org/10.1007/978-3-658-38470-8_5

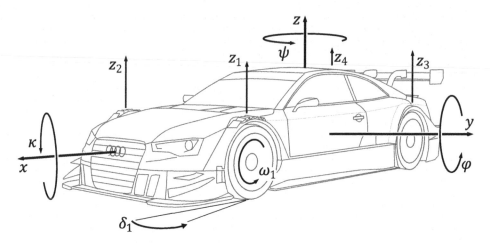

Fig. 5.1 Coordinate system and motion variables of a vehicle

component v_y is closely related to the slip angles (see Sect. 5.2) and the side slip angle of the vehicle. The rotational movement of the body about the y-axis is called pitching movement, which occurs during braking and acceleration. The z-axis is used to describe the vertical or lifting movement of the center of gravity. It is worth mentioning here that the spatial position of the body is defined either by the roll angle, the pitch angle and its vertical movement z or by three different z-coordinates. The rotation around the z-axis is the yaw motion ψ of the vehicle.

In addition to the motionvariables of the vehicle body, the motion variabels of the individual wheels are decisive for the dynamic driving state of a vehicle. The motion quantities of the wheels are:

- $z_1 - z_4$, $\dot{z}_1 - \dot{z}_4$, $\ddot{z}_1 - \ddot{z}_4$: Wheel vertical movement, speed, acceleration
- $\omega_1 - \omega_4$, $\dot{\omega}_1 - \dot{\omega}_4$: wheel rotation speed, acceleration
- $\delta_1 - \delta_4$: Wheel steering angle
- $\gamma_1 - \gamma_4$: Wheel camber angle

The variables z_1–z_4 are the vertical movements of the individual wheels. From them, in conjunction with the chassis geometry, the spring deflections and the relative speeds in the dampers are derived. The wheel slip is determined from the wheel rotational speeds or wheel speeds ω_1–ω_4. The wheel steering angles δ_1–δ_4 result from the driver's steering inputs and the kinematic and elastokinematic properties of the chassis. As a further motion variable, the wheel-individual camber values γ_1–γ_4 should be mentioned; they also depend on the kinematic and elastokinematic properties of the chassis.

Another important motion variable of the vehicle is the side slip angle β, which can be seen from the ratios in Fig. 5.2. The yaw angle ψ describes the rotational motion of the

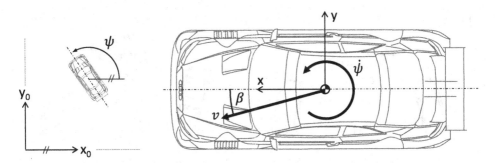

Fig. 5.2 Yaw angle and side slip angle

vehicle's longitudinal axis in an inertial coordinate system "0". However, the velocity vector v *of* the vehicle is generally not along the longitudinal axis of the vehicle, but it is rotated by the side slip angle β with respect to the longitudinal axis of the vehicle. For this reason, the velocity vector v is composed of the previously mentioned components v_x and v_y. However, the side slip angle is not an independent motion variable, but results from a coupling of yaw rate, lateral acceleration and vehicle speed. It holds:

$$a_y = v \cdot \left(\dot{\psi} + \dot{\beta} \right). \tag{5.1}$$

The side slip angle is an important parameter for describing the stability of a vehicle. Figure 5.3 illustrates the significance of the side slip angle. Figure 5.3a shows the author's wife driving through a roundabout. Due to the exemplary driving style for public road traffic, only a small side slip angle occurs. The longitudinal axis of the vehicle follows the radius of the curve almost tangentially. Walter Röhrl drifts through the curve in an Audi quattro Rally in Fig. 5.3b. The side slip angle is about $-40°$. The rear end pushes outwards. This driving condition, which tends to instability, is difficult to control for normal drivers. Figure 5.3c shows the author's attempt to drive through the roundabout like Walter Röhrl. The author loses control and spins backwards off the road, with a side slip angle of about $-180°$. Thus, side slip angles of high magnitude indicate an unstable or strongly oversteering driving condition.

With this selection of motion variables, simplified vehicle dynamics models are set up in the following, which allow a basic analysis of the vehicle dynamics properties of racing vehicles. A comprehensive discussion of the theoretical fundamentals of vehicle dynamics can be found in [1, 2], among others.

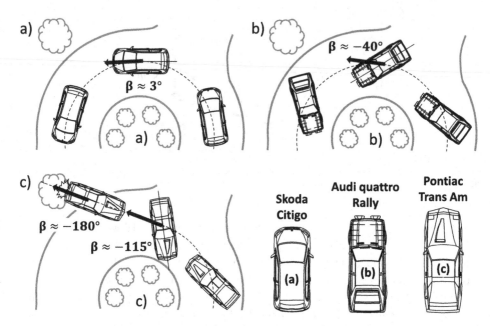

Fig. 5.3 Illustration of the side slip angle: (**a**) The author's wife drives through a roundabout. (**b**) Walter Röhrl drives through a roundabout. (**c**) The author tries to drive through a roundabout like Walter Röhrl

5.2   Single Track Model

The single-track model is the simplest approach to analytically describe the relationship between the forces acting on the vehicle, the vehicle's motion variables, the slip angles and the driver's actions. Characteristic for the single-track model is that the two wheels of an axle are combined to *one* wheel on the longitudinal axis. The resulting kinematic conditions are shown in Fig. 5.4. The velocity vector v rotates around the center of curvature with the sum of the yaw rate $\dot{\psi}$ and the side slip angle velocity $\dot{\beta}$ (time derivative of the side slip angle). The distance between the center of gravity of the vehicle and the center of curvature is the radius of curvature ρ. The following relationship applies:

$$v = \rho \cdot \left(\dot{\psi} + \dot{\beta} \right). \tag{5.2}$$

Multiplying Eq. (5.2) by the magnitude of the velocity v and dividing by ρ yields the formula for the lateral acceleration a_y known from Eq. (5.1)$_y$.

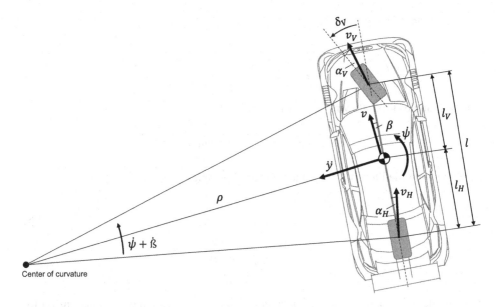

Fig. 5.4 Kinematic variables on the single-track model

$$\frac{v^2}{\rho} = v \cdot \left(\dot{\psi} + \dot{\beta}\right) = a_y \tag{5.3}$$

The relationship between slip angles, movement variables and wheel position is derived from the consideration of the vehicle as a rigid body. A possible wheel steering angle of the rear axle, resulting from the axle kinematics or a rear axle steering, is neglected here. The rigid body properties of the vehicle lead to the illustrative condition that all speed components along the longitudinal axis must be identical. If this were not the case, the vehicle would deform. From this follows:

$$v_x = v \cdot \cos(\beta) = v_{Vx} = v_{Hx}. \tag{5.4}$$

When calculating the velocity components along the y-axis of the vehicle, the rotational movement of the longitudinal axis must be taken into account in order to form the rigid body conditions, so that the following relationships are obtained:

$$\alpha_V = \delta_V - \tan^{-1}\left(\frac{v_{Vy}}{v_{Vx}}\right) = \delta_V - \tan^{-1}\left(\frac{v \cdot \sin(\beta) + \dot{\psi} \cdot l_V}{v \cdot \cos(\beta)}\right). \tag{5.5}$$

The slip angle α_H is:

$$\alpha_H = -\tan^{-1}\left(\frac{v_{Hy}}{v_{Hx}}\right) = -\tan^{-1}\left(\frac{v \cdot \sin(\beta) - \dot{\psi} \cdot l_H}{v \cdot \cos(\beta)}\right). \tag{5.6}$$

For small angles the following approximation can be used:

$$\sin(x) \approx x, \ \cos(x) \approx 1 \Rightarrow \tan(x) \approx 1. \tag{5.7}$$

Thus, the simplified terms are obtained for the slip angles:

$$\alpha_V = \delta_V - \beta - \frac{\dot{\psi} \cdot l_V}{v}, \tag{5.8}$$

$$\alpha_H = -\beta + \frac{\dot{\psi} \cdot l_H}{v}. \tag{5.9}$$

Rearranging the equations to β and then equating provides an important relationship for determining the mean front wheel angle δ_V:

$$\delta_V = \frac{\dot{\psi} \cdot (l_V + l_H)}{v} + \alpha_V - \alpha_H = \frac{\dot{\psi} \cdot l}{v} + \alpha_V - \alpha_H. \tag{5.10}$$

Equation (5.10) allows a basic discussion of the driving behavior, which is carried out in Sect. 5.5. Figure 5.5 shows the front wheel steering angles δ_1 and δ_2, the averaged front wheel steeering angle δ_V and the steering wheel angle δ_L. In the single-track model, the front wheel steering angle δ_V represents the average of the left and right front wheel steering angles δ_1 and δ_2. The exact relationship between steering wheel angle and front wheel steering angle is obtained from various characteristics of the front axle. Further details of the suspension and steering system are discussed in Chap. 7. Elastic properties of the steering system are not considered here, and a constant steering ratio is assumed for simplicity, so that holds:

$$\delta_V = \frac{\delta_1 + \delta_2}{2} = \frac{\delta_L}{i_L}. \tag{5.11}$$

The steering ratio is a measure of the steering wheel angle requirement or the agility of a vehicle. The lower the steering ratio, the lower the steering wheel angle required to negotiate a bend and the more direct the vehicle. Figure 5.5 summarizes typical values for the steering ratio of various vehicles in tabular form.

In the next step, the forces and moments occurring on the single-track model are considered in Fig. 5.6. The axle lateral forces F_{yV} and F_{yH}, which are perpendicular to the wheel center line, and the axle longitudinal forces F_{xV} and F_{xH}, which are on the wheel

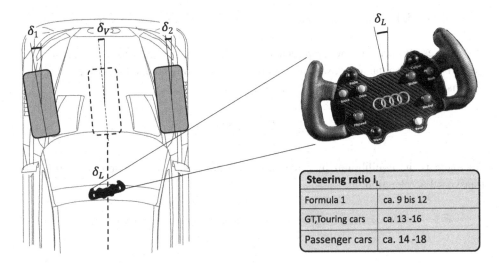

Fig. 5.5 Steering ratio

Steering ratio i_L	
Formula 1	ca. 9 bis 12
GT,Touring cars	ca. 13 -16
Passenger cars	ca. 14 -18

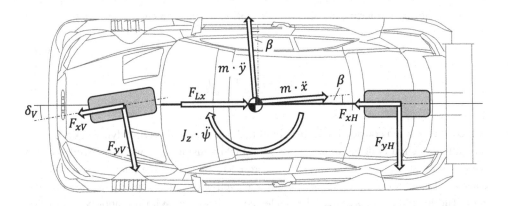

Fig. 5.6 Forces and moments on the single-track model

center line, act on the front and rear axles respectively. At the center of gravity, the inertial forces of longitudinal and lateral acceleration each act in the direction of the associated acceleration. The yaw moment of inertia also acts around the center of gravity. The air resistance force F_{Lx} is taken into account as a further force, whereby the aerodynamic forces only have a longitudinal component here for simplification. The aerodynamic forces occurring on the vehicle are dealt with comprehensively in Chap. 6.

From setting up the equilibria of forces and moments, the three equations of motion of the single-track model are obtained. The equilibrium of forces along the x-axis fixed to the vehicle is:

$$F_{xV} \cdot \cos(\delta_V) + F_{xH} = m \cdot \ddot{x} \cdot \cos(\beta) + F_{Lx} + F_{yV} \cdot \sin(\delta_V) + m \cdot \ddot{y} \cdot \sin(\beta). \quad (5.12)$$

If one takes into account the simplification introduced by Eq. (5.7) for small angles and neglects the sinusoidal terms, which are only very small compared to the cosine terms, one obtains for the longitudinal acceleration of the vehicle:

$$\ddot{x} = a_x = \frac{F_{xV} + F_{xH} - F_{Lx}}{m}. \quad (5.13)$$

Analogue is the equilibrium of forces in y-direction resp. the equation of motion for transversal acceleration:

$$F_{yV} \cdot \cos(\delta_V) + F_{xV} \cdot \sin(\delta_V) + F_{yH} + m \cdot \ddot{x} \cdot \sin(\beta) = m \cdot \ddot{y} \cdot \cos(\beta), \quad (5.14)$$

$$\ddot{y} = a_y = \frac{F_{yV} + F_{yH}}{m}. \quad (5.15)$$

By forming the moment equilibrium around the center of gravity and using the approximation of the angular terms, the equation of motion for the yawing motion is obtained:

$$F_{yV} \cdot \cos(\delta_V) \cdot l_V + F_{xV} \cdot \sin(\delta_V) \cdot l_V = F_{yH} \cdot l_H + J_z \cdot \ddot{\psi}, \quad (5.16)$$

$$\ddot{\psi} = \frac{F_{yV} \cdot l_V - F_{yH} \cdot l_H}{J_z}. \quad (5.17)$$

A special case often considered for the simplified discussion of driving dynamics effects is the stationary circular drive. In this case, a constant radius of curvature ρ is travelled along with a constant driving speed v. Thus, according to Eq. (5.3), the lateral acceleration is constant. Longitudinal and yaw acceleration as well as the side slip angular velocity are zero. The longitudinal forces can be neglected in this case. With the approximation for small angles, the conditions shown in Fig. 5.7 result for steady-state circular motion. With the help of the single-track model, the terms limiting lateral acceleration and balance are introduced. Furthermore, a first analysis of the interactions between tire and vehicle properties is carried out, whereby their influence on the lateral acceleration limit is also considered. For this purpose, the relationship between mass distribution, wheel loads and lateral forces is first considered. Depending on the position of the center of gravity, the axle loads on the front and rear axles are obtained:

$$F_{zV} = \frac{l_H}{l} \cdot m \cdot g, \quad (5.18)$$

$$F_{zH} = \frac{l_V}{l} \cdot m \cdot g. \quad (5.19)$$

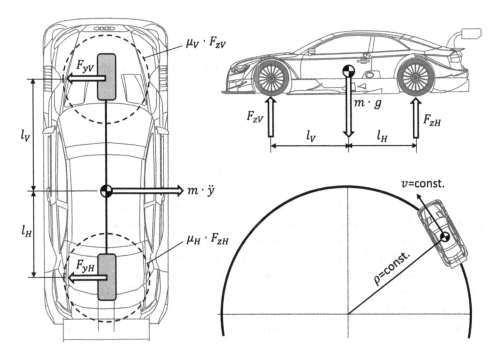

Fig. 5.7 Forces on the single-track model during steady-state cornering

The same applies to the lateral forces required to support a lateral acceleration:

$$F_{yV} = \frac{l_H}{l} \cdot m \cdot \ddot{y}, \tag{5.20}$$

$$F_{yH} = \frac{l_V}{l} \cdot m \cdot \ddot{y}. \tag{5.21}$$

For a stationary circular motion ($\ddot{\psi} = 0$) the following condition must be fulfilled according to Eq. (5.17):

$$F_{yV} \cdot l_V = F_{yH} \cdot l_H. \tag{5.22}$$

The maximum possible lateral acceleration at which this condition is just fulfilled is defined here as the limiting lateral acceleration $a_{y,gr}$. Reaching the limiting lateral acceleration requires that at least one axle reaches its frictional limit and thus generates its maximum lateral force. The maximum lateral forces result from the friction circles at the front and rear axles. This results in the following relationships:

$$F_{yV,\max} = F_{zV} \cdot \mu_V = \frac{l_H}{l} \cdot m \cdot g \cdot \mu_V = \frac{l_H}{l} \cdot m \cdot a_{y,gr,V}, \tag{5.23}$$

$$F_{yH,\,max} = F_{zH} \cdot \mu_H = \frac{l_V}{l} \cdot m \cdot g \cdot \mu_H = \frac{l_V}{l} \cdot m \cdot a_{y,gr,H}. \tag{5.24}$$

In general, the front and rear axles have different limiting lateral accelerations. They are:

$$a_{y,gr,V} = \mu_V \cdot g, \tag{5.25}$$

$$a_{y,gr,H} = \mu_H \cdot g, \tag{5.26}$$

$$a_{y,gr} = \min\left(a_{y,gr,V}, a_{y,gr,H}\right). \tag{5.27}$$

The smaller of the two limiting lateral acceleration values indicates which axle reaches its adhesion limit first. The ratio of the lateral limiting accelerations or the adhesion potentials on the front and rear axles is also referred to as the balance of the vehicle. If the limiting lateral acceleration of the front axle is smaller than that of the rear axle, then the front axle reaches its limit of adhesion first. Such a behavior leads to understeering or stable driving behavior, as outlined in Fig. 5.8. In this case, even by increasing the slip angle, it is no longer possible to increase the front axle lateral force. At the limit, the driver cannot use the steering to impose a turning-in yaw acceleration on the vehicle. The vehicle remains stable at the limit, and the rear axle still has power reserves. These reserves mean that the vehicle also remains stable if, for example, bumps cause a brief increase in the wheel load and thus the lateral force on the front axle, because this can be compensated for by an increase in the rear axle lateral force.

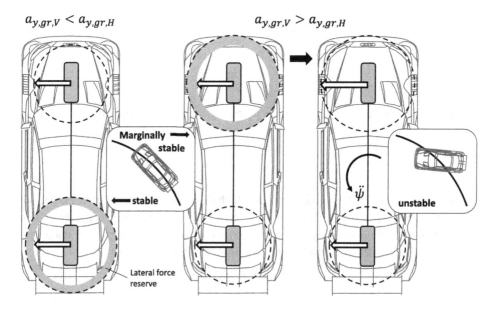

Fig. 5.8 Influence of the balance on the stability behavior

If the rear axle reaches the frictional limit first, then the torque balance of the front and rear axle only represents a limit-stable state. The rear axle's adhesion potential is exhausted, but in this case the driver can increase the lateral force on the front axle by steering further. Since the rear axle has already reached its limit of adhesion, torque balance can no longer occur in this condition. The vehicle experiences yaw acceleration, which causes the vehicle to oversteer. The vehicle is unstable because the yaw acceleration can only be reduced by actively reducing the lateral force on the front axle. The driver has to reduce the steering wheel angle and, if necessary, even countersteer, because side slip angles can build up very quickly under certain circumstances. Production vehicles in particular are therefore designed in such a way that they exhibit a pronounced understeer tendency in the limit range.

The limiting lateral accelerations at the front and rear axles are determined by various factors. A significant influencing factor is the weight distribution of the vehicle. The dependence of the lateral limiting accelerations on the weight distribution results from the wheel load sensitivity of the tires, as shown in Fig. 5.9. Figure 5.9a shows a vehicle with a 50-50 weight distribution on both axles. The axle loads on the front and rear axles are therefore the same and in this example amount to 5000 N. If the tires on the front and rear axles are identical, the adhesion potentials are also identical. The following applies:

$$a_{y,gr,V} = \mu_V \cdot g = a_{y,gr,H} = \mu_{VH} \cdot g = 1.1 \cdot g \Rightarrow a_{y,gr} = 1.1 \cdot g. \tag{5.28}$$

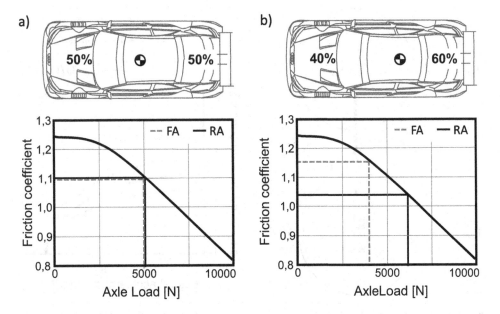

Fig. 5.9 Adhesion potentials of (**a**) a vehicle with medium load and (**b**) a vehicle with rear load

In example (b), a tail-heavy vehicle is considered. The center of gravity shifts towards the rear axle and the rear axle load is greater than that of the front axle. The wheel load sensitivity causes the front axle's adhesion potential to increase, while the rear axle's adhesion potential decreases, with identical tires. The following values for the adhesion potentials or limiting transverse accelerations can be read from Fig. 5.9b:

$$a_{y,gr,V} = \mu_V \cdot g = 1.16 \cdot g, a_{y,gr,H} = \mu_{VH} \cdot g = 1.04 \cdot g \Rightarrow a_{y,gr} = 1.04 \cdot g. \qquad (5.29)$$

In this case, the rear axle limits the limit lateral acceleration, which gives the vehicle an oversteer tendency. The maximum lateral limit acceleration of a vehicle with identical tires on the front and rear axles therefore occurs with a 50-50 weight distribution.

However, a 50-50 weight distribution is seldom the optimum mass distribution, since a vehicle must transmit drive forces in addition to lateral forces. With the help of Fig. 5.10, the influence of the drive force on the lateral acceleration limit of an axle is explained. A rear axle driven vehicle is considered. According to the friction circle, the total force that can be transmitted to the rear axle results in:

$$F_{zH} \cdot \mu_H = \mu_H \cdot \frac{l_V}{l} \cdot m \cdot g = \sqrt{F_{xH}^2 + F_{yH}^2}. \qquad (5.30)$$

At high longitudinal accelerations, the air resistance force can be neglected. For this reason, the following then applies to the longitudinal force on the driven rear axle:

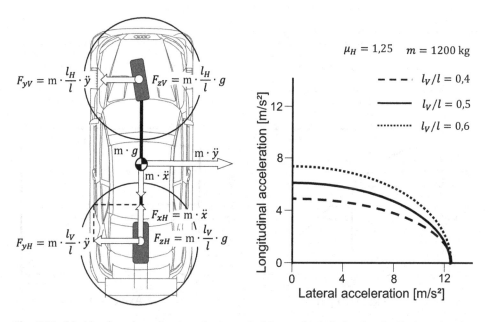

Fig. 5.10 Limiting lateral acceleration of a rear-axle driven vehicle during longitudinal acceleration

$$F_{xH} = m \cdot \ddot{x}. \tag{5.31}$$

For the lateral force applies:

$$F_{yH} = m \cdot \frac{l_V}{l} \cdot \ddot{y}. \tag{5.32}$$

By substituting and then rearranging Eq. (5.30), the dependence of the limiting lateral acceleration of the rear axle on the longitudinal acceleration is obtained:

$$\ddot{y}_{gr,H} = a_{y,gr,H} = \sqrt{(\mu_H \cdot g)^2 - \left(\frac{l}{l_V} \cdot \ddot{x}\right)^2}. \tag{5.33}$$

Plotting the limiting lateral acceleration against the longitudinal acceleration results in the ellipse shown in Fig. 5.10. The maximum longitudinal acceleration is smaller than the maximum lateral acceleration because only the rear axle is available to transmit the longitudinal force in this case. The maximum longitudinal acceleration is:

$$\ddot{x}_{max} = \frac{\mu_H \cdot l_V \cdot g}{l}. \tag{5.34}$$

The maximum transmittable longitudinal force is greater the more of the vehicle weight is on the rear axle. The width of the ellipse therefore depends on the ratio of the wheelbase to the center of gravity. The maximum lateral acceleration is – if the wheel load sensitivity is neglected – independent of the mass distribution. For this reason it makes sense to shift the vehicle weight in favor of the driven axle. This applies in particular if there are many sections with long and strong acceleration phases.

Another way to increase the rear axle's traction reserves is to use mixed tires. This usually involves fitting wider tires on the driven axle than on the non-driven axle. This is a common type of tire combination on racing cars, as the examples in Fig. 5.11 illustrate. Mixed tires are rarely found on production vehicles because it is generally cheaper to purchase as many identical tires as possible. The exception is the less price-sensitive segment of sports and super sports cars, where the use of mixed tires is quite common. Mixed tires are used virtually exclusively on driven rear axles, because wider tires on the front axle restrict steering travel when their insides hit the wheel arches. A famous exception from high-performance racing is the 2015 LMP1 Nissan GT-R LM entered at Le Mans. This was the only LMP1 car at the time to be of front-mid engine design with primary drive on the front axle. From the field of production cars, the all-wheel drive Audi RS3 of 2010 features the rather untypical mixed tires with wider tires on the front axle, which served to reduce the understeer tendency of the vehicle.

Figure 5.12 shows the basic effect of a mixed tire when cornering a rear axle driven racing vehicle. Without mixed tires, the rear axle reaches the adhesion limit first during

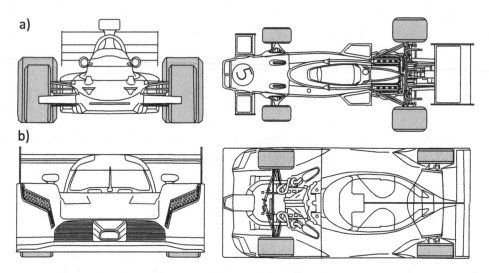

Fig. 5.11 Example of a mixed tire on (**a**) Lotus-Ford 72 (1970) and (**b**) Nissan GT-R LM (2015) LMP1

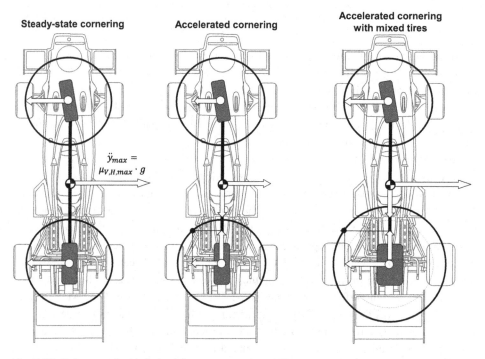

Fig. 5.12 Influence of mixed-sized tires on rear axle stability

acceleration. The remaining potential of the front axle cannot be used, otherwise the vehicle tends to oversteer during acceleration. Mixed tires on the rear axle allow an increase in rear axle lateral force for the same longitudinal acceleration, so that the available frictional potential of the front axle can also be utilized. The vehicle remains neutral for longer during acceleration. The limiting lateral acceleration is now determined by the front axle. The understeer tendency or the stability of the vehicle increases, which also has a positive effect when braking. In addition, higher maximum longitudinal acceleration and braking deceleration can be achieved with mixed tires. However, since the permissible tire dimensions are generally very strictly regulated today, mixed tires cannot be freely configured.

5.3   Two-Track Model

In the case of the two-track model, the kinematic variables and the wheel forces are considered individually. Figure 5.13 summarizes the variables in the x-y plane. Analogous to the single-track model, the equations of motion of the two-track model are obtained by linearizing the angular relationships:

$$\ddot{x} = \frac{F_{x1} + F_{x2} + F_{x3} + F_{x4} - F_{Lx}}{m}, \tag{5.35}$$

$$\ddot{y} = \frac{F_{y1} + F_{y2} + F_{y3} + F_{y4}}{m} = v \cdot \left(\dot{\psi} + \dot{\beta} \right), \tag{5.36}$$

$$\ddot{\psi} = \frac{\left(F_{y1} + F_{y2} \right) \cdot l_V - \left(F_{y3} + F_{y4} \right) \cdot l_H + \left(F_{x2} - F_{x1} \right) \cdot \frac{s_V}{2} + \left(F_{x4} - F_{x3} \right) \cdot \frac{s_H}{2}}{J_z}. \tag{5.37}$$

A difference to the single-track model results from the influence of the drive forces on the yaw acceleration. This is of particular importance if – for example due to the use of a limited slip differential (see Chap. 8) – asymmetrical drive force distributions result. For driving dynamics, the influence of dynamic wheel loads on the lateral and longitudinal force build-up is also a decisive factor. The dynamic wheel loads differ from the static wheel loads due to the leverage effect of inertial forces. In order to provide a basic understanding of the generation of dynamic wheel loads, a simplified chassis and body model is introduced for the interaction of these components.

To this end, the essential components of the chassis are explained in advance with the aid of Fig. 5.14. The rear axle of an Audi R8 of model year 2007 is shown. The wheels are attached to the wheel carrier, which also carries the brake calliper and accommodates the wheel bearing. The wheel carrier is connected to the body via the lower wishbone, the tie rod, the upper wishbone and a strut. The suspension strut consists of a coil spring and a telescopic damper. The coil spring bears the weight of the body. Furthermore, the wheel

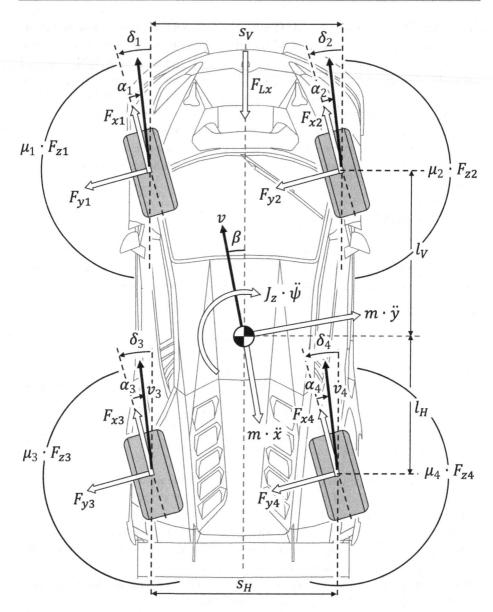

Fig. 5.13 Forces and kinematic variables of the two-track model

carriers are each connected to the stabilizer via a coupling rod. The anti-roll bar is guided on the body by two bearings so that it can rotate.

The function of the stabilizer is shown in Fig. 5.15. If both wheels move upwards (or downwards) in the same direction, the stabilizer rotates without resistance in the bearings. The movement of the body is not affected. In the case of alternating upward

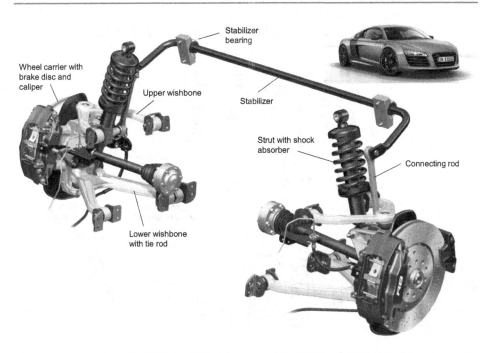

Fig. 5.14 Rear axle of the 2007 Audi R8. (Courtesy of © AUDI AG 2018. All Rights Reserved)

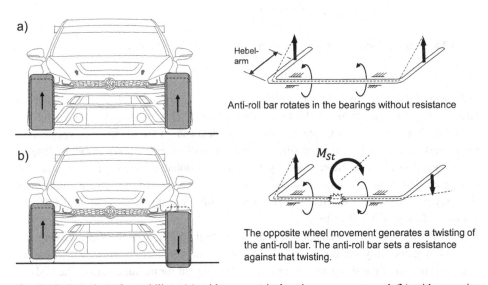

Fig. 5.15 Function of a stabilizer (**a**) with symmetrical spring movement and (**b**) with opposite spring movement

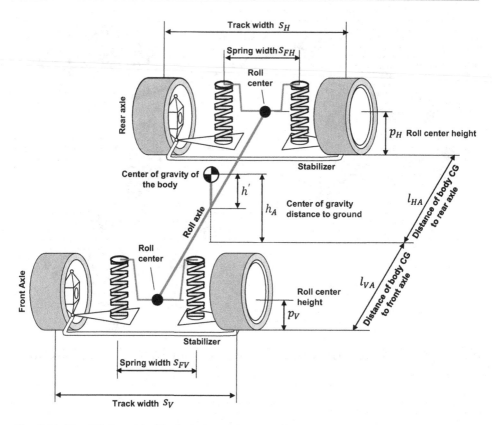

Fig. 5.16 Simplified model of body during stationary straight-ahead travel

and downward deflection, as occurs during rolling or any asymmetrical wheel movement, a twisting angle occurs between the two stabilizer ends. This twisting is resisted by the stabilizer. The resistance is greater the shorter the lever arms are to the force application points and the higher the torsional stiffness of the stabilizer. For a given stabilizer length, the torsional stiffness is determined by the cross-sectional geometry and the material properties. The stabilizer torque M_{St} generated about the x-axis of the vehicle counteracts the body roll. The characteristic property of the stabilizer is that it increases roll stiffness without affecting vertical stiffness.

Figure 5.16 shows the interaction of chassis and body during straight-ahead driving. To simplify matters, only stationary driving conditions are considered below, which means that the function of the shock absorbers can be disregarded for the time being. The body is symbolized by the roll axis and the support in the strut bearings. The roll axis is the connecting line between the roll centers at the front and rear axle. The meaning of the roll axis will be discussed in detail in Chap. 7. At this point it is only important to know that the roll center represents the point around which the body rotates and in which the support of

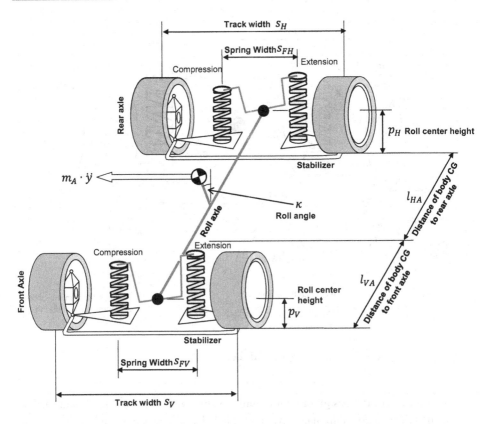

Fig. 5.17 Simplified model of the body during steady-state cornering

the lateral forces on the body can be summarized. With the two-track model, a subdivision is made into the sprung mass of the body (sprung masses) m_A and the unsprung masses m_V and m_H of the suspensions. The position of the body's center of gravity is described by the setback l_{VA}, the projection l_{HA} and its height above the road h_A. The lever arm h' is created between the body center of gravity and the rolling axis. In this state, the body springs only support the weight force of the body.

When cornering, the centrifugal force acting at the center of gravity and the lever arm h' create a moment which turns the body around the roll axis. This rolling motion leads to the roll angle κ. Due to the rolling motion, the vehicle moves downwards (compression of the spring/bump) on the outside of the curve and upwards (rebound) on the inside of the curve. These relationships are shown in Fig. 5.17. With the aid of this simplified model, a mathematical relationship is established between lateral acceleration, roll angle and dynamic wheel loads. In the first step, the geometric relationships are illustrated with the aid of Fig. 5.18. The geometric relationships are shown in Fig. 5.16. Due to the compression and rebound, the connecting line between the upper spring mounts twists by the roll

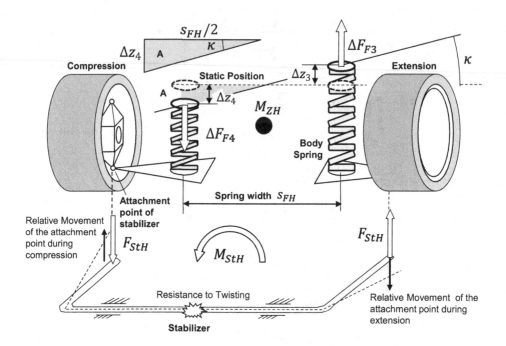

Fig. 5.18 Geometrical conditions on the rear axle during rolling

angle κ compared to the design or equilibrium position when driving straight ahead. It is assumed that there is no additional heave movement. The changes in spring travel Δz_3 and Δz_4 are then equal in amount and result from the triangular relationships between changes in spring travel, spacing of the body springs s_{FH} and roll angle κ. It holds:

$$\Delta z_3 = \Delta z_4 = \frac{s_{FH}}{2} \cdot \sin(\kappa). \tag{5.38}$$

For small wall angles you get:

$$\Delta z_3 = \Delta z_4 = \frac{s_{FH}}{2} \cdot \kappa. \tag{5.39}$$

Due to the changes in spring travel, the forces generated by the body springs change. These deviations from the static spring force are the spring force changes ΔF_{F3} and ΔF_{F4}. They form a moment that counteracts the roll motion. The relationship between changes in spring travel and changes in spring force can be seen from the spring characteristic curve of the body spring, which is shown in Fig. 5.19. In the equilibrium position, the body spring has the absolute spring travel z_0. In this condition, the spring must carry half the rear axle load. It is assumed at this point that there is a linear relationship between spring travel and spring

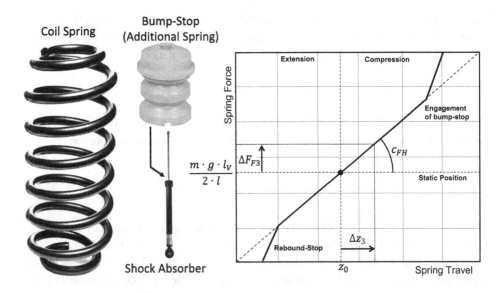

Fig. 5.19 Properties of the body spring

force for the helical body spring. This linear relationship is described by the spring stiffness c_{FH}. It applies to the amounts of the spring force changes:

$$\Delta F_{F3} = \Delta F_{F4} = c_{FH} \cdot \Delta z_{3,4} = c_{FH} \cdot \frac{s_{FH}}{2} \cdot \kappa. \tag{5.40}$$

During compression, the force supported in the spring increases, and during extension, it decreases. The two spring forces counteract the rolling motion with the following moment:

$$M_{FH} = \Delta F_{F3} \cdot \frac{s_{FH}}{2} + \Delta F_{F4} \cdot \frac{s_{FH}}{2} = c_{FH} \cdot \kappa \cdot \frac{s_{FH}^2}{2}. \tag{5.41}$$

In addition, the effect of an additional spring (colloquially also called a bump stop or rubber spring) is sketched in Fig. 5.19. The additional spring is pushed onto the piston rod of the shock absorber. The effect of the additional spring starts when the lower part of the shock absorber has been pushed up so far during compression that the additional spring is braced between the shock absorber tube and the strut mount. From this point on, the spring effects of the body spring and the additional spring add up (parallel connection of two springs), and a progressive spring is created. The additional spring prevents the shock absorber from hitting hard mechanically and thus protects it from overload. A similar mechanism is also used for the rebound stop during extension. In addition to the body springs, the stabilizer is active in Fig. 5.18. At the compressing wheel, the stabilizer end is moved upwards relative to its initial position, at the rebounding wheel, the stabilizer end moves downwards. The reaction forces F_{StH} are thus generated at the ends of the stabilizer. These two forces produce the stabilizer torque M_{StH}, which counteracts the rolling motion. For simplicity, this moment is assumed to be proportional to the roll angle κ, hence:

$$M_{StH} = c_{StH} \cdot \kappa. \tag{5.42}$$

The parameter c_{StH} is the stabilizer stiffness. At the rear axle, the body springs and the stabilizer thus provide the following moment of resistance to the rolling motion:

$$M_H = M_{FH} + M_{StH} = \left(c_{FH} \cdot \frac{s_{FH}^2}{2} + c_{StH} \right) \cdot \kappa. \tag{5.43}$$

Equations (5.38, 5.39, 5.40, 5.41, 5.42 and 5.43) can be applied analogously to the front axle. To determine the wheel loads and the roll angle, the next step is to cut the body free from the chassis. This procedure is shown in Fig. 5.20. On the body, the moments caused by the centrifugal force and the moments caused by the weight force are in equilibrium with the roll resistance moments of the front and rear axles, so that the moment equilibrium about the roll axis applies:

$$m_A \cdot h' \cdot \ddot{y} + m_A \cdot g \cdot h' \cdot \kappa = M_V + M_H, \tag{5.44}$$

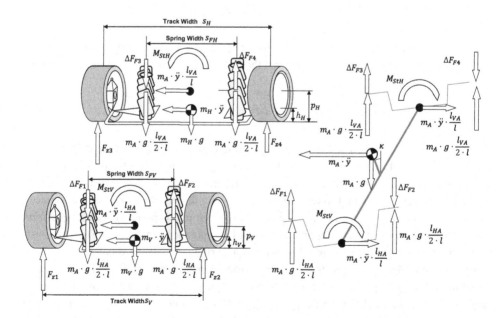

Fig. 5.20 Forces and moments at the free section between body and chassis

$$m_A \cdot \ddot{y} \cdot h' + m_A \cdot g \cdot h' \cdot \kappa = \left(\frac{c_{FV} \cdot s_{FV}^2}{2} + c_{StV} + \frac{c_{FH} \cdot s_{FH}^2}{2} + c_{StH} \right) \cdot \kappa. \quad (5.45)$$

By reshaping, one obtains the roll angle to:

$$\kappa = \frac{m_A \cdot \ddot{y} \cdot h'}{\left(\frac{c_{FV} \cdot s_{FV}^2}{2} + c_{StV} + \frac{c_{FH} \cdot s_{FH}^2}{2} + c_{StH} \right) - m_A \cdot g \cdot h'}. \quad (5.46)$$

The individual wheel loads are obtained by forming the equilibria of forces around the individual wheel contact points. One obtains:

$$F_{z1} = \frac{m_A \cdot g \cdot l_{HA}}{2 \cdot l} + \frac{m_V \cdot g}{2} - \Delta F_{F1} \cdot \frac{s_{FV}}{s_V} - \frac{M_{StV}}{s_V} - \frac{m_A \cdot \ddot{y} \cdot l_{HA} \cdot p_V}{l \cdot s_V}$$
$$- \frac{m_V \cdot \ddot{y} \cdot h_V}{s_V}, \quad (5.47)$$

$$F_{z2} = \frac{m_A \cdot g \cdot l_{HA}}{2 \cdot l} + \frac{m_V \cdot g}{2} + \Delta F_{F2} \cdot \frac{s_{FV}}{s_V} + \frac{M_{StV}}{s_V} + \frac{m_A \cdot \ddot{y} \cdot l_{HA} \cdot p_V}{l \cdot s_V}$$
$$+ \frac{m_V \cdot \ddot{y} \cdot h_V}{s_V}, \quad (5.48)$$

$$F_{z3} = \frac{m_A \cdot g \cdot l_{VA}}{2 \cdot l} + \frac{m_H \cdot g}{2} - \Delta F_{F3} \cdot \frac{s_{FH}}{s_H} - \frac{M_{StH}}{s_H} - \frac{m_A \cdot \ddot{y} \cdot l_{VA} \cdot p_H}{l \cdot s_H}$$
$$- \frac{m_H \cdot \ddot{y} \cdot h_H}{s_H}, \quad (5.49)$$

$$F_{z4} = \frac{m_A \cdot g \cdot l_{VA}}{2 \cdot l} + \frac{m_H \cdot g}{2} + \Delta F_{F4} \cdot \frac{s_{FH}}{s_H} + \frac{M_{StH}}{s_H} + \frac{m_A \cdot \ddot{y} \cdot l_{VA} \cdot p_H}{l \cdot s_H}$$
$$+ \frac{m_H \cdot \ddot{y} \cdot h_H}{s_H}. \quad (5.50)$$

If the relationships given in Eq. (5.40) are used for the change in spring force and the expression given in Eq. (5.46) is used for the roll angle contained therein, the wheel loads can be calculated as a function of the chassis characteristics and the lateral acceleration as follows:

$$F_{z1} = \frac{m_A \cdot g \cdot l_{HA}}{2 \cdot l} + \frac{m_V \cdot g}{2}$$
$$- \left(\frac{\left(\frac{c_{FV} \cdot s_{FV}^2}{2} + c_{StV} \right) \cdot m_A \cdot h'}{c_W - m_A \cdot g \cdot h'} + \frac{m_A \cdot l_{HA} \cdot p_V}{l} + m_V \cdot h_V \right) \cdot \frac{\ddot{y}}{s_V}, \quad (5.51)$$

$$F_{z2} = \frac{m_A \cdot g \cdot l_{HA}}{2 \cdot l} + \frac{m_V \cdot g}{2}$$

$$+ \left(\frac{\left(\frac{c_{FV} \cdot s_{FV}^2}{2} + c_{StV} \right) \cdot m_A \cdot h'}{c_W - m_A \cdot g \cdot h'} + \frac{m_A . l_{HA} \cdot p_V}{l} + m_V \cdot h_V \right) \cdot \frac{\ddot{y}}{s_V}, \qquad (5.52)$$

$$F_{z3} = \frac{m_A \cdot g \cdot l_{VA}}{2 \cdot l} + \frac{m_H \cdot g}{2}$$

$$- \left(\frac{\left(\frac{c_{FH} \cdot s_{FH}^2}{2} + c_{StH} \right) \cdot m_A \cdot h'}{c_W - m_A \cdot g \cdot h'} + \frac{m_A . l_{VA} \cdot p_H}{l} + m_H \cdot h_H \right) \cdot \frac{\ddot{y}}{s_H}, \qquad (5.53)$$

$$F_{z4} = \frac{m_A \cdot g \cdot l_{VA}}{2 \cdot l} + \frac{m_H \cdot g}{2}$$

$$+ \left(\frac{\left(\frac{c_{FH} \cdot s_{FH}^2}{2} + c_{StH} \right) \cdot m_A \cdot h'}{c_W - m_A \cdot g \cdot h'} + \frac{m_A . l_{VA} \cdot p_H}{l} + m_H \cdot h_H \right) \cdot \frac{\ddot{y}}{s_H}, \qquad (5.54)$$

$$\text{with} : c_W = \frac{c_{FV} \cdot s_{FV}^2}{2} + c_{StV} \cdot \frac{c_{FH} \cdot s_{FH}^2}{2} + c_{StH}. \qquad (5.55)$$

It can be seen from the equations that the wheel loads on the outer wheels increase, while they decrease on the wheels on the inside of the curve. Figure 5.21 illustrates the meaning of the formulae Eqs. (5.51, 5.52, 5.53, and 5.54) by means of an example. Two tuning variants of a Golf GTI are considered, which differ only in the stabilizer stiffnesses of the front and rear axle stabilizers, although the sum of both stabilizer stiffnesses remains constant. The roll angle is therefore always the same for both variants. The roll angle related to the lateral acceleration is referred to as the roll gradient. In this example it is 4°/g. The course of the wheel loads as a function of the lateral acceleration is shown for both variants. The static weight is 60% on the front axle and 40% on the rear axle, which corresponds to a typical axle load distribution of a front-wheel drive vehicle. The static wheel loads are identical for both cases, as they are not affected by the stabilizer stiffnesses. In series-production vehicles, the roll moment is usually supported significantly in favor of the front axle, which is expressed in the significantly higher front-axle stabilizer stiffness in case 1.

The wheel load changes on the front axle are therefore significantly greater than those on the rear axle, which means that the front wheels on the inside of the bend are almost completely unloaded in the limit range. In case 2, the stiffness of the rear axle stabilizer is significantly increased and lowered by the corresponding amount at the front axle. The result of this is that the wheel load changes decrease at the front axle and increase at the rear axle. This results in an enormously important law for driving dynamics:

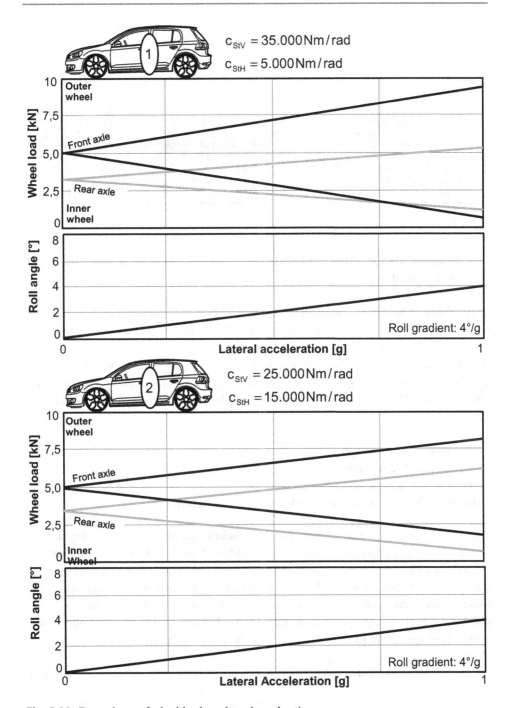

Fig. 5.21 Dependence of wheel loads on lateral acceleration

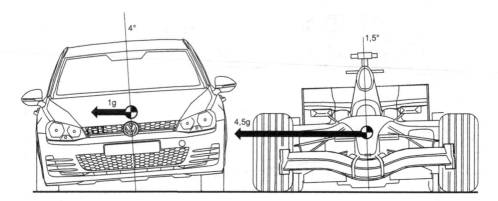

Fig. 5.22 Roll gradients of a Golf VII GTI and an F1 vehicle

▶   Increasing the roll stiffness of an axle leads to an increase in the dynamic wheel load
     changes on that axle.

   By changing the spring and stabilizer stiffness, it is thus possible to influence both the
height of the roll angle and the distribution of the wheel load changes between the front and
rear axles. It should be noted that the body's center of gravity moves against the direction of
the curve during roll. Therefore, a high roll stiffness and a high roll damping or a low roll
gradient lead to a faster response of the vehicle to steering inputs. Figure 5.22 shows the
roll behavior of a Golf VII GTI and an F1 vehicle. Typical roll gradients of different
vehicles are:

- approx. 6°/g for passenger cars
- approx. 3–4.2°/g for sports cars
- less than 1.5°/g for near-series racing vehicles
- approx. 0.2–0.7°/g for racing vehicles with massive downforce

The wheel load changes caused by longitudinal accelerations and longitudinal
decelerations can be treated by analogous procedure. Figure 5.23 shows the conditions
for braking with constant deceleration. The influence of the dampers is also neglected for
this special case. The stabilizers do not generate any forces, since the front and rear axles
compress and decompress in the same direction. Additionally, the influence of the air
resistance is neglected compared to the inertia force. For small angles, one first obtains the
spring travel and spring force changes at the front and rear axle to:

$$\text{Front axle}: \Delta z_V = l_V'' \cdot \varphi, \text{Rear axle}: \Delta z_{VH} = l_H'' \cdot \varphi, \tag{5.56}$$

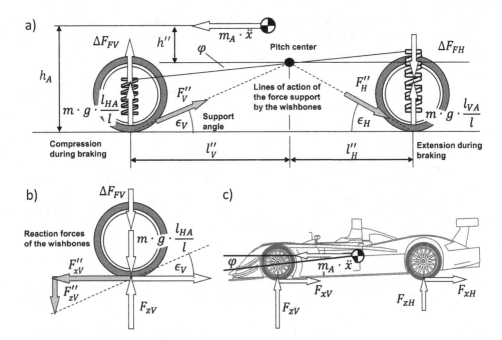

Fig. 5.23 Forces during brake pitching: (**a**) Body cut free from control arms and springs, (**b**) Front wheels cut free from body and roadway, (**c**) Vehicle cut free from roadway

$$\text{Front axle}: \Delta F_{FV} = \Delta F_{F1} + \Delta F_{F2} = 2 \cdot c_{FV} \cdot l_V'' \cdot \varphi, \tag{5.57}$$

$$\text{Rear axle}: \Delta F_{FH} = \Delta F_{F3} + \Delta F_{F4} = 2 \cdot c_{FH} \cdot l_{VH}'' \cdot \varphi. \tag{5.58}$$

The vehicle body rotates around the pitch center, which results from the geometric arrangement of the control arms. The pitch center is the intersection of the forces transmitted to the body by the control arms. These characteristics of the chassis are discussed in Sect. 7.7.2. The moment equilibrium around the pitching pole is:

$$m_A \cdot \ddot{x} \cdot h'' = \Delta F_{FV} \cdot l_V'' + \Delta F_{FH} \cdot l_{VH}''. \tag{5.59}$$

The reaction forces transmitted in the control arms do not contribute to the pitch excitation. By substituting Eqs. (5.57) and (5.58), the pitch angle can be determined from Eq. (5.59). It holds:

$$\varphi = \frac{m_A \cdot \ddot{x} \cdot h''}{2 \cdot \left(c_{FV} \cdot l_V'' + c_{FH} \cdot l_H'' \right)}. \tag{5.60}$$

The dynamic wheel loads result from the static axle loads, the spring force changes and the vertical component from the support forces transmitted in the control arms. Provided that the springs do not absorb any lateral forces, the following applies to this component:

$$F''_{zV} = F_{xV} \cdot \sin(\epsilon_V) \ \text{ and } \ F''_{zH} = F_{xH} \cdot \sin(\epsilon_H). \tag{5.61}$$

For the wheel loads you get:

$$F_{z1} = F_{z2} = \frac{m_A \cdot g \cdot l_{HA}}{2} + \frac{m_V}{2} + c_{FV} \cdot l''_V \cdot \frac{m_A \cdot \ddot{x} \cdot h''}{2 \cdot \left(c_{FV} \cdot l''_V + c_{FH} \cdot l''_H\right)}$$

$$+ \frac{F_{xV} \cdot \sin(\epsilon_V)}{2}, \tag{5.62}$$

$$F_{z3} = F_{z4} = \frac{m_A \cdot g \cdot l_{VA}}{2} + \frac{m_H}{2} - c_{FH} \cdot l''_{VH} \cdot \frac{m_A \cdot \ddot{x} \cdot h''}{2 \cdot \left(c_{FV} \cdot l''_V + c_{FH} \cdot l''_H\right)}$$

$$- \frac{F_{xH} \cdot \sin(\epsilon_H)}{2}. \tag{5.63}$$

According to the equations, the wheel loads on the front axle increase during deceleration, while the rear wheels are relieved. Here too, a higher vertical stiffness of the axle leads to an increase in the wheel load changes occurring on this axle. Under real conditions, longitudinal and lateral accelerations – e.g. when braking into a curve or accelerating out of a curve – occur simultaneously. In such cases, the two effects overlap, and the proportional spring deflections from pitching and rolling add up. If the spring deflections are derived in time, the relative velocities are obtained against which the damper works. The following applies, provided that the heave movement of the body is neglected:

$$\Delta z_1 = -\kappa \cdot \frac{s_{FV}}{2} + \varphi \cdot l''_V \rightarrow \Delta \dot{z}_1 = -\dot{\kappa} \cdot \frac{s_{FV}}{2} + \dot{\varphi} \cdot l''_V, \tag{5.64}$$

$$\Delta z_2 = +\kappa \cdot \frac{s_{FV}}{2} + \varphi \cdot l''_V \rightarrow \Delta \dot{z}_2 = +\dot{\kappa} \cdot \frac{s_{FV}}{2} + \dot{\varphi} \cdot l''_V, \tag{5.65}$$

$$z_3 = -\kappa \cdot \frac{s_{FH}}{2} - \varphi \cdot l''_{VH} \rightarrow \Delta \dot{z}_3 = -\dot{\kappa} \cdot \frac{s_{FH}}{2} - \dot{\varphi} \cdot l''_H, \tag{5.66}$$

$$z_4 = +\kappa \cdot \frac{s_{FH}}{2} - \varphi \cdot l''_{VH} \rightarrow \Delta \dot{z}_4 = +\dot{\kappa} \cdot \frac{s_{FH}}{2} - \dot{\varphi} \cdot l''_H. \tag{5.67}$$

The body dampers basically have the same influence on the wheel load changes as body springs and anti-roll bar. The harder the damper is in its rebound or compression stage, the higher the wheel load change caused by it. Figure 5.24 illustrates the distribution of wheel loads as well as rebound and compression behavior of the suspension in different driving

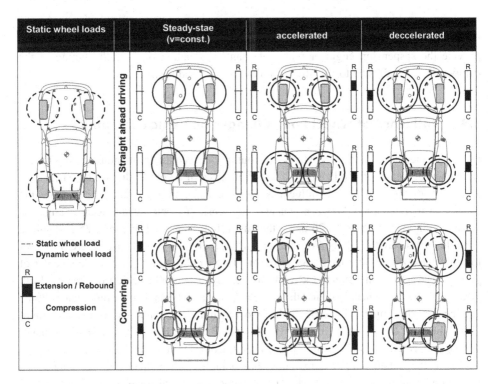

Fig. 5.24 Schematic wheel load distribution and effect of dampers

situations. When cornering, a dynamic wheel load transfer takes place from the inside to the outside wheels. On turn-in, the inside wheels deflect and the inside dampers work in rebound. On the outside of the turn, the wheels moves towards the body and the dampers are in compression. When steering back to straight ahead, the velocity directions in the dampers reverse, but this is not shown in Fig. 5.24. During acceleration a wheel load transfer from the front axle to the rear axle takes place. At the dampers of the front axle the rebound stage is active due to the extension movement, while the dampers of the rear axle are loaded to compression due to the compression. During deceleration, these conditions are reversed. During accelerated or decelerated cornering, these effects overlap. During accelerated cornering, the deflection processes on the wheel on the inside of the bend increase, which is therefore relieved of a great deal of load. For this reason, a driven front wheel tends to spin the most. In this situation the damper of the inside wheel works strongly in the rebound stage. On the outside rear wheel, the two compression processes overlap and the compression stage is effective. At the other two wheels, the spring movements of the longitudinal and lateral accelerations weaken each other. During decelerated cornering, the rebound movement is strongest on the inside rear wheel, and this wheel is most likely to lock up during deceleration. The rebound stage is effective on this damper. On the outer front wheel, the compression movements overlap and lead to a pronounced activation of the compression stage on this damper. Knowledge of the condition of the damper can provide the suspension engineer with important information on how the tuning of the damper must be adjusted to specifically influence the handling.

5.4   Influence of Wheel Load Changes on Vehicle Dynamics

Dynamic wheel load changes and their distribution between the front and rear axles have a significant influence on the handling and performance of a vehicle due to the wheel load sensitivity of the tires. These influences must be taken into account when tuning the chassis. The interaction between wheel load sensitivity and wheel load changes is visualized in Figs. 5.25 and 5.26 using the lateral force potential. Figure 5.25 first shows the axle lateral force resulting from the two tires for the theoretical case where no wheel load changes occur under lateral acceleration. This would require the vehicle's center of gravity to be at the same level as the road surface, which cannot be achieved in reality. Both tires have the same wheel load and therefore also achieve the same maximum lateral force at a certain slip angle. The sum of the two maximum lateral forces is the maximum achievable axle lateral force. In this example it is 8 kN. In Fig. 5.26, on the other hand, the wheel load changes occurring during lateral acceleration are taken into account. The wheel on the outside of the curve now has a higher wheel load and can transmit a correspondingly higher lateral force. The transmissible lateral force on the inside wheel

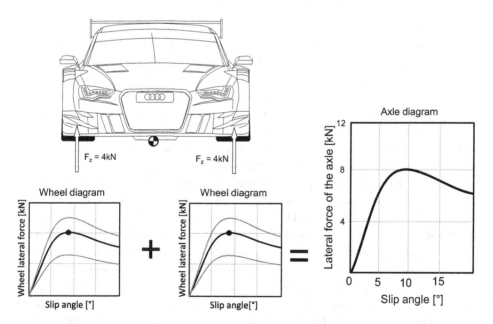

Fig. 5.25 Lateral force potential without dynamic wheel load changes

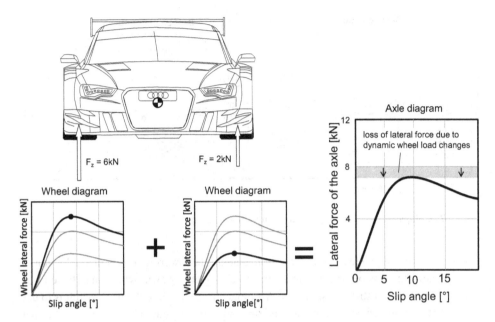

Fig. 5.26 Lateral force potential with dynamic wheel load changes

decreases due to the reduction in load. It has already been explained in Sect. 4.3 that there is a degressive relationship between the wheel load and the coefficient of adhesion or the maximum transmissible lateral force. This means that the increase in lateral force on the outer wheel cannot compensate for the loss of lateral force on the inner wheel. The wheel load changes thus reduce the maximum transmittable axle lateral force and the maximum achievable lateral acceleration. The same consideration can be applied to the maximum longitudinal acceleration and deceleration of a vehicle with identical results.

To achieve the best possible driving performance, it is therefore necessary to keep the wheel load changes that occur as low as possible. It should be expressly noted at this point that the magnitude of the dynamic wheel load changes depends exclusively on the center of gravity heights h_V, h_H and h_A, the track widths and the wheelbase as well as the masses m_V, m_H and m_A, with the body parameters making the dominant contribution. All other chassis parameters discussed in the previous section only affect the distribution of dynamic wheel loads between the front and rear axles. The minimization of the wheel load changes desired for reasons of driving dynamics is thus only possible by reducing the masses, lowering the body and achieving the highest possible track widths and wheelbases, with the wheelbase influencing the wheel load changes from longitudinal accelerations. These measures can be clearly seen in particular in near-series racing vehicles, as Fig. 5.27 shows using the Seat Leon as an example.

The Cup version has been lowered significantly compared to the standard vehicle, which moves the body center of gravity downwards. Another positive side effect is the reduction

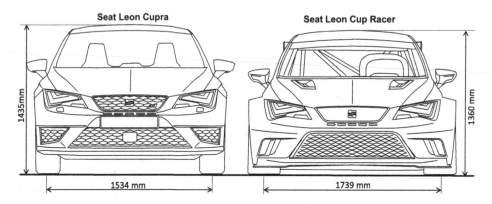

Fig. 5.27 Comparison of a standard Seat Leon with a Seat Leon CUP

in air resistance due to a reduction in the frontal area. The lowering results in a reduction of the maximum suspension travel before the vehicle touches down, which may have to be compensated for by a stiffer suspension. A simple way to increase the track width is to use spacers between the wheel bearing and the rim and to reduce the offset on the rim. Further conceptual possibilities for achieving the lowest possible overall center of gravity are explained in Sect. 9.2.

The unavoidable wheel load changes are distributed to the two axles by tuning the springs, dampers and stabilizers, taking into account the chassis geometry. The basic influence on the driving behavior is explained using the example of a front- and a rear-wheel drive vehicle. Figure 5.28 first shows the influence of three basic design philosophies on the driving behavior of a front-wheel drive vehicle. The starting point is balanced roll support, so that the wheel load changes on the front and rear axles are approximately equal. In this case, the friction circles at both axles are also approximately equal. By redistributing the roll support in favor of the front axle, the front axle's adhesion potential decreases and that of the rear axle increases, as expressed by the size of the friction circles. This type of roll support is achieved by using harder body springs, harder anti-roll bars or harder dampers on the front axle, or by softer tuning of the rear axle.

According to Sect. 5.2 and Fig. 5.8, a vehicle balanced in this way exhibits good stability behavior and a tendency to understeer even under maximum lateral acceleration, because it still has sufficient lateral force reserves on the rear axle. For this reason, production vehicles are generally tuned in such a way that such behavior is ensured. The disadvantage of such a set-up for front-wheel-drive vehicles is that their understeer tendency increases further during acceleration. The maximum transmissible lateral force on the front axle is reduced by the longitudinal force that has to be transmitted at the same time. This phenomenon is referred to as power-on understeer. The power reserves of the rear axle increase, but cannot be used for an increase in driving performance. When roll support is provided in favor of the rear axle, the result is a tendency to oversteer. Due to the increased tendency of the vehicle to enter an unstable driving state, such a design is

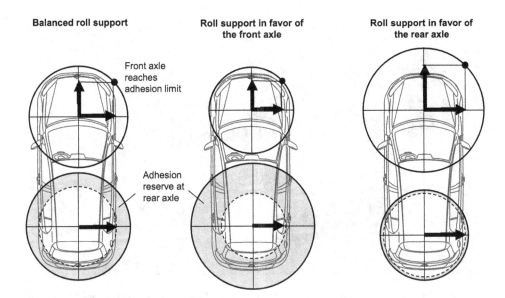

Fig. 5.28 Influence of the roll support with front-wheel drive

unsuitable for production vehicles. In a front-wheel-drive vehicle, however, such a design can improve acceleration behavior in curves. The previously unusable lateral force reserves of the rear axle are used to implement a rear-axle-heavy roll support. The resulting enlargement of the friction circle at the front axle increases the longitudinal force that can be transmitted there.

However, such a roll support is usually only used on front-wheel drive racing vehicles. As Fig. 5.29 shows, the high stabilization of the rear axle can be recognized by a lifting of the inside rear wheel on the bend. The adhesion potential of the front and rear axles is affected by the different design philosophies in rear-wheel drive vehicles (Fig. 5.30) in the same way as in front-wheel drive. The main difference, however, is that the requirements for high stability and good traction are not contradictory in a rear-wheel drive vehicle. The power reserves on the rear axle resulting from a high degree of stabilization of the front axle can be utilized by the drive. Vehicles with rear-axle drive therefore make better use of the available adhesion potential. Roll support in favor of the rear axle is counterproductive for rear-wheel drive vehicles in terms of both vehicle stability and traction.

Another conceptual advantage of rear-wheel-drive vehicles results from the fact that during acceleration the dynamic wheel load changes lead to an increase in the wheel load on the driven axle. In a front-wheel drive vehicle, on the other hand, the driven axle is relieved by the wheel load changes. Thus, for rear-wheel drive vehicles, the transmissible tire forces increase during acceleration, while they decrease for front-wheel drive. Figure 5.31 shows this effect using the front-wheel drive Audi TT Cup (2015) and the rear-wheel drive Bentley Continental GT3 as examples. Front-wheel drive cars compensate for

Fig. 5.29 Roll support of a front-wheel drive Polo Cup car. (Courtesy of © Volkswagen Motorsport GmbH 2018. All Rights Reserved)

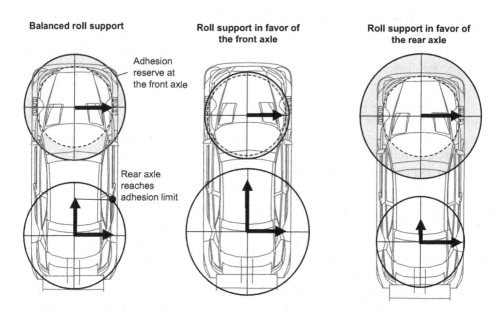

Fig. 5.30 Influence of the roll support for rear-wheel drive

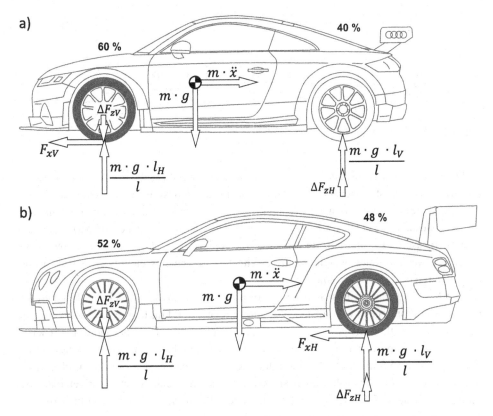

Fig. 5.31 Comparison of (**a**) front-wheel drive and (**b**) rear-wheel drive during acceleration

some of this loss by having about 60% of their weight on the front axle. At low friction coefficients, such as those encountered in snow or very heavy rain, front-wheel-drive vehicles can even have a traction advantage over rear-wheel-drive vehicles at low longitudinal accelerations. However, at high longitudinal accelerations, such as those achieved in high-performance racing in particular, the advantage due to favorable wheel load transfer outweighs this. This results in a higher maximum longitudinal acceleration and a traction advantage when accelerating from the apex of the curve. Another side effect is that a more balanced static weight distribution can be realized, which leads to a more effective utilization of the tire potential according to Fig. 5.9. Due to their front-heavy weight distribution, front-wheel drive vehicles also have a higher slip angle requirement at the front axle than at the rear axle, which increases the tendency to understeer. These reasons are key to why sports cars and high-performance race cars are almost invariably rear-wheel drive vehicles (or rear-wheel drive based all-wheel drive). The key advantages of front-wheel-drive vehicles, such as the lower cost of the powertrain topology and the space available in the interior, are not decisive in racing.

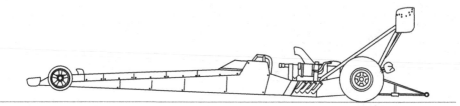

Fig. 5.32 Top Fuel Dragster during acceleration

In racing series with production-based racing vehicles, such as the FIA WTCC, in which the drive type must correspond to that of the production vehicle and rear-axle driven vehicles therefore compete against front-axle driven vehicles, the conceptual advantages of rear-axle driven vehicles are often compensated for by a "balance of performance". Typical measures for this are, for example, the replacement of standing starts by flying starts, a lower minimum weight for front-wheel drive vehicles or greater leeway in the aerodynamic design of the vehicles.

Top Fuel Dragsters represent a special form of racing vehicles, which are designed for pure acceleration races over a distance of approximately 300 m. Figure 5.32 shows a Top Fuel Dragster in the start phase of such an acceleration race. Top Fuel Dragsters reach a speed of 160 km/h after about 0.8 s, which corresponds to an average longitudinal acceleration of 55.5 m/s^2. In addition to the tires, which can have a coefficient of friction of up to 6.5, Top Fuel Dragsters are designed so that all the weight is on the driven rear axle during acceleration. Trim wings on the front axle prevent the vehicle from rolling over.

5.5   Driving Behavior as a Function of Balance, Brake Force Distribution and Drive Type

The basic driving behavior of vehicles can be described in terms of simple circular movements. A variant of this is the stationary circular motion. In the case of steady-state circular travel, the vehicle is accelerated very slowly on a constant radius of curvature, and the driving state variables are recorded as a function of the lateral acceleration. This procedure is shown as an example in Fig. 5.33. In this case, the course of the steering wheel angle, the side slip angle and the roll angle are plotted against the lateral acceleration as an example. The steering wheel angle requirement is a measure of the vehicle's agility and defines its steering tendency. With the help of the side slip angle, the stability of the vehicle can be inferred. As a rule, the aim is for the vehicle to have a low side slip angle in the driving dynamics limit range. The condition for this is a stiff rear axle with corresponding adhesion reserve. Side slip angles of high magnitude are an indication of unstable driving conditions. The development of the steering wheel and side slip angle curves as well as the influence of various chassis parameters are considered analytically in

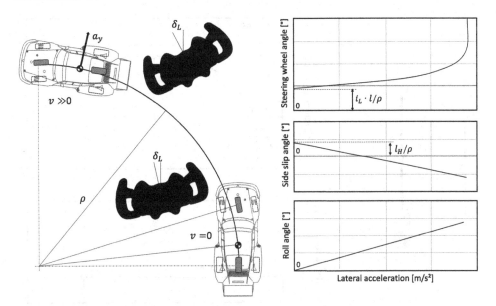

Fig. 5.33 Performing a steady-state cornering

the following with the aid of the single-track model. Due to the low longitudinal acceleration, the drive forces can be neglected compared to the lateral forces. The starting point for
the considerations is a base vehicle with the following properties:

- Vehicle mass $m = 1200$ kg
- Medium-heavy weight distribution, i.e. $l_V = l_H = \frac{l}{2}$
- Identical tires on front and rear axle with $\mu_V = \mu_H = 1.25$
- Neglect of wheel load changes
- Neglect of elastic and kinematic properties of the chassis

In a stationary circular motion, the side slip angular velocity $\dot{\beta}$ assumes the value zero, thus
resulting in a simplified relationship between yaw rate and lateral acceleration:

$$a_y = \ddot{y} = \frac{v^2}{\rho} = v \cdot \dot{\psi}. \tag{5.68}$$

With this simplification it follows for the steering wheel angle according to Eq. (5.10):

$$\delta_L = i_L \cdot \left(\frac{l}{\rho} + \alpha_V - \alpha_H \right). \tag{5.69}$$

The slip angles depend on the lateral forces, for which the following applies:

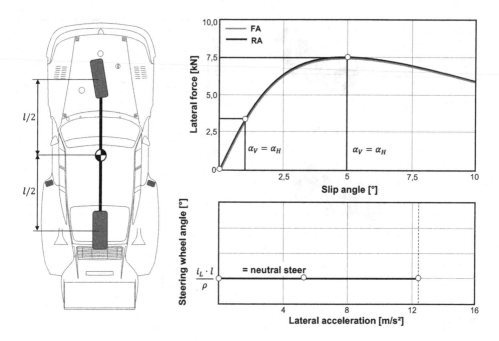

Fig. 5.34 Steady-state cornering of the neutral-steering base vehicle

$$F_{yV} = m \cdot \ddot{y} \cdot \frac{l_H}{l} \quad \text{and} \quad F_{yH} = m \cdot \ddot{y} \cdot \frac{l_V}{l}. \tag{5.70}$$

After calculating the lateral forces, the slip angles can be read from the lateral force slip diagram as shown in Fig. 5.34. Equation (5.69) is then used to determine the steering wheel angle. The steering wheel angle at the beginning of the circular drive results from the steering ratio, wheelbase and the radius of curvature to be driven over. The course over the lateral acceleration results from the slip angle difference. Due to the characteristics of the base vehicle, this is zero over the entire lateral acceleration curve. Thus, the steering wheel angle is constant until the maximum lateral acceleration is reached.

This behavior defines a neutral steering vehicle. The side slip angle of the base vehicle is calculated from Fig. 5.6 and Eq. (5.68):

$$\beta = \frac{l_H}{\rho} - \alpha_H. \tag{5.71}$$

The amount of the side slip angle thus results exclusively from the properties of the rear axle for a given vehicle geometry. Vehicles that are perceived as stable require low side slip angles and thus a stiff rear axle that responds without distortion. A stiff rear axle can be achieved, among other things, by using a more powerful tire on the rear axle. The influence of this variation compared to the basic variant is shown in Fig. 5.35.

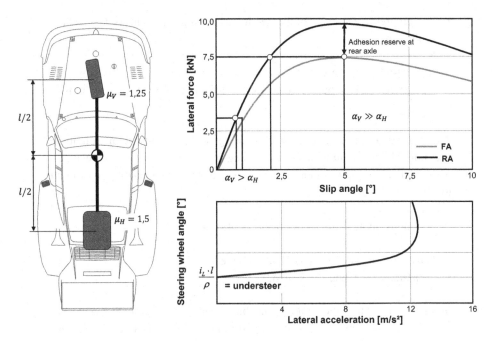

Fig. 5.35 Steady-state cornering with increased cornering stiffness and rear tire adhesion potential

In this case, the wider rear tire has both a higher slip stiffness and a higher adhesion potential than the front axle tires. A higher slip angle is therefore required on the front axle than on the rear axle to achieve a certain lateral acceleration, which leads to an increase in the steering wheel angle above the lateral acceleration. This behavior defines an understeering vehicle. Strictly speaking, the terms "neutral steering, understeer and oversteer" are defined only for steady-state circular driving, but in common usage they are also applied to dynamic driving situations. In the range of linear tire behavior, the steering wheel angle initially increases approximately linearly. With increasing lateral acceleration, the front axle then enters saturation, which leads to an increasing progression in the rise of the steering wheel angle. The limit lateral acceleration is reached when the maximum adhesion at the front axle is reached. A further increase of the slip angle by "overtightening" the steering wheel leads to a decrease of the lateral force at the front axle, which decreases the lateral acceleration of the vehicle and increases the drivable circle radius. In principle, both production and racing vehicles tend to understeer in a stationary circle. However, there are significant differences in the steering wheel angle requirement or in the extent of the understeer tendency and the maximum lateral acceleration.

▶　　The essential characteristics of an understeering design are:

- The understeering driving condition is stable and remains so even without driver intervention.

- The progressive increase of the steering wheel angle announces the reaching of the limit range.
- Stability reserves on the rear axle can be used for propulsion.
- In front-wheel drive vehicles, understeer is amplified by the drivetrain.
- Understeer cannot be corrected by the driver, or only to a very limited extent.

The basic advantage of an understeering design is the stability of this condition. The normal driver can actually do nothing wrong in this state. However, an unpopular characteristic among racers is that understeering handling is difficult to correct.

A famous quote from Walter Röhrl on this is:

- *"Oversteer* is when the passenger is scared."
- *"Understeer* is when *I'm* scared."

In order to make this quote understandable, the use of adhesion on a vehicle in an understeering and an oversteering driving situation is considered. Figure 5.36 shows the conditions on an understeering vehicle when maximum lateral acceleration is reached. At the front axle, the friction circle is fully utilized. The vehicle is in a state of equilibrium and the yaw acceleration of the vehicle is zero, or the vehicle has a constant yaw rate. Since the yaw rate can no longer be increased by the driver, the side slip angle and therefore the lateral force on the rear axle can also no longer be increased. The vehicle remains stable without any further driver action, and the rear axle still has stability reserves. The vehicle tends to leave the ideal line to the outside of the curve, it

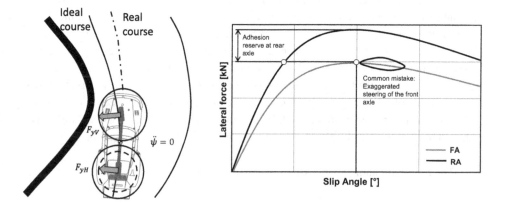

Fig. 5.36 Tire utilization with adhesion understeer

"pushes" over the front axle. The racing driver also refers to driving the vehicle "on the nose" [3].

A common mistake when understeer occurs is that the inexperienced driver intuitively steers in harder. This reduces the lateral force potential of the front axle even further and the tendency to understeer increases.

▶   Possible driving measures for understeer:

- Opening of the steering and increasing sensitively the steering wheel angle in the event of too high slip angles
- Initiate wheel load transfer to the front axle by changing the load (foot off the accelerator, tap the brake).

Moderate understeer is relatively easy to correct, but severe understeer is virtually impossible to correct effectively.

As Fig. 5.37 shows, an oversteering vehicle first reaches the adhesion limit at the rear axle. In contrast to the understeering vehicle, the driver can increase the lateral force at the front axle by further steering, resulting in a turning-in yaw acceleration. This condition is unstable because without driver intervention, the yaw rate increases continuously and can cause the rear of the vehicle to break away. In technical jargon, this is referred to as a "loose back" or "loose rear" [3].

▶   Stabilization of this driving condition requires:

- Fast countersteering and, if necessary, compensation of the counter swing

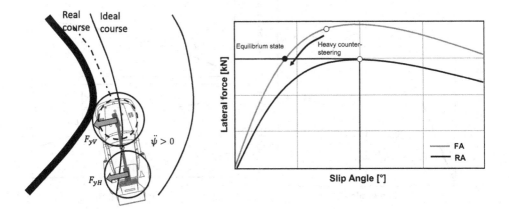

Fig. 5.37   Tire utilization in case of adhesion oversteer

- In the case of front-wheel drive vehicles, an operation of the accelerator pedal

Pressing the brake pedal to regain control is counterproductive and should only be used to reduce speed in hopeless situations. Both the required countersteering movement and the avoidance of the brake do not correspond to the intuition of a normal driver, which is why an oversteering vehicle at the performance limit is difficult to control without extensive training. Professional racing drivers correct intuitively, and intervene at an early stage before critical driving conditions arise.

With a vehicle that is set up to understeer, the risk of leaving the ideal line is higher and then always associated with a relatively high loss of time. Over a long race distance, however, such a set-up can reduce the stress on the driver and his error rate.

An important means of influencing the understeer and oversteer characteristics of a vehicle is, as already mentioned, the distribution of dynamic wheel load changes between the front and rear axles. Figure 5.38 summarizes again the influence of wheel load on cornering stiffness c_α and coefficient of adhesion μ for explanation. Both the cornering stiffness and the coefficient of adhesion show a degressive dependence on the wheel load. The essential information of this picture is that the cumulative lateral stiffness and cumulative coefficient of adhesion of the inner and outer wheel decrease by increasing dynamic wheel load change.

For the understeering vehicle in Fig. 5.39, the dynamic wheel load changes are now also taken into account. First, the case of a front-axle-heavy roll support is considered. In a stationary circular drive, such a behavior is achieved by relatively higher spring stiffness

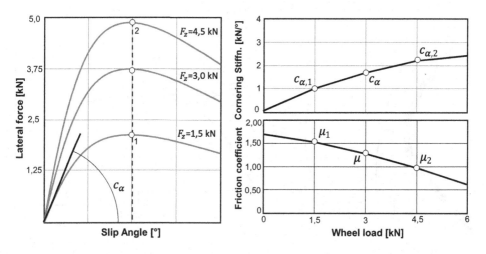

Fig. 5.38 Effect of wheel load sensitivity on cornering stiffness and adhesion

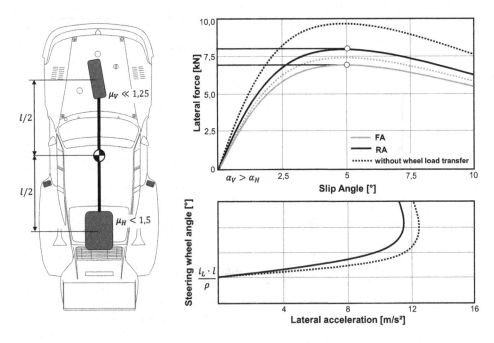

Fig. 5.39 Influence of a roll support in favor of the front axle during steady-state cornering

and stabilizer stiffness at the front axle. The dampers have no effect on the steering behavior during a steady-state circular drive. As shown in Fig. 5.38, such tuning reduces the cornering stiffness and coefficient of adhesion at the front axle more than at the rear axle. The understeer tendency of the vehicle is increased as a result. The steering wheel angle requirement rises significantly steeper even from zero lateral acceleration. Due to the lower maximum front axle lateral force, the maximum achievable lateral acceleration also decreases.

In Fig. 5.40, the reverse case of a rear axle-heavy roll support is considered. This is achieved by a relative increase of the spring and stabilizer stiffnesses at the rear axle. The dynamic wheel load changes now influence the slip stiffness and adhesion potential on the rear axle more than on the front axle. Compared to the front-axle-heavy roll support, understeer is reduced. The steering wheel angle demand is reduced and the vehicle becomes more agile. At the same time, the vehicle achieves higher maximum lateral acceleration. However, it still understeers due to its mixed tires, but is less understeery or "more oversteery" than a front-axle-heavy roll support. The maximum lateral acceleration is also lower than in the case of neglected wheel load changes, since despite rear-axle-heavy roll support, the small wheel load changes on the front axle also lead to a drop in the maximum transmissible front axle side force.

The tendency to understeer or oversteer is also evident in dynamic situations. Typical of this are so-called load change reactions, in which the accelerator pedal is either abruptly

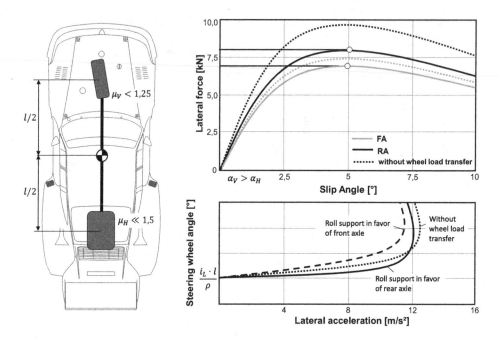

Fig. 5.40 Influence of a roll support in favor of the rear axle during steady-state cornering

depressed or abruptly released when cornering. If the accelerator pedal is depressed abruptly, this is referred to as a power-on load change; if the accelerator pedal is released abruptly, this is referred to as a power-off load change.

Equally typical of a dynamic driving situation is braking in a curve. The effects occurring in these driving situations and their influence on the driving behavior are discussed below. Figure 5.41 shows the behavior of a front-wheel drive vehicle during a power-on load change from a stationary circular drive. Here, two main effects influence the driving behavior. The drive slip development at the front axle reduces the ability of the front axle to transmit lateral forces according to the friction circle. The inertial forces that occur cause a dynamic wheel load transfer from the front axle to the rear axle. The power transfer potential of the front axle is reduced, and the power transfer potential of the rear axle increases. Both effects, traction slip on the front axle and dynamic wheel load transfer, cause the vehicle to tend to understeer. This "power-on-understeer" is typical of front-wheel drive vehicles.

In Fig. 5.42, the power-off load change of a front-wheel drive vehicle is considered. The vehicle is initially in an accelerated circular drive, and the driver completely releases the accelerator pedal during this. The slip required to transmit the driving force is reduced and, correspondingly, the lateral force generated at the front axle increases for a given slip angle. At the same time, a relative wheel load transfer from the rear axle to the front axle takes place due to the reduction in longitudinal acceleration. Due to the air resistance, the vehicle

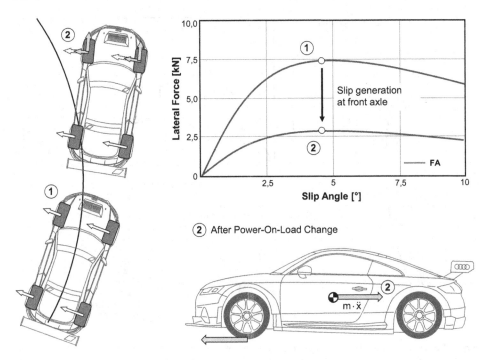

Fig. 5.41 Power-on load change with front-wheel drive

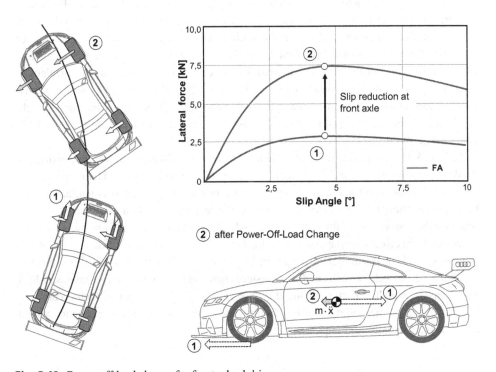

Fig. 5.42 Power-off load change for front-wheel drive

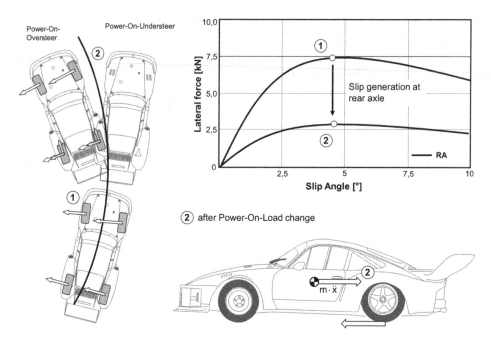

Fig. 5.43 Power-on load change with rear-wheel drive

is slightly decelerated, which increases the wheel load transfer to the front axle. The wheel load transfer to the front axle further increases the lateral force on the front axle. Both effects create an oversteer tendency of the vehicle. This is also referred to as load transfer oversteer. Since front-wheel drive vehicles tend to understeer when accelerating, a moderate oversteer load change response is desirable, since a defined release of the accelerator pedal allows a course correction to be made. However, the vehicle must be tuned in such a way that load change reactions remain predictable and controllable. However, for a vehicle that is already on the verge of instability or loss of control due to oversteer, the power-off load change should be avoided.

In Fig. 5.43, the power-on load change of a rear-wheel drive vehicle is considered. Analogous to the front-wheel drive vehicle, the resulting longitudinal acceleration causes a wheel load transfer to the rear axle. The increase in rear axle adhesion potential tends to have an understeering effect. However, driving the rear axle simultaneously causes a slip-related reduction in lateral force on the rear axle. This effect tends to cause the vehicle to oversteer. Both effects overlap, so that the load change reaction of a rear axle driven vehicle is not always clear. If sufficient drive power is available to cause the rear wheels to spin, the rear-wheel drive vehicle will tend to oversteer. However, in higher speed ranges, wheel load transfer, which is partly enhanced by the aerodynamic behavior of racing cars (see Sect. 6.9), can be the dominant effect. In this case, acceleration understeer occurs, which racing drivers usually find unpleasant.

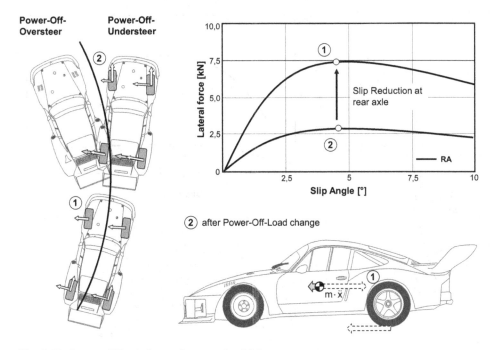

Fig. 5.44 Power-off load change for rear-wheel drive

The power-off load change of a rear axle driven vehicle is shown in Fig. 5.44. Also in this case two effects overlap in opposite directions. The first effect is the relative wheel load transfer from the rear to the front axle, which leads to an increase in the lateral force on the front axle and a reduction in the lateral forces on the rear axle. This causes – just as with the front-wheel drive vehicle – a tendency to oversteer. At the same time, however, the drive slip on the rear axle is reduced, which in turn leads to an increase in lateral force. Generally, when drive slip is low, wheel load transfer during a power-off load change is the dominant effect. However, the vehicle is stabilized by throttle removal if the load change is preceded by significant acceleration oversteer.

In addition to the transmission of drive forces, the dynamic change in the lateral force potential can also be caused by the generation of braking forces. A distinction is made between the case of an overbraked front axle and an overbraked rear axle. Which case occurs depends in particular on the selected brake force distribution (see also Sect. 7.9). Figure 5.45 shows the case of an overbraked front axle. In this case, at least one of the front wheels locks and loses its ability to transmit lateral forces. The loss of lateral force caused by a locked wheel cannot be compensated even by the wheel load transfer to the front axle. The vehicle understeers, but remains basically stable.

Due to the overbraking of the rear axle, shown in Fig. 5.46, at least one of the rear wheels locks, which significantly reduces the lateral force transfer at the rear axle. This has an even more severe effect because the tendency to oversteer is exacerbated due to the wheel load transfer caused by longitudinal deceleration. The vehicle is in danger of

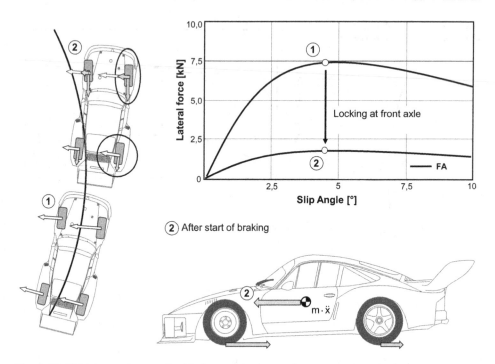

Fig. 5.45 Vehicle reaction to wheel-lock up at the front axle

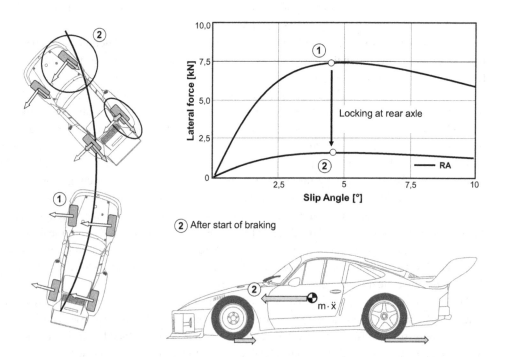

Fig. 5.46 Vehicle reaction to wheel-lock up at the rear axle

becoming unstable "on the brakes". However, a high degree of stability on the brakes is an important prerequisite for the driver to retain confidence in his vehicle and to be able to move it at the limit accordingly quickly. In modern production vehicles, braking stability is ensured by the electronic brake control systems that have become standard today in the form of electronic brake force distribution and the anti-lock braking system. Both systems ensure that the adhesion ratios at the individual wheels remain in the optimum range without locking and without the occurrence of sudden yawing moments. However, electronic brake and traction control systems are only permitted in mass sports or in isolated racing series. In high-performance racing they are prohibited without exception. The required level of braking stability must therefore be achieved by conventional means, i.e. through appropriate suspension tuning, a differential lock and, if necessary, aerodynamics. It remains to be added that, of course, even with production vehicles, a coherent conventional set-up of the vehicle cannot be replaced by control systems and always forms the basis of good driving dynamic characteristics.

The wheel load changes that occur during braking and acceleration, in conjunction with the type of drive, also determine the optimum weight distribution of a vehicle. This results, as summarized in Fig. 5.47, from the requirements with regard to traction, braking and driving stability. High traction requires a corresponding weight on the driven axle. Due to the wheel load transfer to the rear axle, the proportional front axle weight must be greater for front-wheel drive than for rear-wheel drive. The proportional front axle weight is limited by the requirement for sufficiently high braking stability. A rear axle weight that is too low, in conjunction with the wheel load transfer to the front axle, increases the oversteer tendency of the vehicle during braking, since without a corresponding wheel load the rear axle cannot build up sufficient lateral forces. For a front-wheel drive system, therefore, the optimum front axle weight ratio is between 59% and 61%. For a rear-wheel drive with unit tires, the optimum front axle weight percentage is between 49% and 51%. With the same or better traction, this weight distribution has the advantage of better braking

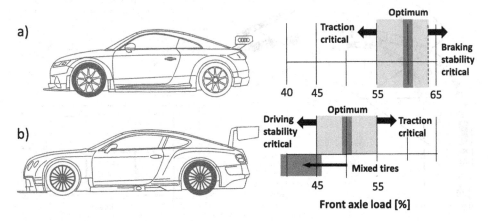

Fig. 5.47 Optimum weight distribution (**a**) for front-wheel drive and (**b**) for rear-wheel drive

stability and, especially on the front axle, a lower influence of wheel load sensitivity. These are two further conceptual advantages of a rear axle driven vehicle. The proportionate rear axle load is limited by the requirement for high driving stability. By using a mixed tire, this limit can be shifted to higher rear axle loads that promote traction and braking stability. In this case, the optimum weight share of the front axle is around 40–45%.

5.6   Basic Driving Techniques

For a better understanding of the requirements for the dynamic behavior of a vehicle, as judged by the driver if necessary, it is useful to look at the basic driving technique of a professional driver. Figure 5.48 shows an example of this in driving through a curve. This process can be divided into the phases "Approach", "Entry", "Midcorner" with Apex and "Exit" [4].

During the "Approach", the vehicle essentially drives straight ahead and is braked down to the speed for the curve entrance. At the turn entrance or "Entry", the driver steers in. He remains on the brakes during this time. This has the effect of continuing to transfer wheel load to the front axle. This increase in the tendency to oversteer is used specifically to make the vehicle more agile or to increase the willingness to turn in. Until the apex is reached, which is in the middle part of the curve ("midcorner"), the brake pedal is continuously reduced. This technique is known as "trail braking".

From the start of the curve to the apex, lateral acceleration and roll angle build up. The suspension on the inside of the curve compresses and the suspension on the outside of the curve rebounds. The dampers on the inside wheels are in rebound and the outside wheels are in compression. However, it should be noted that during this phase the negative pitch angle caused by braking must also be reduced. In the area of the apex, the vehicle then only

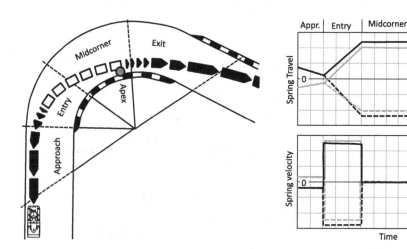

Fig. 5.48 Curve sections and spring behavior

has to transmit lateral forces for a short period of time, so that the maximum lateral acceleration and the maximum roll angle are achieved in this phase.

Behind the apex is the corner exit or "exit". In this phase, the driver presses the accelerator pedal to accelerate the vehicle. The lateral acceleration decreases, as the vehicle now has to transmit drive forces. The accelerator pedal is progressively depressed as lateral acceleration decreases. By the time the front axle wheels are straight again, the driver has fully depressed the accelerator pedal. This phase is characterized by a decrease in roll angle, but the roll is superimposed by a pitching motion caused by longitudinal acceleration. Nevertheless, one can initially assume that the velocity directions are reversed in the damper compared to the curve entrance: The curve outer side now goes into the rebound stage, and the curve inner side is in the compression stage. For the tuning of the suspension and especially its dampers it is important to know how the speed of the dampers behaves. For this reason it is advisable to measure the suspension travel. In addition, the current height of the body above the road can be determined via the spring travel, from which conclusions can be drawn about the aerodynamic behavior of the vehicle.

This driving technique places high demands on the modulation of the brake pressure and on the "driveability of the engine". Engine drivability is the relationship between accelerator pedal application and actual power delivery to the wheels. The accelerator pedal curve is usually progressive, allowing the driver to sensitively influence the yaw response of the vehicle under partial load by initiating a load change. Outside these ranges, there is then a steep increase to prevent the pedal travel from becoming too long. The dynamic behavior of the engine or drive train also plays a major role here. A high time lag between accelerator pedal actuation and the build-up of drive forces, such as can be caused by a turbocharger or a converter gearbox (not used in high-performance racing), makes it difficult to find the right time for accelerator pedal actuation. It is important that the driver can precisely assign the behavior of his vehicle to the individual phases of the curve. Only then can the race engineer decide on the right action to correct the set-up. The choice of racing line and apex depends on various factors, including above all the geometry of the corner and the following section of the track, as well as the weather conditions. Furthermore, the racing line is influenced by the driving style of the driver and the driving dynamic characteristics of the vehicle. The course shown in Fig. 5.48 is only an example. Professional driving techniques and the choice of the ideal line are described in [4–7], among others. It should be noted that inexperienced drivers who are driving on race tracks for the first time or rather seldom and who are driving in the dynamic limit range are instructed to separate the steering process from the braking process and sometimes also from the acceleration process. The reason for this is that a steering movement during braking can cause the vehicle to oversteer very quickly. This applies in particular to front-wheel-drive vehicles due to the front-axle-heavy mass distribution, as the author knows from his own experience (without enemy contact and contact with the track boundary). The disadvantage of this driving technique is shown in Fig. 5.49a. During the transition phases from deceleration to lateral acceleration and back to longitudinal acceleration, the potential of the tires is not fully exploited. In contrast, a professional racer is always at the limit of the friction circle, which gives a corresponding advantage on the stopwatch. The lower utilization of the

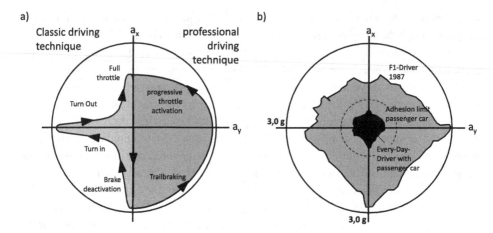

Fig. 5.49 (**a**) g-g diagram with different driving techniques, (**b**) g-g diagram with the range of experience of a normal driver and an F1 driver

friction circle in the direction of acceleration results from the fact that the vehicles shown are driven on "only" two wheels but braked on all four wheels.

As Fig. 5.49b shows, the range of experience of a normal driver – represented by acceleration values measured in everyday driving – is significantly below the performance potential of his vehicle. For this reason, good-natured vehicle behavior in the limit range is required, which is also supported by driving dynamics control systems in today's production cars. A professional racing driver, on the other hand, constantly balances his vehicle at the physical limits, which in the case of high-performance racing cars are also far above the limits of a production car due to the aerodynamic downforce. These differences become clear in the g-g diagram, which also shows the acceleration values measured for an F1 car from 1987. The longitudinal and lateral accelerations achieved as well as the braking decelerations are two to three times higher than the maximum values of a production car. The driver's "trail braking" is also clearly visible.

References

1. Mitschke, M., Wallentowitz, H.: Dynamik der Kraftfahrzeuge. Springer, Wiesbaden (2015)
2. Pacejka, H.: Tire and Vehicle Dynamics. Butterworth-Heinemann, Oxford (2012)
3. Berger, G.: Zielgerade. Wiener, Himberg (1997)
4. Bentley, R.: Ultimate Speed Secrets – the Complete Guide to High-Performance and Race Driving. Motorbooks, Minneapolis (2011)
5. Frére, P.: Sports Car and Competition Driving. Bentley Publishers, Cambridge (1992)
6. Krumm, M.: Driving on the Edge – the Art and Science of Race Driving. Icon Publishing, Worcestershire (2011)
7. Prost, A., Rousselot, P.-F.: Competition Driving. Hazleton Publishing, Richmond (1990)

Aerodynamics

6

6.1 Aerodynamic Forces on the Complete Vehicle

Aerodynamics has become the dominant competitive factor in high-performance racing. This section introduces the most important parameters for describing the aerodynamic properties of a vehicle and explains their influence on driving dynamics. With the help of these parameters, the dimensions in which the aerodynamic behavior of a racing vehicle differs from that of a production vehicle are clarified. In the further course of the chapter, the function and interaction of the aerodynamic components of a racing vehicle are described.

The aerodynamic parameters are usually determined in a wind tunnel. As an example of a modern wind tunnel, Fig. 6.1 shows the aeroacoustic wind tunnel of Audi AG. Its turbine (2) has an output of 2.6 megawatts, with which it can simulate incident flow speeds of up to 300 km/h in the measuring section. Deflection corners (3) and rectifiers (4) minimize turbulence and ensure a smooth flow of air to the vehicle in the test section. The wind tunnel is of the "closed-return" design, in which the air circulates in a closed circuit. As a result, the energy of the air flow that is still retained when it leaves the measuring section is not lost, which reduces the energy requirement of the system.

The vehicle stands on a turntable (1) with which the angle of attack τ_L can be varied. Each of the four wheels stands on a measuring scale which records the aerodynamic forces occurring in the x, y and z directions individually for each wheel (Fig. 6.2). The typical result of a measurement run with a constant angle of attack τ_L is the course of the aerodynamic forces as a function of the velocity of attack v. This course is usually approximated by a quadratic equation. Figure 6.3 illustrates this procedure for the aerodynamic forces in the longitudinal and vertical directions. These aerodynamic forces are

L. Frömmig, *Basic Course in Race Car Technology*, https://doi.org/10.1007/978-3-658-38470-8_6

Fig. 6.1 Audi AG aeroacoustic wind tunnel. (Courtesy of © Audi AG 2018. All Rights Reserved)

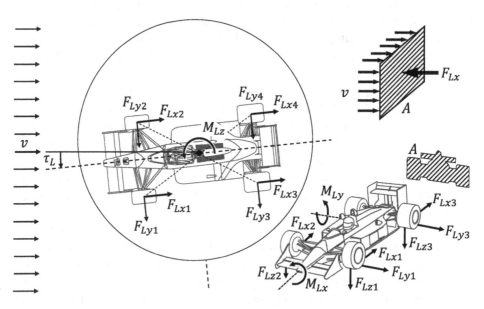

Fig. 6.2 Determination of aerodynamic forces in the wind tunnel

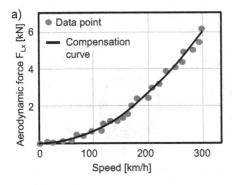 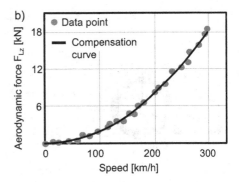

Fig. 6.3 Evaluation of wind tunnel measurements: (**a**) in longitudinal direction, (**b**) in vertical direction

called drag (Fig. 6.3a) and lift or down force (Fig. 6.3b). The approximate equations are defined as:

$$F_{Lx} = F_{Lx1} + F_{Lx2} + F_{Lx3} + F_{Lx4} = c_x(\tau_L) \cdot A \cdot \frac{\rho_L}{2} \cdot v^2, \tag{6.1}$$

$$F_{Lz} = F_{Lz1} + F_{Lz2} + F_{Lz3} + F_{Lz4} = -c_z(\tau_L) \cdot A \cdot \frac{\rho_L}{2} \cdot v^2. \tag{6.2}$$

In these equations, c_x is the drag coefficient, which for an angle of attack of $\tau_L = 0°$ is called the c_W value, and c_z is the lift coefficient for positive values and the downforce coefficient for negative values. The parameter A is the frontal area of the vehicle obtained from a frontal projection, and ρ_L is the air density. The same procedure can be used to determine the coefficient c_y.

$$F_{Ly} = F_{Ly1} + F_{Ly2} + F_{Ly3} + F_{Ly4} = c_y(\tau_L) \cdot A \cdot \frac{\rho_L}{2} \cdot v^2 \tag{6.3}$$

This coefficient is of importance if the angle of attack differs from zero or the vehicle is asymmetrical. In addition to the aerodynamic forces, their moment effects about a reference point 0 are of interest. These moments are described by the following equations:

$$M_{Lx} = c_{Mx}(\tau_L) \cdot A \cdot l \cdot \frac{\rho_L}{2} \cdot v^2, \tag{6.4}$$

$$M_{Ly} = c_{My}(\tau_L) \cdot A \cdot l \cdot \frac{\rho_L}{2} \cdot v^2, \tag{6.5}$$

$$M_{Lz} = c_{Mz}(\tau_L) \cdot A \cdot l \cdot \frac{\rho_L}{2} \cdot v^2. \tag{6.6}$$

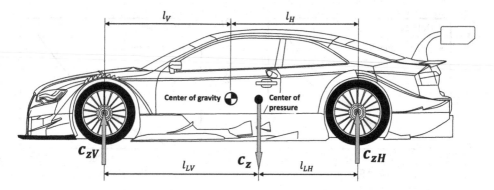

Fig. 6.4 Aerodynamic balance of a vehicle

The dependence of the coefficients on the angle of attack is obtained by repeating the measurement for different turntable positions. These properties will be discussed separately later in the corresponding sections. In the following, the focus will be on the drag coefficients and the lift and downforce coefficients in order to discuss their influence on the driving dynamics. From the point of view of driving dynamics, not only the resulting lift or downforce at the pressure point is of importance, but also in particular its distribution between the front and rear axles of the vehicle (Fig. 6.4). Therefore, the axle lift or axle downforce coefficients are introduced at this point. It applies:

$$F_{Lz1} + F_{Lz2} = -c_{zV}(\tau_L) \cdot A \cdot \frac{\rho_L}{2} \cdot v^2, \tag{6.7}$$

$$F_{Lz3} + F_{Lz4} = -c_{zH}(\tau_L) \cdot A \cdot \frac{\rho_L}{2} \cdot v^2. \tag{6.8}$$

In the typical case of racing cars, where downforce is generated at both axles, the aerodynamic balance is calculated from these variables:

$$A_B = \frac{F_{Lz1} + F_{Lz2}}{F_{Lz1} + F_{Lz2} + F_{Lz3} + F_{Lz4}} = \frac{c_{zV}}{c_{zV} + c_{zH}}. \tag{6.9}$$

Aerodynamic balance has a significant influence on a vehicle's handling characteristics and is a key design objective in the development and tuning of racing and production vehicles. Basically, an increase in aerodynamic balance reduces the understeer tendency of a vehicle, while a reduction in aerodynamic balance increases understeer tendency and thus stability. This is because an increase in aerodynamic balance increases the lateral forces that can be transmitted to the front axle, while a reduction increases the lateral force potential of the rear axle.[1]

---

[1] If lift is generated at both axes, these statements are reversed accordingly.

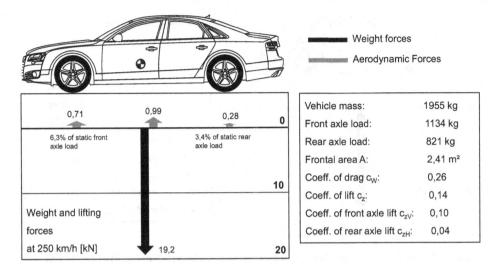

Vehicle mass:	1955 kg
Front axle load:	1134 kg
Rear axle load:	821 kg
Frontal area A:	2,41 m²
Coeff. of drag c_W:	0,26
Coeff. of lift c_z:	0,14
Coeff. of front axle lift c_{zV}:	0,10
Coeff. of rear axle lift c_{zH}:	0,04

Fig. 6.5   Weight forces and aerodynamic forces on the Audi A8 (2010), calculated with data from [1]

In order to establish the relationship between the lift and downforce coefficients of a vehicle and its handling characteristics and performance, its axle loads are first calculated. These are obtained by taking the aerodynamic coefficients into account:

$$F_{zV} = \frac{m \cdot g \cdot l_H}{l} - c_{zV} \cdot A \cdot \frac{\rho_L}{2} \cdot v^2, \tag{6.10}$$

$$F_{zH} = \frac{m \cdot g \cdot l_V}{l} - c_{zH} \cdot A \cdot \frac{\rho_L}{2} \cdot v^2. \tag{6.11}$$

Figures 6.5, 6.6, 6.7, 6.8 and 6.9 clearly illustrate the significance of the lift or downforce coefficient by considering the static wheel loads and the lift or downforce forces for various vehicles. The data of the mass geometry and the aerodynamic coefficients are listed in tabular form. The weight force acting at the center of gravity, the resulting lift or downforce and the lift or downforce at the front and rear axles resulting from the aerodynamic balance are shown graphically. The aerodynamic forces are given for a vehicle speed of 250 km/h.

With the Audi A8 (model year 2010), a production vehicle is considered in advance. The typical silhouettes of production vehicles usually generate lift, so that in the case of the Audi A8 the front axle is unloaded by 6.3% at 250 km/h, while the rear axle is only unloaded by 3.4%. The relatively lower unloading of the rear axle is a typical design for production vehicles in order to ensure the best possible driving stability even at maximum speed.

In production sports cars, attempts are made to reduce axle lift to a minimum or even to generate moderate downforce on the rear axle. Generating downforce is usually accompanied by an increase in aerodynamic drag and fuel consumption. In order to resolve

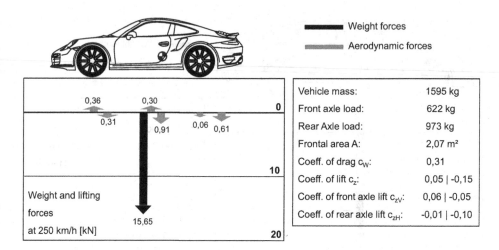

Vehicle mass:	1595 kg
Front axle load:	622 kg
Rear Axle load:	973 kg
Frontal area A:	2,07 m²
Coeff. of drag c_W:	0,31
Coeff. of lift c_z:	0,05 | -0,15
Coeff. of front axle lift c_{zV}:	0,06 | -0,05
Coeff. of rear axle lift c_{zH}:	-0,01 | -0,10

Fig. 6.6 Weight forces and aerodynamic forces on the Porsche 911 Turbo (2014), calculated with data from [2]

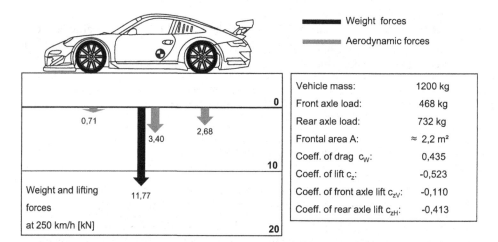

Vehicle mass:	1200 kg
Front axle load:	468 kg
Rear axle load:	732 kg
Frontal area A:	≈ 2,2 m²
Coeff. of drag c_W:	0,435
Coeff. of lift c_z:	-0,523
Coeff. of front axle lift c_{zV}:	-0,110
Coeff. of rear axle lift c_{zH}:	-0,413

Fig. 6.7 Weight forces and aerodynamic forces on the Porsche 997 GTR RSR, calculated with data from [3]

this conflict of objectives, active aerodynamic elements are used in the Porsche 911 Turbo (model year 2014) under the designation "Porsche Active Aerodynamics" (PAA), the setting of which can be varied by the driver in three stages [2]. The aerodynamic behavior in the "PAA Start" and "PAA Performance" modes is shown in Fig. 6.6.

Figure 6.7 shows the Porsche 997 GTR RSR derived from the production car, which was used in various one-make cups and GT series. The downforce achieved in this racing vehicle corresponds to almost one third of the weight force of the vehicle at 250 km/h. The

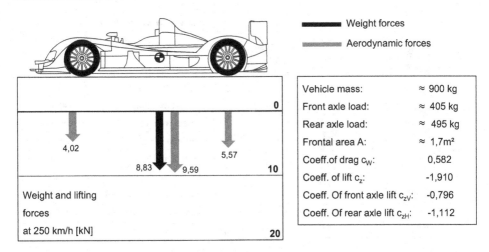

Vehicle mass:	≈ 900 kg
Front axle load:	≈ 405 kg
Rear axle load:	≈ 495 kg
Frontal area A:	≈ 1,7 m²
Coeff. of drag c_W:	0,582
Coeff. of lift c_z:	-1,910
Coeff. Of front axle lift c_{zV}:	-0,796
Coeff. Of rear axle lift c_{zH}:	-1,112

Weight and lifting forces at 250 km/h [kN]

Values shown: 0, 4,02, 8,83, 9,59, 5,57, 10, 20

Fig. 6.8 Weight forces and aerodynamic forces on the Zytek Z11SN (LMP2) calculated with data from [4]

distribution in favor of the rear axle shows that the focus of this vehicle was on increasing high-speed and braking stability.

The LMP2 category vehicles used at Le Mans already generate downforce forces at 250 km/h that are higher than the weight force of the vehicle. Thus, the Zytek Z11SN shown in Fig. 6.8 can theoretically drive upside down on the ceiling at this speed. The distribution of aerodynamic forces between the front and rear axles approximates the mass balance of the vehicle. This aerodynamic balance is accompanied by an increase in lateral force potential on the front and rear axles, which allows significantly higher lateral accelerations to be achieved. The higher the ratio of downforce to vehicle weight, the higher the lateral accelerations that can be achieved.

In a Formula 1 car, such as the Honda RA107 from 2007 (Fig. 6.9), the downforce forces at 250 km/h already correspond to 1.5 times its weight force. This results from the high downforce coefficient and the low total weight of an F1 vehicle.

The high downforce coefficients of high-performance racing cars can only be achieved by specially designed components, such as front and rear wings or diffusers in particular. Figure 6.10 summarizes the contributions of various aerodynamic components to total drag and downforce using the example of the 2000 Ferrari F-2000. The function and design of these and other aerodynamic components will be explained in the following sections. First, the relationship between the aerodynamic properties of a racing car and its achievable driving performance is derived. Figure 6.10 already shows that the downforce-generating components – especially the front and rear wings – also contribute significantly to aerodynamic drag. For this reason, racing cars generally have significantly higher aerodynamic drag than production vehicles. In Formula 1 cars, drag is significantly increased by the free-standing wheels. This explains the large difference between the c_W value of an F1 vehicle and that of an LMP vehicle.

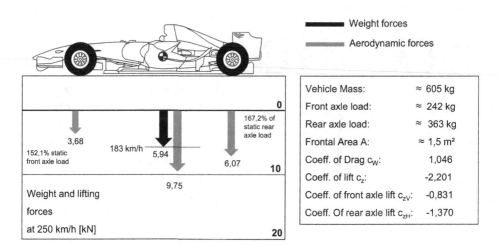

Fig. 6.9 Weight forces and aerodynamic forces on the Honda RA107 from the 2007 season, calculated with data from [5]

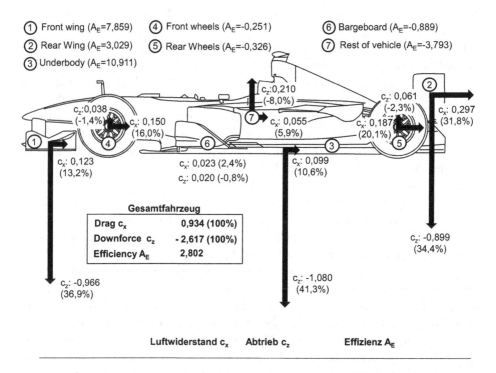

Fig. 6.10 Contribution of different aerodynamic components to drag and downforce on the Ferrari F-2000 [6]

6.2   Driving Performance

Maximum lateral acceleration, maximum speed, maximum longitudinal acceleration and braking deceleration are used here to characterize the driving performance of passenger cars and racing vehicles. Figure 6.11 shows the dependence of the maximum transmissible tire forces on the downforce with the aid of the friction circles, although the dynamic wheel load changes are not taken into account.

Taking into account the aerodynamic share of the wheel load, the following relationship results for the maximum achievable lateral forces:

$$F_{yV,\max} = \mu_V \cdot \left(\frac{m \cdot g \cdot l_H}{l} - c_{zV} \cdot A \cdot \frac{\rho_L}{2} \cdot v^2 \right) = \frac{m \cdot \ddot{y}_{gr,V} \cdot l_H}{l} \qquad (6.12)$$

$$F_{yH,\max} = \mu_H \cdot \left(\frac{m \cdot g \cdot l_V}{l} - c_{zH} \cdot A \cdot \frac{\rho_L}{2} \cdot v^2 \right) = \frac{m \cdot \ddot{y}_{gr,H} \cdot l_V}{l} \qquad (6.13)$$

The basic principle of aerodynamic downforce is therefore to increase the wheel load without increasing the mass to be supported and thus to maximize the achievable lateral accelerations. The coefficients of adhesion of the front and rear tires are often referred to as the "mechanical grip" of the vehicle, the contribution by the downforce forces as "aerodynamic grip". According to the considerations in Sect. 5.2, the axle with the lower limiting lateral acceleration determines the driving behavior at the adhesion limit. For this reason, a

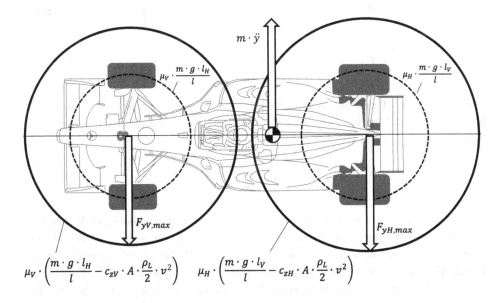

Fig. 6.11 Influence of the output on the Kamm circles

high aerodynamic balance favors the tendency to oversteer, and a low aerodynamic balance favors the tendency to understeer. In order to simplify the relationships, we will assume here that the vehicle has a 50:50 weight distribution with an aerodynamic balance of $A_B = 0.5$ and identical tires on the front and rear axles. For this special case, the maximum achievable lateral acceleration is given by:

$$\ddot{y}_{max} = \mu \cdot \left(g - \frac{c_z \cdot A \cdot \rho_L \cdot v^2}{2 \cdot m} \right) \tag{6.14}$$

This equation states that the lateral acceleration potential increases quadratically with the driving speed. The open sports car prototype achieves a maximum lateral acceleration of about 3.5 g at its top speed. The downforce coefficients of the Formula 1 car, combined with its much lower mass, allow maximum lateral acceleration values of about 5.5 g at top speed. Figure 6.12 shows that the lateral acceleration for these two vehicles increases by a factor of about 2.0 and 2.5, respectively, between 100 and 300 km/h, while for production vehicles it is constant over the entire speed range if lift is neglected. The pilots of high performance racing cars have the challenging task of optimally exploiting the limit of their vehicle in all speed ranges and under racing conditions. The complexity of this task becomes even more apparent when longitudinal accelerations and decelerations are also included in this consideration. Analogous to the maximum transmissible lateral forces, the maximum transmissible longitudinal forces are obtained:

$$F_{xV, max} = \mu_V \cdot \left(\frac{m \cdot g \cdot l_H}{l} - c_{zV} \cdot A \cdot \frac{\rho_L}{2} \cdot v^2 \right), \tag{6.15}$$

$$F_{xH, max} = \mu_{VH} \cdot \left(\frac{m \cdot g \cdot l_V}{l} - c_{zH} \cdot A \cdot \frac{\rho_L}{2} \cdot v^2 \right). \tag{6.16}$$

These forces describe the adhesion limit. When the vehicle accelerates, the transmittable force is additionally limited by the available engine power. This is referred to as the power limit. The following applies to this:

$$(F_{xV} + F_{xH})_{max} = \frac{P_{max}}{v}. \tag{6.17}$$

The curve from this equation is called the "ideal tractive force hyperbola". In the case of a single-axle driven vehicle, one of the two longitudinal force components is zero. For the sake of clarity, the following formulae are formulated for the case of a rear-axle driven vehicle ($F_{xV} = 0$). The intersection of the adhesion limit and the power limit is the longitudinal force at which the maximum longitudinal acceleration is achieved. The maximum longitudinal acceleration from the adhesion of a vehicle is thus obtained as:

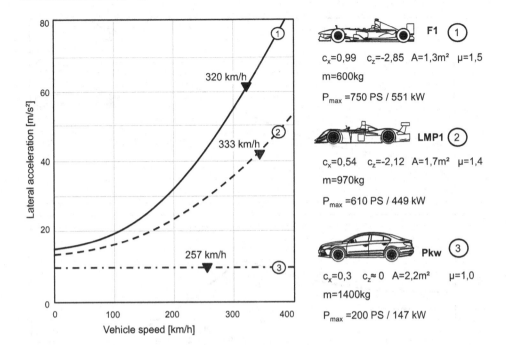

Fig. 6.12 Maximum lateral acceleration and maximum speed

$$\ddot{x}_{max} = \frac{\mu_H \cdot F_{zH} - F_{Lx}}{m} = \mu_H \cdot \frac{l_V}{l} \cdot g + \frac{(-c_{zH} \cdot \mu_H - c_x) \cdot A \cdot \frac{\rho_L}{2} \cdot v^2}{m}. \tag{6.18}$$

For longitudinal acceleration limited by engine power:

$$\ddot{x}_{max} = \frac{\frac{P_{max}}{v} - F_{Lx}}{m} = \frac{\frac{P_{max}}{v} - c_x \cdot A \cdot \frac{\rho_L}{2} \cdot v^2}{m}. \tag{6.19}$$

Figure 6.13 shows in the upper part the course of the maximum driving force resulting from the friction circle ("adhesion limit") or the traction force hyperbola ("power limit"). The F1 vehicle and the production car from Fig. 6.12 are compared. The drive wheels can be made to spin up to the point of intersection of these two lines. The Formula 1 vehicle under consideration is therefore limited up to 200 km/h by the adhesion of its tires. This limit is already reached at around 80 km/h in the production vehicle, which is also quite well powered. The adhesion potential of the production car remains constant relative to the driving speed. In the F1 car, it increases due to downforce. The maximum longitudinal acceleration, as can be seen in the lower diagram of Fig. 6.13, is about 13 m/s² when the frictional limit is reached. This compares with a maximum longitudinal acceleration of just under 5 m/s² of the production car. At the point at which the F1 achieves its maximum longitudinal acceleration, the standard car can only accelerate moderately.

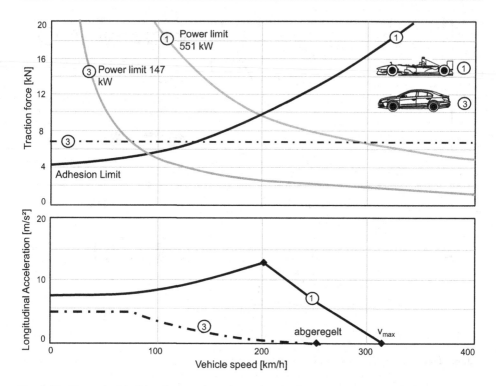

Fig. 6.13 Transmittable drive force and maximum longitudinal acceleration of the F1 vehicle and the production car from Fig. 6.12

In contrast to driving (exception: all-wheel drive), braking uses the tires of both axles. The maximum longitudinal deceleration is achieved when both axles operate simultaneously at the respective adhesion limit without locking. The following applies in this case:

$$\ddot{x}_{max} = -\frac{g}{l} \cdot (\mu_V \cdot l_H + \mu_H \cdot l_V) - \frac{(-c_{zV} \cdot \mu_V - c_{zH} \cdot \mu_H + c_x) \cdot A \cdot \frac{\rho_L}{2} \cdot v^2}{m}. \qquad (6.20)$$

It should be noted here that air resistance now contributes to deceleration. The power during braking does not have to be provided by the engine, but is dissipated in the brakes or converted into heat. A power limit does not exist for this case. Figure 6.14 shows in the upper diagram the maximum braking deceleration as well as the share of air resistance. The standard passenger car achieves a maximum braking deceleration of about 10 m/s². The proportion due to air resistance is relatively low and is only noticeable at high speeds.

The Formula 1 vehicle would be able to achieve longitudinal decelerations beyond 80 m/s² in this consideration. In reality, due to the wheel load sensitivity, the dynamic wheel load changes, a non-existent brake force amplification and a brake force distribution ensuring stability, maximum decelerations of about 5–6 g are achieved. It is worth mentioning that at high speeds, the deceleration due to air resistance can be as much as

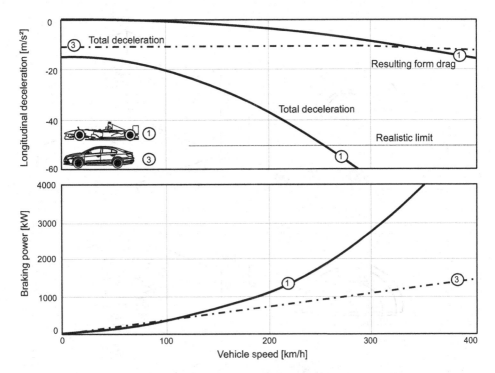

Fig. 6.14 Maximum longitudinal deceleration and braking power of the F1 car and the production car from Fig. 6.12

1 g. Thus, at high speeds, lobbing the accelerator pedal or interrupting traction to change gear is equivalent to full braking of a standard passenger car.

The power required by a vehicle to reach a certain speed depends essentially on its aerodynamic drag. If we neglect the losses due to rolling resistance and the efficiency losses in the drive train, we obtain the following relationship:

$$P = F_{Lx} \cdot v = c_x \cdot A \cdot \frac{\rho_L}{2} \cdot v^3. \tag{6.21}$$

The maximum speed of the vehicle is obtained from the above equation by substituting the available maximum power. The lower part of Fig. 6.14 shows the total power absorbed by the brakes at maximum deceleration, which is important for the thermal design of the braking system, among other things. Part of this energy is recovered today in Formula 1 and in the sports car prototypes of the LMP1 category by the KERS (Kinetic Energy Recovery System).

In summary, the achievable lateral accelerations and braking decelerations are thus directly dependent on the driving speed. The longitudinal acceleration is additionally determined by the engine power. These relationships can be shown very clearly in a g-g-v diagram. Figures 6.15 and 6.16 show the g-g-v diagrams of four vehicles of different

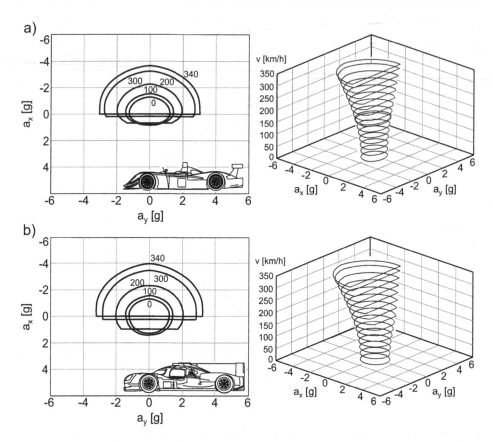

Fig. 6.15 g-g-v diagrams. (**a**) Volkswagen Golf VII GTI Clubsport S (2017), (**b**) Williams-Renault FW 14B (1992)

categories. Particularly concise is the comparison between a Formula 1 car and a production car, which is performed in Fig. 6.15 with a Williams-Renault FW 14B and a VW Golf VII GTI Clubsport S. The g-g-v diagram of the Clubsport S is essentially an ascending column as the adhesion potential is almost constant over vehicle speed. At a driving speed of $v = 0$ km/h, the g-g-v diagram corresponds to a friction circle truncated on the drive side. The truncation results from the fact that the vehicle is driven only via the front axle. Although the GTI Clubsport S has downforce-generating components in the form of a front splitter and a rear wing on the roof edge that measurably increase the vehicle's performance, downforce has only a minor influence on the wheel loads compared to the total vehicle weight. The maximum lateral acceleration and the maximum longitudinal deceleration of about 1.2 g are determined primarily by the coefficient of adhesion of the semislicks. The longitudinal acceleration potential drops to zero in line with the engine power when the maximum speed of 264 km/h is reached. For the other vehicles, the friction circles at $v = 0$ km/h show the increased adhesion potential of racing tires, which for the

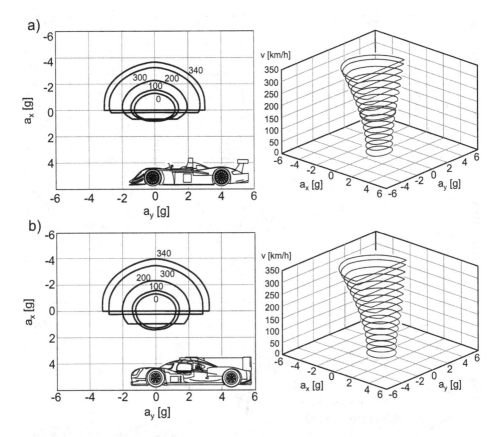

Fig. 6.16 g-g-v diagrams. (**a**) Audi R8 (2000), (**b**) Porsche 919 Hybrid (2017)

Williams FW 14B was assumed to be $\mu_V = 1.5$ for the front axle and $\mu_H \approx 1.65$ for the rear axle.

The coefficients of adhesion of the Audi R8 and Porsche 919 Hybrid sports car prototypes were estimated to be $\mu_V \approx \mu_H \approx 1.3$. For the high-performance race cars considered, the g-g-v diagram visibly widens in the direction of lateral acceleration and longitudinal deceleration with increasing road speed. The Williams-Renault FW 14B achieves a maximum lateral acceleration of about 5 g in the high-speed range. For the Audi R8 and the Porsche 919 Hybrid, the maximum lateral acceleration is about 3.2 g. The characteristic intercept due to engine power is also evident in these vehicles. The Porsche 919 Hybrid is the only one of the vehicles considered to feature an all-wheel drive system in which the rear axle is driven by an internal combustion engine and the front axle by an electric motor. From a standstill, the Porsche 919 Hybrid therefore has similar deceleration as acceleration potential. These images illustrate how demanding the driving task is for the pilot of a high-performance racing car. Under great physical strain, he must be able at all times to correctly assess the performance potential of his vehicle in all speed ranges. Due to

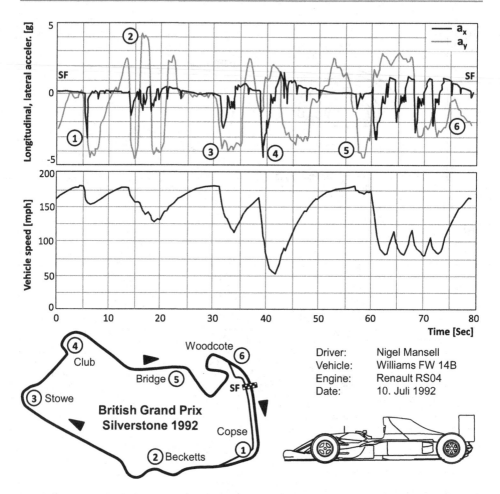

Fig. 6.17 Nigel Mansell's qualifying lap for the 1992 British Grand Prix at Silverstone Circuit [7]. (Courtesy © Williams Grand Prix Engineering Ltd. 2018. All Rights Reserved)

the regulations, driving dynamics control systems, which are standard in modern production cars, are not available. The pressure applied during braking must be tracked via the pedal as the speed drops in order to prevent the wheels from locking. At the same time, overtaking maneuvers have to be carried out or attacks by the opponent have to be fended off.

Figure 6.17 demonstrates the performance of a modern Formula 1 car using some measured data from Nigel Mansell's qualifying lap for the 1992 British Grand Prix. With a time of 1 min 18.965 s, Mansell put his 730 hp. Williams-Renault FW14B on pole position. Shown are the time histories of longitudinal and lateral acceleration as well as vehicle speed during a flying lap on the Silverstone Circuit.

The start and finish line is in the right-hand "Woodcote", which Mansell passes at around 164 mph and a lateral acceleration of 2.5 g. The car is then braked to "Copse" at the end of the straight. At the end of the straight, the brakes are applied at up to 3 g on "Copse," which is then passed with a lateral acceleration of 4.3 g. In the multiple chicane around "Becketts", lateral accelerations of between 4.2 and 4.4 g are achieved. The maximum longitudinal deceleration of 4.5 g is achieved by the vehicle as it brakes into "Club", reducing speed from 162 to 53 mph in less than 4 s. Full acceleration is then achieved from "Club". The maximum acceleration value here is 1.5 g. During the acceleration phase, this value decreases continuously for the reasons previously stated. At 174 mph, "Bridge" is then passed, where the maximum lateral acceleration of 4.5 g occurs. After the final corners, "Woodcote" and the timing on start-and-finish are passed again. Mansell won the race and became world champion at the end of the season.

Achieving optimum driving performance or gaining a competitive advantage requires efficient use of the available engine power. One of the essential goals of aerodynamic development is therefore to optimize the ratio of downforce to aerodynamic drag. This criterion is expressed by aerodynamic efficiency:

$$A_E = \frac{-c_z}{c_x}.$$ (6.22)

In Fig. 6.18, the drag coefficients for various vehicle categories are plotted against their downforce coefficients. The aerodynamic efficiency A_E can be assumed to be

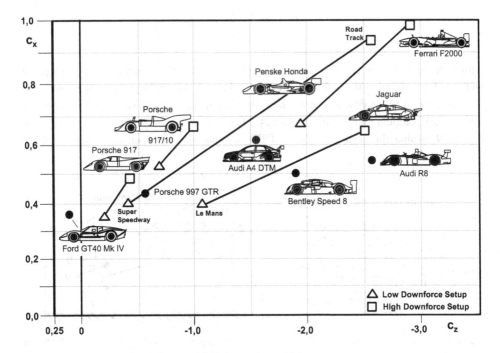

Fig. 6.18 Aerodynamic efficiency of different racing vehicles

approximately constant within a racing vehicle category, so that the values of a low-downforce and a high-downforce set-up can be connected by a straight line. The generation of downforce is therefore always associated with a proportional increase in aerodynamic drag and thus with a reduced top speed. This is the central conflict of goals for which the race engineer must find the track-optimal compromise. Even a small gain in aerodynamic efficiency can mean the decisive advantage over the competition.

6.3   Basic Description of Flow Processes

As illustrated in Fig. 6.19, complex three-dimensional and unsteady flow processes take place on a racing vehicle. The dependence on time results from the constantly changing longitudinal and lateral accelerations. If one also takes into account that air as a gas is a compressible fluid, the general case of a flow process is present. The flow state is known if the state variables of the fluid, in this case air, are known at any time and at any relevant location. The state variables are the local prevailing pressure p and the local temperature T, the density ρ_L of the fluid, and the local velocity vector $\vec{v}$. In summary, the following state variables are obtained:

$$\begin{aligned}
p &= p(t, x, y, z), \\
T &= T(t, x, y, z), \\
\rho_L &= \rho_L(t, x, y, z), \\
\vec{v} &= \vec{v}(t, x, y, z).
\end{aligned} \tag{6.23}$$

Although these state variables can be determined by complex measurements, analytical methods for simulating these flow processes are increasingly being used in the context of modern racing vehicle development. These simulation methods are summarized under the term CFD (Computational Fluid Dynamics). The representation of the flow processes in Fig. 6.19 is also the result of such a CFD simulation. The main advantage of CFD

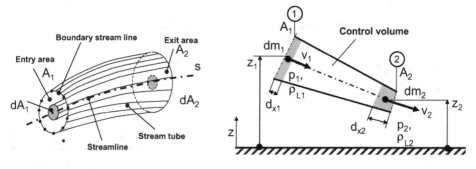

Fig. 6.19 Flow process on a BMW M4 DTM. (Courtesy of © BMW AG 2018. All Rights Reserved)

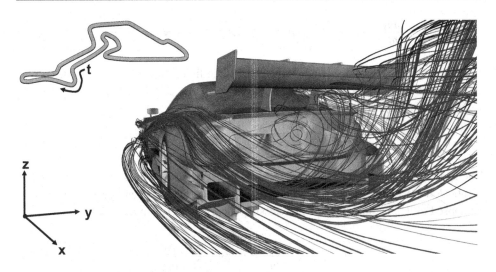

Fig. 6.20 Stream line and control volume

calculations is that concept studies can be carried out on the vehicle or its components before the first component is even manufactured. In order to obtain meaningful results, a high simulation quality is necessary, which requires massive computer capacities. For the analytical solution of the above system of equations, which consists of six unknowns and four independent variables, the continuity equation, the thermal equation of state, the energy conservation laws and the equations of motion are available. The mesh of these equations leads to the "Navier-Stokes equations".

At this point we reduce the complexity to a level that is necessary for the basic understanding of flow processes on racing vehicles. For this purpose, we introduce the terms current filament and control space (Fig. 6.20). The stream thread essentially consists of an inlet surface A_1, an outlet surface A_2, the boundary streamlines and the stream thread axis. The entrance and exit surfaces, in conjunction with the edge streamlines, form a tubular control space. In this control space, the mass is maintained so that the masses entering and leaving the control space are identical. On the streamline axis, the state variables are summarized as an average value. The thought model behind this corresponds to a plane pipe flow. For this simple flow, the continuity equation and the law of conservation of energy can be established. The continuity equation describes just the fact that incoming and outgoing mass flow are identical. Therefore, it holds:

$$dm_1 = dm_2 = A_1 \cdot dx_1 \cdot \rho_{L1} = A_2 \cdot dx_2 \cdot \rho_{L2}. \qquad (6.24)$$

From the temporal derivation it follows:

$$\frac{dm_1}{dt} = \dot{m}_1 = A_1 \cdot v_1 \cdot \rho_{L1} = \dot{m}_2 = A_2 \cdot v_2 \cdot \rho_{L2} \quad \text{with} : \frac{dx}{d_t} = v. \tag{6.25}$$

In the following, air is considered an incompressible fluid. This means that its density is constant and thus identical at the inlet and outlet of the control space. This gives the following relationship between the inlet and outlet velocities:

$$v_1 \cdot A_1 = v_2 \cdot A_2 \quad \text{with} : \rho_{L1} = \rho_{L2}. \tag{6.26}$$

The law of conservation of energy states that the energy of the fluid at entry and exit is identical if no energy is added to the control space. The total energy of the fluid is composed of its potential, kinetic and pressure energy. This is expressed by:

$$g \cdot z_1 + \frac{v_1^2}{2} + \frac{p_1}{\rho_{L1}} = g \cdot z_2 + \frac{v_2^2}{2} + \frac{p_2}{\rho_{L2}}. \tag{6.27}$$

In this representation it is already taken into account that incoming and outgoing mass flow are identical and can accordingly be eliminated from Eq. (6.27). The potential energy of air can be neglected compared to its kinetic and pressure energy. It follows:

$$\frac{v_1^2}{2} + \frac{p_1}{\rho_{L1}} = \frac{v_2^2}{2} + \frac{p_2}{\rho_{L2}}. \tag{6.28}$$

An incompressible fluid is also postulated when using the law of conservation of energy. By rearranging Eq. (6.28), the pressure at the outlet can then be determined as a function of the inlet and outlet velocities.

$$p_2 = p_1 + \frac{v_1^2 - v_2^2}{2} \cdot \rho_L \quad \text{with} : \rho_{L1} = \rho_{L2} = \rho_L. \tag{6.29}$$

The ratio of inlet and outlet velocity is determined by the ratio of inlet and outlet cross section according to the continuity equation. Thus, substituting Eq. (6.26) into Eq. (6.29), we obtain:

$$p_2 = p_1 + v_1^2 \cdot \frac{1 - \frac{A_1^2}{A_2^2}}{2} \cdot \rho_L. \tag{6.30}$$

The following law results from these equations: An increase in the flow velocity along the streamline axis leads to a reduction in pressure; conversely, a reduction in the flow velocity leads to an increase in pressure. This effect is of central importance for the function of the aerodynamic components of a racing vehicle. The pressure distribution along a streamline or surface is often described by the dimensionless pressure coefficient C_p, which is defined as:

$$C_p = \frac{p_2 - p_1}{\frac{1}{2} \cdot \rho_L \cdot v_1^2}. \tag{6.31}$$

In relation to the static pressure p_1, a negative pressure coefficient C_p means a pressure reduction and a positive value means a pressure increase. In the further considerations, the static pressure p_1 always corresponds to the atmospheric pressure and the speed v_1 to the vehicle speed. Simple examples are used to illustrate the significance of the pressure distribution along a surface for the aerodynamic forces. The following basic relationship applies: the resulting force on a surface corresponds to the multiplication of the surface by the pressure prevailing there. It follows from this:

$$F = p \cdot A. \tag{6.32}$$

In Fig. 6.21, the pressure along the top and bottom of a simple flat surface is considered and shown both in terms of absolute pressure and in terms of pressure coefficient. In case a, the ambient atmospheric pressure prevails on both the top and bottom surfaces. The pressure coefficient at the upper and lower side over the entire length is $C_p = 0$. The pressure forces acting on the upper and lower side cancel each other out, so that the resulting force on the surface is zero. In case b, constant negative pressure is applied to the entire underside, which accordingly leads to a negative pressure coefficient over the entire length. The

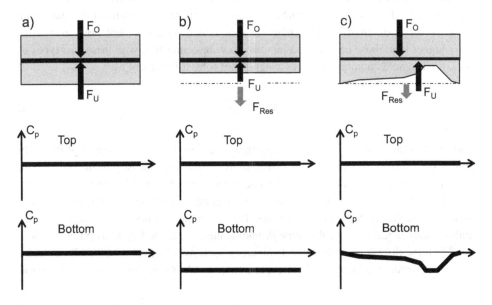

Fig. 6.21 Relationship between pressure coefficient and aerodynamic forces

pressure force on the underside decreases, and a downward resultant force is produced. According to the considerations in Sect. 6.1, this generates a downward force. Case c shows a more realistic distribution of the pressure coefficient. In this case, too, there is negative pressure on the underside on average, which results in an output force. However, the asymmetry of the pressure distribution simultaneously causes a shift in the point of force application. On a racing vehicle, this effect is equivalent to a shift in the aerodynamic balance. Knowledge of the pressure distribution or how it can be specifically influenced is a crucial skill in the development of racing vehicles.

The pressure coefficient can be made visible as a result of a CFD simulation by different coloring on the surface of the body flowing around it, from which the associated aerodynamic forces can also be concluded. Figure 6.22 shows this procedure using the example of a Lamborghini Gallardo LP570. Blue areas mark an area with strong negative pressure, while red areas mark strong positive pressure. Ambient pressure prevails in green areas. High overpressure therefore prevails in front of the front end, resulting in a corresponding drag force. Above the bonnet and the whole roof area however, negative pressure prevails. This negative pressure creates a suction effect, which results in body lift. Without further aerodynamic measures, this results in positive axle lift values, as already shown on the Audi A8 in Fig. 6.5. Lift-reducing measures on the Gallardo are the splitter at the front end and the rear wing. The functionality of these components will be discussed in detail later on.

When considering flow processes, it must be taken into account that frictional forces are transmitted between the surface and the air, which withdraw energy from the flow. Due to this withdrawal of energy, a laminar flow can change into a turbulent flow. In a laminar flow, all flow processes take place parallel to each other, so that individual layers do not mix with each other. In contrast, a turbulent flow has a three-dimensional flow field with a random velocity distribution. The friction losses caused by a turbulent flow are significantly higher than those caused by a laminar flow. In connection with these two types of flows, the Reynolds number Re plays a significant role. It is defined as:

$$\mathrm{Re} = \frac{v \cdot l}{\vartheta_L}. \tag{6.33}$$

In addition to the flow velocity v, this equation takes into account the length of the body being flowed around or the distance l covered as well as the viscosity of the air ϑ_L. The Reynolds number is also of great importance in the context of wind tunnel tests, namely when the investigations are not carried out on the original vehicle but with the aid of scaled-down models. The Reynolds number states that the turbulence behavior of geometrically similar bodies is identical for the same Reynolds number. If the Reynolds number is less than 2320, the flow is assumed to be laminar; as soon as it is greater than 2320, the flow begins to separate and increasingly changes into a turbulent flow form. The consideration

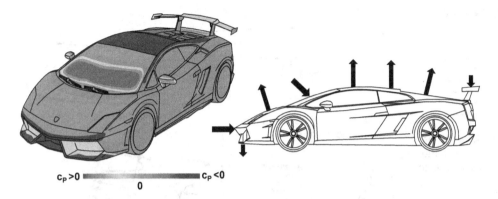

$c_P > 0$ ▬▬▬▬▬▬▬▬▬▬ $c_P < 0$
0

Fig. 6.22 Pressure coefficient distribution and aerodynamic forces on a Lamborghini Gallardo LP570 Supertrofeo Stradale 2011

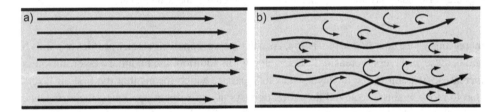

Fig. 6.23 (a) Laminar flow, (b) Turbulent flow

of these two flow forms, illustrated in Fig. 6.23, also plays an important role in connection with the boundary layer and the phenomenon of separation.

The concept of boundary layer is introduced with the help of Fig. 6.24. The flow conditions on a wing profile are shown, as they occur in a wind tunnel. The boundary layer theory is based on the assumption that those air particles which are in direct contact with the wing surface virtually stick to the surface. The relative velocity between these air particles and the wing is therefore zero. At a certain distance from the contact surface, on the other hand, the air particles again assume the flow velocity. The distance between these air particles is the boundary layer thickness. Due to the viscous properties of air, the air layers of different air particles rub against each other, which leads to a degressive, but homogeneously decreasing velocity profile of the air layers in the boundary layer at point a. The velocity profile of the air layers in the boundary layer decreases in the direction of the flow. Along the direction of flow, especially the air layers close to the wing surface

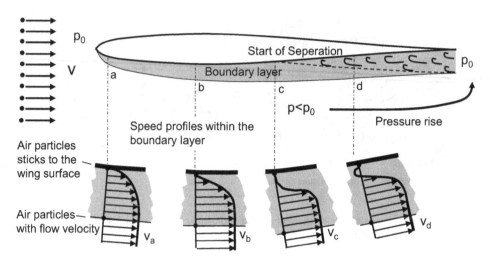

Fig. 6.24 Boundary layer and separation on a wing profile in the wind tunnel

increasingly lose energy due to friction. As can be seen from the velocity profiles at points b and c, these air layers increasingly lose velocity. In addition, the boundary layer thickness increases. At point d, the air layers start to separate, which means that the first air layers start to move against the actual flow direction. As the movement continues, the turbulence of the flow in the boundary layer increases. The separation of the boundary layer occurs especially when the air particles are flowing against a pressure rise, which is the case on the undersides of wing profiles in racing cars. Under the wing profile there is negative pressure, while behind the wing profile there is ambient pressure. This pressure increase promotes the loss of speed in the boundary layer. Detaching boundary layers reduce the ability of the airfoil to generate upward or downward forces, and they increase drag. A major goal in the development of wing profiles for both aircraft and racing vehicles is therefore the best possible avoidance of flow separation.

The occurrence of boundary layers also has a major influence on the accuracy of the test results when designing or using wind tunnels. The flow conditions when a vehicle is driving freely and when a stationary vehicle is flowing into the wind tunnel differ significantly due to the relative velocity between the vehicle and the road. The flow conditions for different scenarios are shown in Fig. 6.25. In free running, the relative velocity between the vehicle and the roadway corresponds to the vehicle velocity. The air particles adhere to the contact surface of the moving vehicle. Therefore, the particles adhering to the underbody move with the vehicle speed, and a boundary layer is formed there, at the end of which the speed of the air particles is zero again. The air particles on the ground also have zero velocity. Between the ground and the end of the boundary layer of the vehicle, the relative velocity is zero, so that no boundary layer is formed at the ground.

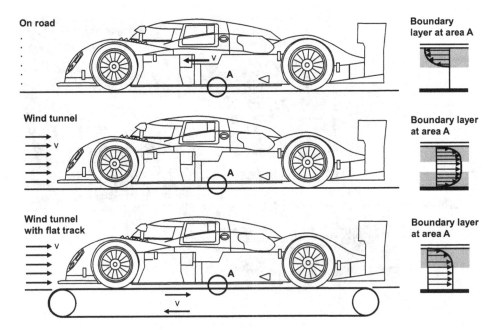

Fig. 6.25 Boundary layers and velocity profiles for a vehicle on the road and for different wind tunnel concepts

On a vehicle standing in the wind tunnel, if no additional design measures have been taken, the relative velocity between the vehicle and the road surface is zero. Both the air particles adhering to the underbody of the vehicle and the air particles adhering to the road thus exhibit a difference in velocity compared to the incident flow velocity. For this reason, a boundary layer is formed on both the road surface and the underbody of the vehicle. The formation of this second boundary layer can significantly falsify the measurement results. In modern wind tunnels, measures are therefore taken to avoid the formation of this second boundary layer. A typical measure is the insertion of one or more flat belts in the floor of the wind tunnel so that the correct relative speed is applied between the vehicle and the road and correct boundary layer conditions are established.

Figure 6.26 shows an example of the measurement section in the wind tunnel of Dr. Ing. h.c. F. Porsche AG at the development center in Weissach. This wind tunnel has a flat belt that extends over approximately 1.5 times the width of the vehicle. The design challenge is to make the flat belt so rigid that the aerodynamic forces do not cause any deformation of the flat belt. In addition to achieving correct boundary layer conditions, this design also has the advantage of taking into account the influence of the rolling wheels. Stationary wheels are another source of error in wind tunnel measurements. The drums that drive the wheels

Fig. 6.26 Wind tunnel of Dr. Ing. h.c. F. Porsche AG with flat track and vehicle restraint. (Courtesy of © Dr. Ing. h.c. F. Porsche AG 2018. All Rights Reserved)

also serve as precision measuring carriages for measuring the aerodynamic forces. The measurement accuracy is about 1 N. With this measurement concept, a compact design is sufficient to restrain the vehicle, thus avoiding further disturbing influences on the flow around the vehicle. The wind tunnel has seven megawatts of power, so that vehicle speeds of up to 300 km/h can be simulated. This design is very complex and requires high investments. Many wind tunnels therefore use simpler measures to minimize the influence of errors caused by boundary layers. For example, the flat belts are only arranged between the wheels or in several partial areas, as is also implemented in the aeroacoustic wind tunnel of Audi AG (see Fig. 6.1). Measurement data recorded in wind tunnels with different configurations are generally not directly comparable. Moreover, it is often the case that precise forecasts are only possible by comparing the wind tunnel measurements with a real driving test. The development of methods that enable wind tunnel measurements to be transferred to the real vehicle is the core competence of the aerodynamics department of a professional racing team.

6.4 Drag

In Sect. 6.2 it has already been shown how drag affects the maximum attainable speed and how it can be described by the drag coefficient c_x. In general, the following applies to the drag force of a body:

$$F_{Lx} = c_x(\tau_L) \cdot A \cdot \frac{\rho_L}{2} \cdot v^2. \tag{6.34}$$

Accordingly, there are two strategies to minimize drag. The first strategy is to reduce the frontal area A of the vehicle. The second strategy is to reduce the drag coefficient c_x, which is made up of frictional resistance, pressure resistance and flow resistance. In vehicles, up to a drag coefficient of $c_x = 0.25$, pressure resistance dominates over frictional resistance. Only below this limit does frictional resistance also play a significant role. Another component of the air resistance of vehicles is the flow resistance, which arises from the need to supply the combustion engine with air and to cool various components and operating fluids.

Figure 6.27 shows the drag coefficients of various bodies. The highest c_x value is exhibited by a flat plate standing vertically in the air flow. Its air resistance consists almost exclusively of pressure resistance. The air flow hitting the front side of the plate creates a strong high pressure area there, which is effective over almost the entire plate surface, since the air can only flow around the plate poorly. Behind the plate, the flow separates and a wake area is created, in which strong negative pressure acts. This suction effect also

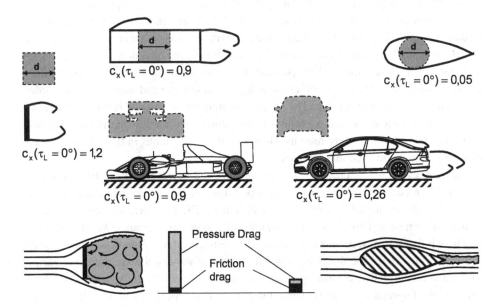

Fig. 6.27 Drag coefficients of different bodies. (Based on [8, 9])

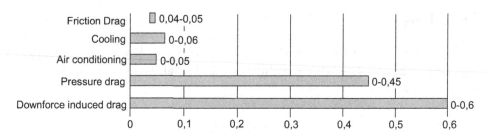

Fig. 6.28 Shares of different contributors to the drag coefficient

contributes significantly to the high pressure difference between the front and rear sides, which counteracts the movement of the plate in the air flow and leads to the very high drag coefficient of 1.2.

In contrast, a streamlined body has a drag coefficient of only 0.05 with an identical frontal area. The drop shape allows the air layers to flow over the upper and lower sides with relatively little resistance. The air flows on the upper and lower side are brought together again behind the body with almost no separation phenomena. Only very small areas of high pressure or low pressure occur in front of and behind the body. The air resistance results almost exclusively from the friction between the body surface and the surrounding air layers. One can deduce from this: In order to achieve a low air resistance, the inflow of blunt surfaces and detachment phenomena should be avoided as far as possible. It is noticeable that both typical production cars and F1 cars are positioned between these two extreme examples. Furthermore, it is noticeable that the drag coefficient of a Formula 1 car is virtually identical to that of a quadricycle. This significant difference from a production car results partly from the open wheels and partly from the drag induced by downforce, which will be discussed in the following sections of this chapter. Figure 6.28 shows the typical proportions of various forms of drag in the total drag. It also shows the dominance of pressure drag and the possibly enormous share of downforce-induced drag. Cooling and air conditioning together form the flow resistance and have a not inconsiderable share of the total resistance. The air conditioning of the interior is indispensable for closed racing vehicles, as the maximum temperatures permitted in the interior are now rigorously regulated for safety reasons. However, the explanations in this section concentrate on pressure and frictional resistance.

Figure 6.29 shows a pressure distribution and the correspondings streamline of a Porsche Cayenne. At the front, two stagnation zones with excess pressure are created in the area of the front end and the lower windshield. The high pressure counteracts the vehicle movement and thus contributes significantly to the air resistance. What is unfavorable here is that the overpressure can act on large vertical surfaces. The flow detaches at the rear edge of the roof. The detachment results in backflows, and a wake area is created in which a strong negative pressure prevails on average. Again, large backward surface is unfavorable. Especially this suction effect significantly determines resulting air resistance. In this case, this results in a drag value of $c_x = 0.36$ with a frontal area of 2.8 m^2. The wake

Fig. 6.29 Stagnation zones and wake area of a Porsche Cayenne

area consists of various vortex forms, whose formation and properties are examined in more detail below.

In the case of vortices arising due to detachment, a distinction is made between vortex shapes which run transversely to the direction of inflow and those which run obliquely or longitudinally to the direction of inflow. Figure 6.30 shows an example of the formation of transverse vortices at a blunt tail. The kinetic energy of the vortex field is dissipated and irreversibly converted into heat. In the vortex core there is a strong negative pressure, which has a considerable share in the suction effect in the wake region (cf. [9]). The strong suction effect increases the power requirement for overcoming the air resistance.

Vortices whose axes are essentially parallel to the direction of flow are formed when the detachment edge is inclined with respect to the direction of flow, as is the case, for example, on a hatchback as shown in Fig. 6.31. As with the transverse vortex, a strong negative pressure region is formed along the vortex axis, so that large negative pressure coefficients with pronounced pressure peaks occur on the surfaces swept by the vortices. The high negative pressures at the rear slope contribute significantly to the air resistance and rear axle lift of a car.

The negative pressure effect of such vortices is specifically used on the upper side of the wings of combat aircraft to generate lift. Figure 6.32 shows this procedure using the example of the pressure coefficient distribution above the wings of a Boeing F-18 Hornet. The vortices generated by the leading edge extensions serve to stabilize the flight attitude at high angles of attack ($>20°$). The negative pressures generated by vortices can also be exploited in the field of racing vehicle aerodynamics, which will be discussed explicitly in connection with the underbody design (Sect. 6.6).

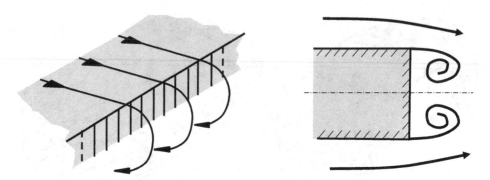

Fig. 6.30 Formation of transverse vortices. (Adapted from [9]; courtesy of © Springer Fachmedien Wiesbaden 2018. All Rights Reserved)

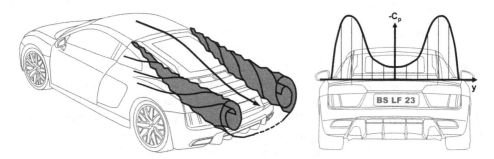

Fig. 6.31 Formation of longitudinal vortices on the hatchback of a sports car

In connection with the goal of minimizing the air resistance of a vehicle or racing vehicle, the preceding considerations mean that the direct inflow and outflow over blunt surfaces must be avoided. This principle is applied in numerous areas of a racing vehicle. Figures 6.33, 6.34 and 6.35 summarize typical measures for reducing the drag of various racing vehicles.

Figure 6.33 shows the Audi R18 and the Audi R10 TDI, each a typical example of an enclosed and an open wheeled prototype. Characteristic of both vehicles is the intensive use of streamlined teardrop shapes, which can be clearly seen in the area of the driver's cabin or cockpit and the wheel arches. Additionally, in order to avoid detachment molds, most components are shaped with a relatively long runout. These parts include exterior mirrors, turbo snorkels, and in the case of the Audi R10, both roll bars. The "boat tailing principle" is also used on both cars: The rear end is slightly retracted, similar to boats and ships. This reduces the width of the wake area created behind the vehicle and thus the air resistance.

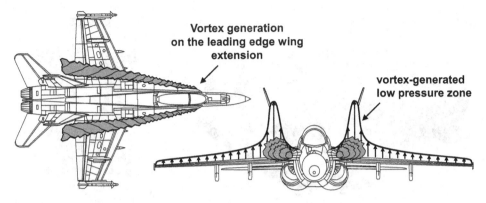

Fig. 6.32 Lift generation by longitudinal vortices on the wings of a Boeing F-18 Hornet fighter aircraft

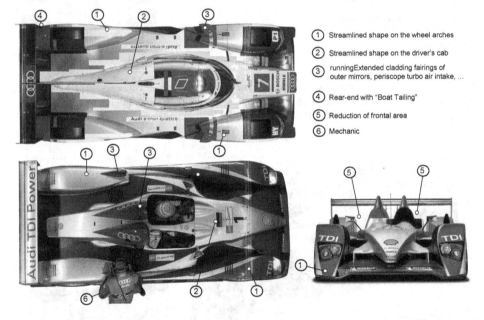

1. Streamlined shape on the wheel arches
2. Streamlined shape on the driver's cab
3. runningExtended cladding fairings of outer mirrors, periscope turbo air intake, ...
4. Rear-end with "Boat Tailing"
5. Reduction of frontal area
6. Mechanic

Fig. 6.33 Typical aerodynamic drag reduction measures on prototype sports cars. (Courtesy of © Audi AG 2018. All Rights Reserved)

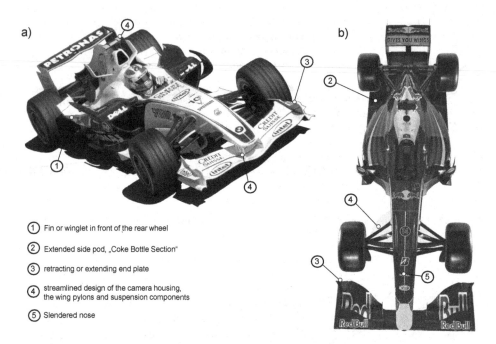

a)

1. Fin or winglet in front of the rear wheel

2. Extended side pod, „Coke Bottle Section"

3. retracting or extending end plate

4. streamlined design of the camera housing,
   the wing pylons and suspension components

5. Slendered nose

b)

Fig. 6.34 Typical aerodynamic drag reduction measures for Formula One cars. (**a**) BMW Sauber F1.07, 2007. (Courtesy © BMW AG 2018. All Rights Reserved), (**b**) Red Bull RB5, 2009. (Courtesy © Getty Images/Red Bull Content Pool 2018. All Rights Reserved)

1. Streamlined wheel fairing

2. Narrow tires

3. Open roll bar

Fig. 6.35 Measures to reduce the aerodynamic drag of a Formula E vehicle. (Courtesy of © Audi AG 2018. All Rights Reserved)

The Audi R10 also features a so-called flow-through concept: Virtually all areas not occupied by vehicle components are exposed, visibly reducing the vehicle's transverse clamping surface.

The Formula 1 cars shown in Fig. 6.34 also have measures that are basically identical to those of the two prototypes. For example, the cameras prescribed by the regulations are housed in aerodynamically shaped housings, and the front wing pylons and the chassis suspension components also have aerodynamically shaped cross-sections.

Typical of modern Formula 1 cars are the long tapered and early retracting sidepods, which form the "coke-bottle" section. On the one hand, this design helps to avoid separation effects behind the car and, at the same time, allows air to flow over the outside and inside of the rear tires, thus avoiding the formation of a stagnation zone with unnecessarily high pressure. In monoposti, the exposed front and rear wheels contribute significantly to drag, so further measures are taken to avoid direct airflow to the tires. On older F1 cars, therefore, a fin or winglet with an upward sweep is found at the end of the side pod, which directs the air over the rear wheel. The advantage of this component is that downforce is generated at the same time. According to today's regulations the use of such components is no longer allowed. Depending on the width of the front wing, the end plates of the front wings are designed to guide the flow around the inside or outside of the tire (see also Sect. 6.5.6).

The loss of aerodynamic efficiency is particularly critical in the monoposti of Formula E. This championship has been contested by the FIA with purely electrically powered vehicles since September 2014. The batteries used for electric vehicles have a significantly lower energy density than conventional fuels for petrol and diesel engines. In order not to unnecessarily limit the range of these vehicles, at least partial fairing of the front and rear wheels is permitted for these vehicles, as Fig. 6.35 shows. In addition, the use of relatively narrow tires is mandatory in this racing series. This measure, as well as the air-permeable roll bar, contributes to the reduction of the frontal area.

The influence of the frontal area A and the enormous differences between individual vehicle classes are impressively demonstrated by Fig. 6.36. There, the Porsche 911 GT3 with about 2.1 m^2 frontal area and the Porsche 919 Hybrid with about 1.6 m^2 frontal area are compared. The sports car prototype appears almost dainty compared to its series-produced brother. Its small frontal area supports the efficient use of the installed power potential.

Figure 6.37 shows the projected front view of various racing vehicles and compares the author's estimates of the frontal areas of different racing vehicles. According to this, the frontal areas of monoposti are approximately between 0.8 and 1.5 m^2. For sports car prototypes from Group C to today's LMP cars, the frontal areas are approximately between 1.6 and 1.7 m^2. In the case of the Nissan Deltawing, an experimental vehicle for demonstrating efficient technologies for motorsport, the frontal area is even only about 1.47 m^2. Near-production racing vehicles or vehicles with at least a near-production

Fig. 6.36 Porsche 911 GT3 and Porsche 919 Hybrid, 2014. (Courtesy of © Dr. Ing. h.c.. F. Porsche AG 2018. All Rights Reserved)

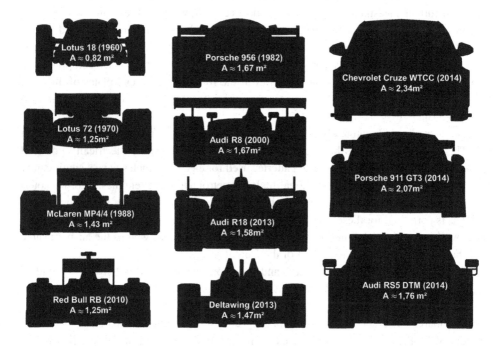

Fig. 6.37 Frontal areas of various racing vehicles. (Own estimates)

silhouette (e.g. DTM Class 1) have frontal areas of 2.0 to about 2.4 m², which corresponds to the dimensions known from production passenger cars.

In contrast to previous considerations, air resistance was used to optimize braking performance on the Mercedes-Benz 300 SLR, which competed in the 24 Hours of Le Mans in 1955. In this vehicle, the driver was able to deploy a flap measuring approximately 0.7 m² during braking. The flap replicated the vehicle contour at the rear and ran the full width of the vehicle (Fig. 6.38). When actuated, the flap increased the c_W value from 0.44 to

Fig. 6.38 Air brake of the Mercedes 300 SLR, 1955. (Courtesy of © Daimler AG 2018. All Rights Reserved)

1.09 [9]. The air brake operated trouble-free and, in conjunction with a frontal area of 1.28 m^2 and a vehicle mass including driver and operating fluids of about 1000 kg, provided an additional braking deceleration of about 3 m/s^2 at 280 km/h [10]. Such solutions are no longer possible today due to the ban on moveable aerodynamic devices.

6.5 Wing Profiles

6.5.1 Basic Characteristics of Wing Profiles

The most conspicuous and at the same time, together with the underbody, the most effective components for generating downforce are wing profiles when used as front or rear wings. The basic properties of wing profiles were known long before they first appeared in motor sports. In aviation, they were used to generate dynamic lift at the beginning of the twentieth century. For example, the Wright brothers achieved the first successful powered flight in 1903. In the decades that followed, further milestones in civil and military aviation were achieved. It was not until 1956 that the Swiss Michael May made the first documented use of a wing profile to generate downforce in motor sports. This is all the more astonishing because May did nothing other than mount a NACA (National Advisory Committee for Aeronautics) airfoil designed for aviation in upside down on his vehicle-a Porsche 550 Spyder (Fig. 6.39). He used this vehicle in the 1956 sports car race at the Nürburgring. Even in rainy conditions and with no experience at the Nürburgring, May managed to overtake the famous drivers Fangio and Behra and set superior best times within his class. Under the pretext that the high wing construction would obstruct the view of the other drivers and probably also at the insistence of the Porsche works team, the wing profile was finally deemed too dangerous [11]. As a result, the idea of generating downforce to increase driving performance was forgotten for another 10 years.

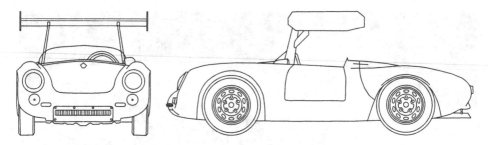

Fig. 6.39 Michael May's Porsche 550 Spyder from 1956

It was not until 1966 that the idea of using a wing profile to generate downforce was taken up again on the Chaparral 2E. The designer Jim Hall attached this airfoil with upright struts directly to the wheel carriers of the rear axle (Fig. 2.14b). To reduce drag, the wing position could be changed with a foot pedal. However, the function of downforce generation was still overlooked by the competition, rather it was assumed to increase braking efficiency and stability. The same principle was used on the Chaparral 2F in 1967. Aerodynamically, these vehicles were ahead of their time, but the chassis in particular was not yet designed for the transmission of high downforce (see also Sect. 2.1.7).

At the 1968 Monaco Grand Prix, the Lotus-Ford 49B made its first appearance with short wing sections attached to the side of the nose (Fig. 6.40). In addition, the body featured a large upsweep at the rear, which acted as a spoiler and created downforce over the rear axle. The subsequent Belgian Grand Prix marked the first use of continuous rear wings on Formula One cars. Inspired by the Chaparral 2E and 2F, it was Ferrari in 1968 that became the first team to use such wing designs in Formula One. On the Ferrari 312 (see Sect. 2.1.7) the wing profile was fixed directly above the engine. Comparable solutions were immediately copied by competitors. For the French Grand Prix, a high-mounted rear wing was used on the Lotus 49B, which was attached directly to the wheel carrier by thin struts. Due to the avoidance of turbulence and the lack of influence of body movements on the angle of attack, this solution was significantly more efficient than the competitors' approaches. However, due to the higher load on the wheel carrier and the vibrations in the struts caused by the wheels, this design was also very susceptible to damage. The variant with high profiles established itself nevertheless also with the competition. On the Brabham BT26 such a construction was used for the first time on both the rear and the front axle, which gave this car the nickname "biplane" (Fig. 2.15a, Sect. 2.1.7). Figure 6.40 summarizes essential steps in the application of wing profiles between the years 1967 and 1969 using the Lotus 49 as an example. To resolve the conflict between downforce and drag, some of the airfoils were coupled to a mechanism that could vary the angle of attack during while driving. Matra used an electric motor to adjust the rear wing, which was controlled by the brake pedal and the gear engaged. In the Ferrari version, the airfoil was adjusted hydraulically via the brake pressure [12].

The use of adjustable wing elements was banned for the 1969 Monaco Grand Prix. Even before these rule changes could come into effect, a series of serious accidents occurred that year at the Spanish Grand Prix at the Circuit de Montjuïc in Barcelona due to failing wing

Lotus-Ford 49
1967

Lotus-Ford 49B
1968

Lotus-Ford 49B
1968

Lotus-Ford 49B
1969

Lotus-Ford 49D
1969

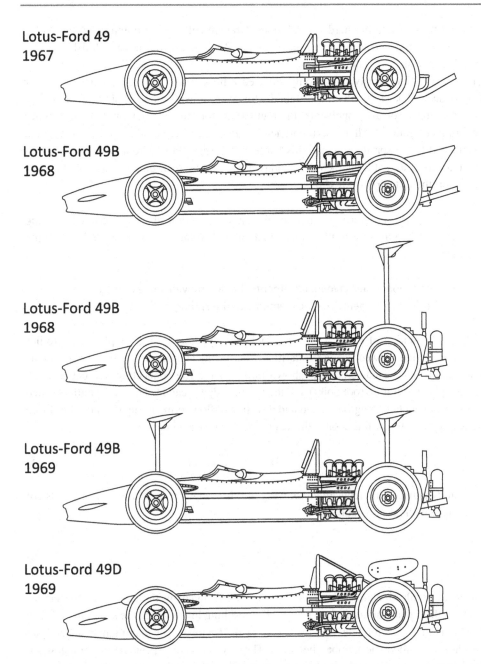

Fig. 6.40 Variants of the Lotus-Ford 49 from 1967 to 1969

struts. First, at the end of the third lap, third-placed Graham Hill crashed his Lotus-Ford 49B, which was used in this race with high wing profiles on the front and rear axles. On the start-finish straight, a strut of the rear wing buckled at about 230 km/h. The uncontrollable

Lotus hit the crash barrier hard several times. Graham Hill remained unhurt. In lap 20 the same defect happened to his team mate Jochen Rindt – also on the start-finish straight. Rindt's car crashed into the wreck of Graham Hill, which had not yet been recovered, and rolled over. Miraculously, Rindt only suffered a broken nose and concussion. The rear wing of Jacky Ickx's Brabham also failed. However, he was initially able to continue the race. As a result of these incidents, the regulations for the construction of wing profiles were further tightened. Their maximum height and width were limited and they now had to be attached to the sprung masses. This led to the arrangement of front and rear wings on formula cars that is still characteristic today and could be seen on the Lotus 49D and its successor, the Lotus 63 from 1969.

▶   In summary, two basic regulations for the construction of wing profiles for racing vehicles originate from this time, which are still valid today in practically all racing series:

- Prohibition of aerodynamic elements that are moved while driving
- Attachment of aerodynamic elements to the sprung masses

At the beginning of modern racing vehicle aerodynamics, quite simple wing profiles were used, which showed strong similarities to aeronautical profiles. Such an airfoil with its basic geometrical characteristics is shown in Fig. 6.41. The connecting line between the leading edge (with nose root point) and the trailing edge is called chord, its length is called profile depth or chord length t. Chord and direction of flow form the angle of attack α. From the width b and the airfoil depth t the wing plan area A_F is obtained.

$$A_F = b \cdot t \tag{6.35}$$

The aerodynamic coefficients given for wing profiles usually refer to this span area. Also of importance is the aspect ratio λ_F.

$$\lambda_F = \frac{b}{t} = \frac{b^2}{A_F} \tag{6.36}$$

The local profile thickness along the chord is determined by the centers of the circles touching the top and bottom of the profile. The resulting line is called the skeleton line. The deviation of the skeleton line from the chord is called camber. The maximum distance from the chord is called the camber height f. These values are especially important when attempting to systematically study different profiles and catalog the results, as has been done by NACA and others [13]. Figure 6.42 shows the CFD simulation of such a NACA profile. The dimensionless pressure coefficient C_p and the lines of constant mass flows (streamlines) are shown.

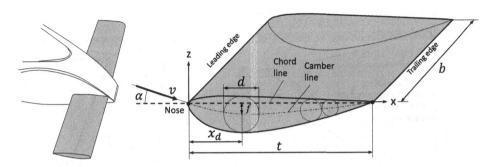

Fig. 6.41 Geometry of a wing profile

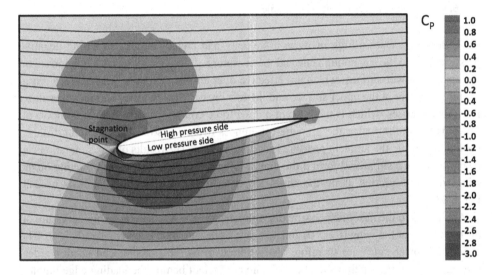

Fig. 6.42 Flow around and pressure distribution on a wing profile (NACA 0015) at $\alpha = 10°$

Air as a fluid always occupies the entire available space. In its movement, it follows any surface shape, such as that of a wing surface [14]. When flowing around a wing profile, the air must change its direction of motion in order to remain in contact with the wing surface. This results in acceleration forces that are supported on the wing surface in the form of pressure changes. The simple basic rule applies, according to which the pressure increases when air flows towards a surface (as is the case at stagnation zones or concave surfaces), and according to which the pressure decreases when air flows away from a surface (convex surface courses). Due to the angle of attack of $10°$, a stagnation zone is formed on the upper surface of the wing, just behind the root of the nose, where the air particles are strongly decelerated. This decrease in velocity causes a strong increase in pressure. At the stagnation point (Fig. 6.42) the flow velocity is zero. There the airflow divides into a portion which flows off onto the upper side and a portion which flows on over the lower side. The strongly convex curved surface of the leading edge, which must be overcome when flowing down to

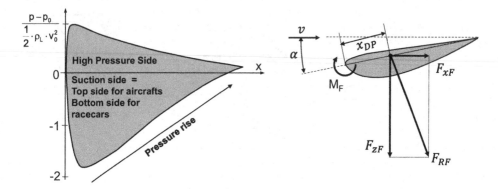

Fig. 6.43 Resulting forces on the wing profile

the underside, causes a high increase in velocity and creates an area of strong negative pressure. Due to the stagnation zone and the curvature, pressure differences are imposed on the upper or lower surface, which are effective over almost the entire wing surface. From the high pressure at the upper surface and the low pressure at the lower surface, a resulting force finally arises, which is made up of a downforce component and an drag component. Towards the trailing edge the pressure differences decrease, so that at a certain distance behind the profile the ambient pressure is assumed again. The results of this simulation can be presented in a simplified form if the dimensionless pressure coefficient C_p acting on the wing surface is plotted along the chord length. This form of representation is used in Fig. 6.43.

Analogous to the absolute pressure distribution in Fig. 6.42, there is an high pressure on the upper side of the wing ($C_p > 0$) and a negative pressure on the lower side ($C_p < 0$). The upper and lower sides are therefore also referred to as the pressure and suction sides, respectively. On the suction side, the pressure coefficient behind the leading edge initially drops steeply to a minimum value, and then rises again at the trailing edge to zero (reassumption of ambient pressure). This means that the air flows against a significant pressure increase at the bottom edge. According to Sect. 6.3, the suction side thus tends towards boundary layer separation, which will play a decisive role in the design of the airfoil in the following.

From the pressure curves over the wing surface, a resulting force is obtained by integrating the pressure curves along the wing surface, which acts on the wing profile and acts at the pressure point DP. This force can be divided into a part parallel to the direction of motion F_{xF} and a part perpendicular to the direction of motion F_{zF}. These force components thus correspond to the drag and the downforce of the wing profile. Analogous to the overall vehicle, the aerodynamic coefficients can also be defined for the wing profile. The following applies:

$$F_{xF} = c_{xF} \cdot A_F \cdot \frac{\rho_L}{2} \cdot v^2, \tag{6.37}$$

$$F_{zF} = -c_{zF} \cdot A_F \cdot \frac{\rho_L}{2} \cdot v^2. \tag{6.38}$$

The pressure at each point of the surface and thus the resulting forces or the aerodynamic coefficients depend on the inflow velocity and angle or on the surface curvature. Higher velocities, larger angles of attack and stronger curvatures require higher pressure changes and thus lead to larger forces on the wing profile.

Figure 6.44 shows the behavior of the coefficients as a function of the angle of attack α in the form of the so-called wing polars. In such a diagram, the downforce coefficient c_{zF} is plotted against the drag coefficient c_{xF}. The angle of attack is varied at a constant approach velocity, which means that the Reynolds number also remains constant. When using such diagrams, it should be noted that the aerodynamic coefficients are generally dependent on the Reynolds number and thus on the flow speed.

In the case of a symmetrical airfoil, which does not interact with the ground, only a drag force is generated at an angle of attack $\alpha = 0°$, but no lift force. This component of the drag is called the airfoil drag. The stagnation point coincides with the leading edge of the airfoil. By varying the angle of attack as shown, the stagnation point moves to the wing surface, requiring another portion of the convex airfoil nose to be overcome as the air flows onto the underside. The negative pressure on the underside increases. High pressure and low pressure are no longer distributed symmetrically over the airfoil, so that the pressure distribution results in a downforce and increased drag. This increase in drag represents the drag induced by downforce.

Up to an angle of attack of $\alpha = 10°$, there is a linear relationship between downforce and drag. However, due to the increase in negative pressure on the suction side, the positive pressure gradient also increases up to the trailing edge of the wing. This increased pressure rise increasingly favors the phenomenon of detachment. Due to the separation starting at the trailing edge, the drag increases more steeply than the lift coefficient, which results in the degressive course.

At an angle of attack of $22.5°$, the flow on the underside is practically completely detached. The downforce can no longer be increased. The phenomenon of separation is one of the central problems in the wing design of racing cars, since one is forced to generate large pressure gradients due to the strictly regulated installation space for wing profiles. The various measures to avoid detachment are considered in the following sections.

6.5.2 Influence of the Wing Geometry

From the wing polar, it can be seen that for a given wing profile, the increase in downforce is accompanied by an increase in drag and that a certain maximum downforce coefficient

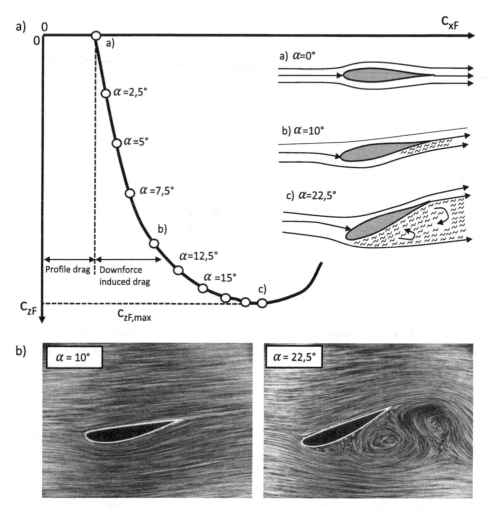

Fig. 6.44 Influence of the angle of attack (**a**) in the wing polar diagram and (**b**) as streamlines in the water channel. (Courtesy of © Deutsches Zentrum für Luft- und Raumfahrt e. V., Central Archive 2018. All Rights Reserved)

$c_{zF,max}$ is achieved. A typical problem in racing vehicle aerodynamics is to increase the maximum downforce while achieving the best possible aerodynamic efficiency.

Two essential parameters of the airfoil design are airfoil thickness and camber respectively their course along the wing chord. Figure 6.45 shows the basic influence of these two parameters on the maximum downforce coefficient. The simple wing profile shown in a) reaches its maximum downforce coefficient at a profile thickness which is about 12% of the chord length. The camber of an airfoil initially has a similar effect to the inclination of a symmetrical airfoil in that it shifts the zero downforce angle to non-zero values. Basically,

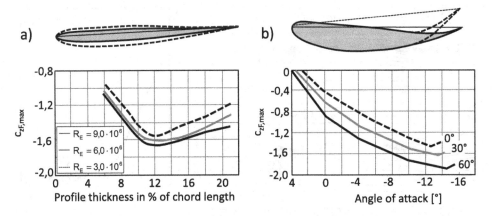

Fig. 6.45 Influence of (**a**) the profile thickness and (**b**) the profile camber

the camber increases the pressure difference between the upper and lower side, so that higher maximum downforce coefficients are also achieved.

In addition to the maximum downforce coefficient, the airfoil geometry also determines the separation behavior of the wing, which is illustrated in Fig. 6.46. It compares the streamlines and the angle of attack dependent downforce behavior of two wing profiles of different thickness but identical camber line. Wing A has a maximum airfoil thickness of 12% related to the chord length, while this parameter is 22% for wing B.

According to Fig. 6.45 a higher maximum downforce coefficient is achieved with wing A than with wing B. The maximum downforce coefficient is reached at an angle of attack of 15°. Shortly before this angle of attack the flow is still attached until far behind the leading edge, but already at a slightly higher angle of attack the flow separates completely at the leading edge. This is called "leading edge stall". The result is a steep drop in the coefficient of lift after reaching the maximum. In the vicinity of the maximum lift coefficient, this airfoil reacts very sensitively to an increase in the angle of attack.

Wing B, on the other hand, has a detachment behavior in which the detachment point moves continuously from the trailing edge towards the leading edge as the angle of attack increases. This behavior is called "trailing edge stall". The drop in the coefficient of drag is much flatter. The airfoil is clearly more robust against variations of the angle of attack in the aerodynamic limit range. Simple airfoils, which are designed for a maximum downforce value, tend to "leading edge stall". Such behavior is favored by sharp curvatures at the leading edge of the airfoil. A higher curvature and a higher profile thickness counteract this [15].

The behavior of a wing profile is directly related to the pressure distribution on its surface. Due to the tendency of boundary layer separation, the suction side of the wing, which is located on the underside of the airfoil in racing vehicles, is particularly important. The influence of airfoil thickness and camber on the pressure distribution on the airfoil surface as well as the resulting wing polars are shown in Fig. 6.47. On the left side the

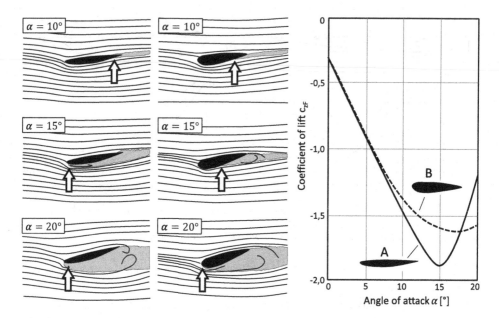

Fig. 6.46 Influence of the profile thickness on the detachment behavior [8]

behavior of three airfoils is compared, which differ only by the relative airfoil thickness of five, ten and 15%. The angle of attack in this example is 5°. It can be seen that with increasing airfoil thickness, the negative pressure level on the suction side increases and thus the main pressure increase in the direction of the trailing edge is stronger. This can be explained by a higher displacement effect of the airfoil nose, which leads to higher flow velocities on the lower side of the airfoil. However, the suction tip at the leading edge of the wing is reduced by the airfoil thickness. The detachment of the boundary layer over the trailing edge of the wing is favored by these effects. From the wing polars it can be seen that thinner airfoils have a lower drag at small angles of attack, which is due to the lower main pressure rise towards the trailing edge of the airfoil. For thin airfoils with low angles of attack, the boundary layer flow is therefore laminar for longer. At higher angles of attack, however, the pressure increase due to the pronounced suction peak at the leading edge favors the change to turbulent boundary layer flows, which causes a significant reduction in aerodynamic efficiency. The suction tip of thin airfoils results in early boundary layer separation at the leading edge of the airfoil at higher angles of attack, limiting the maximum downforce coefficient. This confirms the conclusion from Fig. 6.45 that the maximum downforce coefficient of symmetrical airfoils is only reached at relative airfoil thicknesses of significantly more than 10%.

The profile camber, shown on the right side of the example, is between zero and 4% at an angle of attack of 5°, leads to a higher pressure rise on the pressure side and to a higher pressure drop on the suction side. The downforce coefficient increases accordingly with

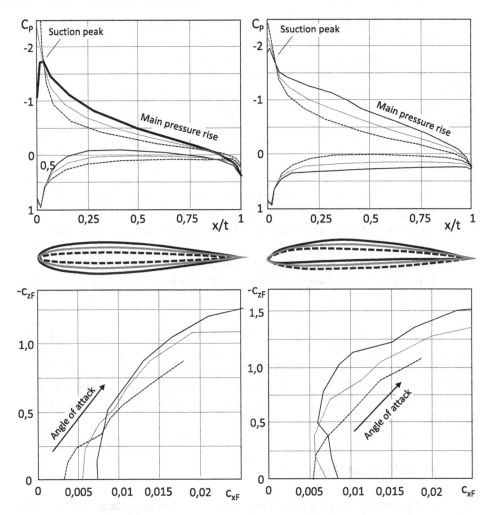

Fig. 6.47 Influence of airfoil thickness and camber on pressure distribution and wing polars. (Adapted from [16]; courtesy of © Institute of Aerodynamics and Gas Dynamics, University of Stuttgart 2018. All Rights Reserved)

higher camber. In the main pressure area on the suction side, the pressure coefficients run almost parallel to each other. This means that the pressure increase in the direction of the trailing edge is initially not significantly influenced by the camber. Turbulent boundary layer separation therefore only takes place at higher downforce values, which is why curved profiles achieve higher maximum downforce coefficients than symmetrical profiles. Increased camber also results in a reduction of the suction tip, which reduces the tendency to separation at the leading edge ("leading edge stall"). From the wing polars it can be seen that airfoils with higher camber have lower drag coefficients at higher downforce coefficients. Therefore, airfoil camber is also an important parameter for improving the aerodynamic efficiency of a wing that needs to achieve high downforce coefficients. However, at low downforce coefficients, camber increases drag.

The wing geometry therefore influences the pressure distribution on the wing surface, which in turn determines the lift and drag behavior of the wing. The aerodynamicist's task is to find an optimum airfoil geometry for the respective application, which in high-performance racing can only be achieved by using powerful CFD simulation programs.

▶   The following values are given in [17] as guide values for the design of simple
        profiles:

- *Angle of attack:* Small for low downforce and low drag, large (up to 14–16°)
  for high downforce with higher drag
- *Curvature:* 4–6% for low output and low resistance, approx. 9% for high
  output
- *Thickness:* Thin for low downforce and low drag, 14–16% for good efficiency
  over a wide range of angles of attack, 18–20% for high downforce.
- *Nose radius:* 1–3%.

The diagrams used in this book generally originate from aviation, so that at this point a fundamental consideration of various effects takes place. General statements are therefore not possible due to the special boundary conditions that exist when using wing profiles in motorsport. The special boundary conditions include in particular the significant influence of the vehicle body on the airflow around the wing as well as the ground effect explained in Sect. 6.5.4.

6.5.3   Multi-Element Wings

The potential of simple airfoils for downforce generation is limited, because the angle of attack and camber cannot be increased arbitrarily due to the separation starting at the suction side. The separation starts at the trailing edge of the wing when the pressure increase towards the trailing edge becomes too high. For this reason, multi-element wings are used – if permitted by the regulations. Multi-element wings consist of a main airfoil, a slat and one or more flaps. Some typical arrangements for multi-element profiles are shown in Fig. 6.48.

In the design of the flaps, a distinction is made between surface-mounted flap (Junkers flap), embedded flap (Fowler flap) and double and triple flaps. The basic principle of the multi-element wing is to use part of the energy on the pressure side to accelerate the air on the suction side. In this way, the detachment process is delayed. For this purpose, the flap and slat form a gap which narrows in the direction of flow. Figure 6.49 shows this using the example of the rear wing profile of an F1 vehicle.

The narrowing gap geometry accelerates the air flowing in on the pressure side. The gap is designed in such a way that the air flowing through it hits the underside of the wing

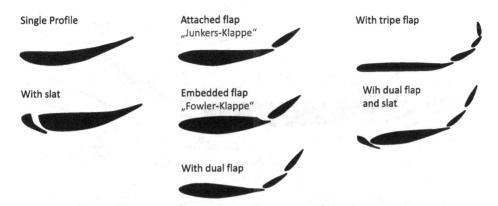

Fig. 6.48 Typical slat and flap arrangements on multi-element wings

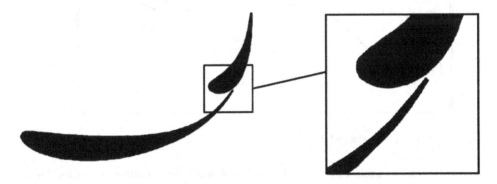

Fig. 6.49 Gap geometry between main profile and flap of an F1 rear wing

approximately in a tangential direction. There, as shown in Fig. 6.50, the gap flow adds new energy to the low-energy boundary layer of the main profile, which tends to detach. By delaying the detachment process, multi-element airfoils can be operated with significantly higher camber and angles of attack than single-element airfoils. However, the higher camber achieved by the flaps simultaneously increases the flow velocity at the leading edge of the airfoil. As a result, suction peaks can occur at the leading edge, whose trailing pressure increase can possibly lead to separation at the leading edge. A slat shifts the suction tip from the main airfoil to the slat, preventing premature separation at the main airfoil and extending the usable angle of attack range [18]. Figure 6.50 illustrates the function of the slat by means of the pressure distribution on a three-element airfoil.

In contrast to the flaps, the slat has practically no influence on the camber and therefore does not increase the downforce directly, but only via the increase in the angle of attack possible with it, as can be seen in Fig. 6.50. The typical effect of a camber-increasing flap can also be seen, which, by increasing the flap angle, on the one hand shifts the zero downforce angle and, on the other, causes an increase in the maximum downforce. The effect of the flap is also shown in Fig. 6.51 in the form of the downforce and drag

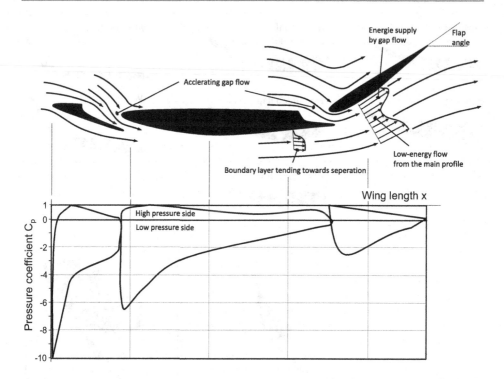

Fig. 6.50 Flow through slat and flap

coefficients plotted against the angle of attack and in the form of the wing polars. These values are contrasted with the behavior of a single-element airfoil. The increase of the maximum downforce coefficient due to the flap is clearly visible. However, the wing polar also shows that the use of a flap is associated with a loss of aerodynamic efficiency. At lower downforce values, the single wing profile generates significantly less drag than the dual-element profile. Only when a correspondingly higher maximum downforce coefficient is to be achieved do the positive effects of the flap come into play.

Multi-element wings serve to increase the downforce. As a rule, they have no positive influence on the aerodynamic efficiency of a wing profile. Therefore, the simple rule of thumb applies: For a drag-optimal wing design, one-piece wings are used, for downforce-optimal applications multi-element wings. This is underlined by a comparison of different rear wing profiles on a McLaren-Mercedes MP4/22. Figure 6.52 shows the respective wing configuration for the Italian Grand Prix at the Autodromo die Monza and for the Canadian Grand Prix at the Circuit de Gilles Villeneuve. Typical for the Monza circuit are its high full-throttle percentage and the low number of corners. Above all, this circuit requires a high maximum speed. Accordingly, a single-element wing profile is used. The Circuit de Gille Villeneuve, on the other hand, requires higher downforce coefficients and thus the use of a dual-element wing profile.

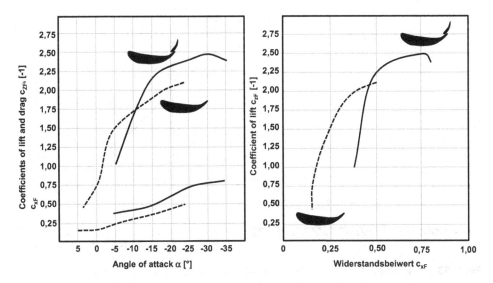

Fig. 6.51 Coefficients of a single and a multi-element profile. (Data from [19, 20])

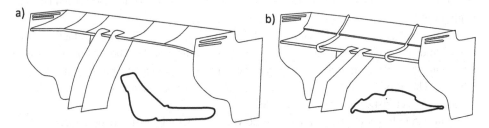

Fig. 6.52 Rear wing configuration of a McLaren MP4/22 (2007) for (**a**) the Italian Grand Prix and (**b**) the Canadian Grand Prix

6.5.4 Ground Effect of Wing Profiles

In connection with wing profiles, another concept is introduced which is of enormous importance for the design of racing cars: the ground effect.

An unadjusted symmetrical airfoil generates no lift and downforce at high altitude above the ground. However, near the ground, the same airfoil generates a down force. This phenomenon is called the ground effect.

Figure 6.53 shows the influence of the ground clearance on the downforce coefficient of a symmetrical airfoil. In the right part of the figure, the curves of the downforce coefficient for two constant altitudes as a function of the angle of attack α are compared. At a very high

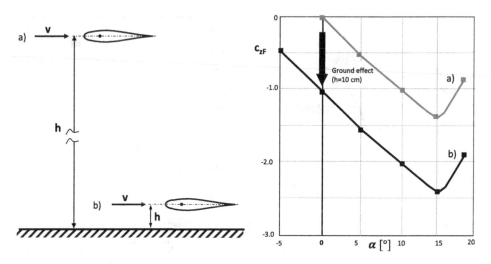

Fig. 6.53 Dependence of the lift coefficient on the ground clearance (see also [15])

altitude above the ground, the downforce coefficient is zero at $\alpha = 0°$ (cf. Figure 6.44). At a height of about 10 cm, i.e. with a typical ground clearance of the front wing, a downforce coefficient of about -1.1 is obtained. Simplified, this can be explained by the fact that the underside of the wing and the ground now form a Venturi nozzle. The interaction of the wing with the ground therefore increases the downforce when the ground distance is reduced. The ground effect contributes significantly to the aerodynamic and driving dynamic characteristics of a racing vehicle.

Figure 6.54 shows the behavior of the downforce coefficient of a two-element airfoil with continuous reduction of the ground clearance under constant angle of attack and constant Reynolds number. In both diagrams the downforce and drag coefficients for two different flap positions are plotted against the ground clearance h related to the chord length c. The curves of the aerodynamic coefficients are strongly dependent on the inflow velocity and Reynolds number. However, this is not considered further here. The chord length of the measured wing profile is 380 mm [21]. The typical working range of the front wing of a 2000 F1 car is approximately 70–110 mm above the ground, which corresponds to a ground clearance-chord length ratio of 0.18–0.29. In this range, the downforce coefficient c_{zF} is approximately linearly dependent on the ground clearance h, and the downforce coefficient increases continuously with decreasing ground clearance. The drag coefficient also increases due to the longitudinal force component induced by the downforce. Changes in ground clearance are experienced by the wing profile of a racing vehicle due to the heaving, pitching and rolling motions of the body that occur during acceleration, deceleration and cornering. Body movements of the vehicle therefore change the aerodynamic balance of the vehicle. This behavior is called "pitch sensitivity" (Sect. 6.9). The ground

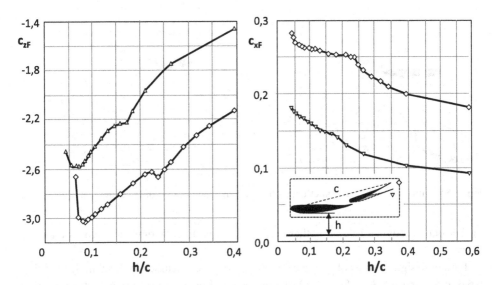

Fig. 6.54 Dependence of downforce and drag coefficient on ground clearance. (Adapted from [22]; courtesy of © Prof. Xin Zhang 2018. All Rights Reserved)

effect therefore also leads to a strong interaction between the aerodynamic characteristics of the vehicle and its chassis.

A central task of the chassis is to keep the aerodynamic balance within a defined working range at all times. The maximum downforce coefficient for the example shown is reached at a ground clearance of approx. 3 cm. If the ground clearance is reduced further from this point, the downforce coefficient drops abruptly. This drop is visibly stronger with a higher angle of attack of the flap. Approaching this working point can lead to "porpoising" in racing vehicles. Until the working point is reached, the spring forces compensate for the downforce forces. If the downforce then drops, the spring relaxes and the body moves upwards. The upward movement now causes the downforce to increase again, and the spring is compressed once more. If this process repeats itself in the context of a low-frequency oscillation, it is referred to as "porpoising" or "bouncing". Such porpoising is observed, for example, on the current generation of DTM cars and was observed in 2014 during the 24 Hours of Le Mans on the Porsche 919 Hybrid. It could also be observed on the new generation of F1 cars introduced in 2022.

The ground effect also has a significant influence on the design of the front and rear wings of a racing car. In a monoposto, the effect of the ground effect is significantly stronger on the low-slung front wing than on the higher-slung rear wing. However, in order to achieve an aerodynamic balance, the front and rear wings must achieve comparable downforce levels (cf. Figure 6.10). To compensate for the lack of interaction with the ground, the rear wing, as Fig. 6.55 shows, is operated with much higher angles of attack and significantly higher camber. For this reason, a front wing has an aerodynamic efficiency of about 8–10, while this is only about 3–4 for the rear wing. The rear wing's share

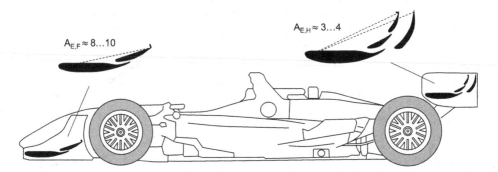

$A_{E,F} \approx 8...10$

$A_{E,H} \approx 3...4$

Fig. 6.55 Influence of the ground effect on the efficiency and design of front and rear wings

of the total aerodynamic drag is therefore about three to four times as high as that of a front wing.

Further design features of racing car wing profiles are discussed in detail in Sects. 6.5.5, 6.5.6, 6.5.7 and 6.5.8. In Sect. 6.12, the influence of the technical regulations on various trends in aerodynamic development and wing design is also considered.

6.5.5   Front and Rear Wings on Racing Vehicles

The two main challenges in the development of front and rear wings for racing vehicles are the strictly limited installation space imposed by the regulations and the strong interaction of all aerodynamic components of the vehicle with each other. In contrast to wing profiles on aircraft, the wing area of racing vehicles is very small compared to the fuselage or body, which leads to a significantly lower and thus less favorable length-to-width ratio λ_F. In order to realize high downforce coefficients despite the limited installation space, high pressure changes and thus high pressure gradients must be achieved, which requires appropriate solutions for the avoidance of separation effects. In this context, wing elements cannot be considered in isolation, but must always be viewed as a component of the overall aerodynamic concept. The flow behind the front wing, for example, significantly influences the incident flow and thus the effectiveness of the underbody. In the same way, the flow of the rear wing is influenced by the body in front of it, while the rear wing in turn influences the flow conditions behind the underbody. In addition, the aerodynamic balance of the vehicle changes permanently due to the ground effect in combination with roll, pitch and heave movements. The drivability of a racing vehicle must nevertheless be guaranteed in all driving conditions. Likewise, the conflict of goals between downforce generation and drag must be solved in the best possible way. In contrast to aircraft wings, there are no movable wing elements available for this purpose in racing technology – with the exception of the drag reduction system (Sect. 6.5.8).

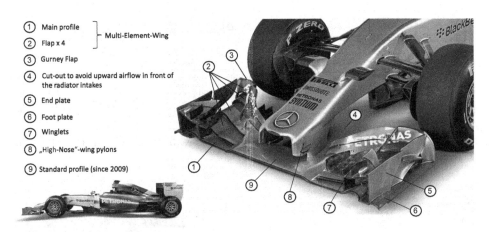

Fig. 6.56 Components of a front wing using the Mercedes F1 W05 Hybrid (2014) as an example. (Courtesy of © Daimler AG 2018. All Rights Reserved)

For these reasons, the front and rear wings of racing vehicles are visibly more complex in design than aircraft wings. Figure 6.56 shows the front wing structure of the Mercedes-Benz F1 W05 Hybrid of the 2014 season. Characteristic of the design of wing profiles on racing cars is above all the use of multi-element wings and relatively large-area end plates. A horizontal footplate is also attached to the outer ends of the front wing. The downforce generation of the front wing is assisted by a so-called Gurney flap, which in this case extends over the outer third of the rearmost wing profile. The function of these components is discussed in Sects. 6.5.6 and 6.5.7. In the central area, the use of a neutral unit profile has been mandatory since 2009, which also explains the trend towards an increase in the number of profile elements on the outer front wing: on the Mercedes-Benz F1 W05, the front wing is divided into five profile elements, while six profile elements were used on the Red Bull RB10 of the same year. Another typical feature of the front wing is the recess in front of the radiator intakes, which ensures an adequate supply of cooling air while avoiding any upwash that might otherwise be generated by the wing. The front wing is suspended from the high nose by two streamlined pylons, which is known as a "high-nose" configuration. The trend towards their use took hold in the early 1990s. Figure 6.57 compares the "high-nose" concept of the 1992 Benetton Ford B192 with a low-slung nose on the 1988 McLaren Honda MP4/4, which had been common until then. The low-slung nose divides the front wing into two airfoils interrupted by the nose, which means that the airfoil does not extend beyond the theoretically permissible width. The "high-nose" configuration allows an increase in the usable wing area of the main airfoil as well as an optimization of the underbody inflow, which increases downforce. Such increases in the efficiency of aerodynamic components are often a consequence of previous changes to the technical regulations, such as those made for the Formula 1 World Championships of the 1990s (Sect. 6.12).

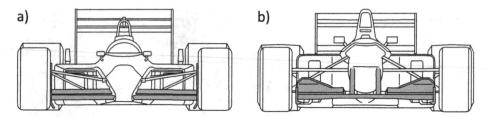

Fig. 6.57 (**a**) McLaren-Honda MP4/4 (1988) with low nose, (**b**) Benetton-Ford B192 (1992) in "high nose" configuration

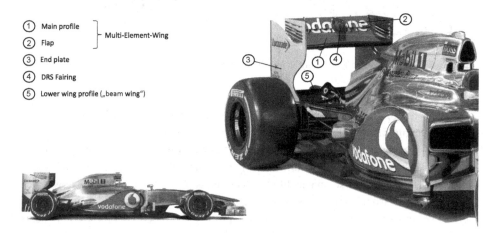

Fig. 6.58 Components of a rear wing using the McLaren-Mercedes MP4–27 (2012) as an example. (Courtesy of © McLaren Technology Group 2018. All Rights Reserved)

The rear wing of a modern F1 car is shown in Fig. 6.58, using the McLaren-Mercedes MP4–27 from 2012 as an example. The large end plates, which also serve as a suspension for the airfoil elements, are striking. The slots above the wing profile are used to specifically influence the drag vortices that are created. Due to the regulations, the rear wing consists of a two-element profile in the upper area and has a further wing profile, the "beam wing", in the lower area. Rear wings on F1 cars of earlier generations sometimes had significantly more wing elements, but today's regulations no longer permit this. Also visible is the fairing for the linkage of the DRS actuator, which can rotate the wing flap or flap around its trailing edge to reduce the angle of attack. Further details of this system are described in Sect. 6.5.8.

Figure 6.59 shows two further variants of the rear wing profile suspension using the example of an Audi R10 TDI and an Audi R8 LMS ultra. In Fig. 6.59a, the rear wing profile is attached to two vertical struts on its underside. An alternative variant is suspension of the airfoil on its upper side, the swan neck suspension. This arrangement has the advantage that the struts do not disturb the sensitive flow on the suction side. Such designs were first seen

a)

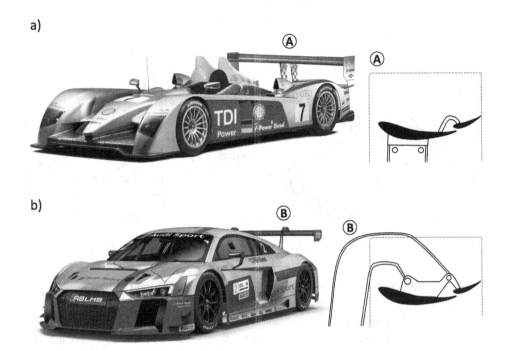

b)

Fig. 6.59 (**a**) Standard rear wing suspension on an Audi R10 TDI, (**b**) Swan neck rear wing suspension on an Audi R8 LMS ultra. (Courtesy of © Audi AG 2018. All Rights Reserved)

on sports car prototypes in the LMP1 category after the permissible rear wing width of these cars was reduced. In the meantime, it has also become established in the DTM and in near-series racing cars.

The interaction of the wing components described here increases the complexity of the optimization task that the aerodynamics department has to deal with.

6.5.6 Wake Vortices and End Plates

As discussed in detail in Sect. 6.5.1, the upper surface of the airfoil of a racing vehicle is highpressurized, while the lower surface is lowpressurized. This pressure difference causes air to flow from the highpressure side to the lowpressure side, causing the air to attempt to return to a state of equilibrium. On a wing airfoil, as Fig. 6.60 shows, this equilibrium movement occurs primarily at the wing tips. The actual inflow process takes place transversely to this equalizing movement. The inflow velocity v and the rotational compensation movement lead to the formation of a wake vortex. Due to the equalizing movement, the pressure difference between upper and lower surface decreases when approaching the wing tips. For wings with a small span ratio λ_F, as found in racing cars, this leads to a noticeable loss of efficiency, while this effect is negligible for very wide

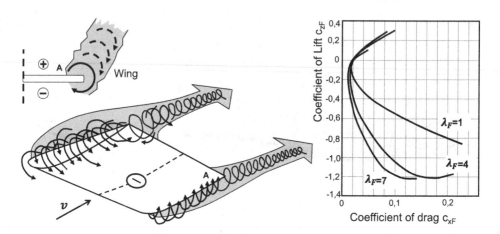

Fig. 6.60 Formation of wake vortices and their influence on the wing polar diagram

wings. The influence of the span ratio on the downforce coefficient is illustrated by the wing polars in Fig. 6.60.

The main task of end plates is to prevent the compensating movement by separating the suction side from the pressure side. As Fig. 6.61 shows, the end plates cause an approximately constant pressure distribution over the span. The available wing area is used more efficiently and higher downforce coefficients can be achieved. Since the end plates are very narrow, their use does not lead to any appreciable increase in the frontal area. The vortex size (cf. Figure 6.63) and thus the energy dissipated in the vortex is reduced. Thus, the downforce can be increased by end plates without an increase in drag. In Fig. 6.61 the curves of the downforce and drag coefficients are plotted for a wing configuration with and without end plates as wing polars. From this diagram it can be seen that the downforce can be increased by approx. 30% with end plates without increasing the aerodynamic drag.

The behavior of drag and downforce depends on further design elements of the end plate. Figure 6.62 illustrates the influence of the end plate area. In simple terms, it can be seen that the larger the end plate area, the greater the increase in downforce. At the same time, the use of the end plate is accompanied by a moderate reduction in drag. However, this effect is significantly smaller than the downforce increase. In contrast to the downforce, the jump between the first two endplate sizes shows only a small decrease in drag relative to the endplate area. The decrease in drag results from the decrease in wake vortex size outlined in Fig. 6.63.

End plates do not completely prevent the formation of wake vortices. Therefore, further measures are taken on the wings which mitigate the negative effects caused by vortex formation. A complete prevention is also not possible because the technical regulations limit the permissible end plate areas. A further element for the targeted influencing of vortex formation on the rear wing are slots above the wing profile (Fig. 6.64). These were used for the first time on the Renault R24 in the 2004 season. Due to the strong

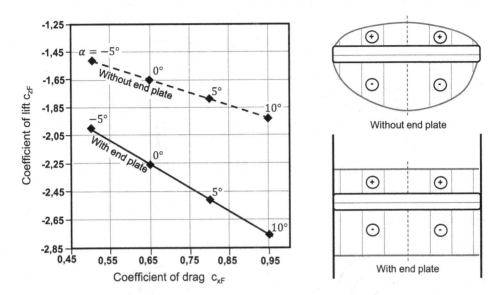

Fig. 6.61 Effect of end plates on pressure distribution and drag and downforce coefficients [15]

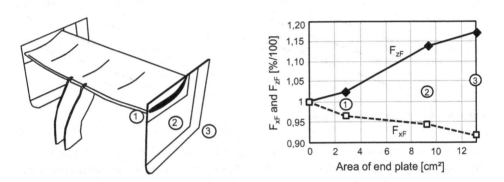

Fig. 6.62 Influence of end plate area on downforce and drag [23]

highpressure on the upper side of the airfoil, the air flows through these slots and, if the shape is appropriately upward, a flow movement is created which counteracts the actual balancing movement and thus the vortex formation. The vortex size and thus the drag of the rear wing are reduced by this measure.

On front wings of F1 cars it is noticeable that the wing profile does not extend over the entire permissible width. Instead, a horizontal foot plate is arranged next to the outside of the end plate. The relatively small end plates of a front wing cannot completely stop the flow movement from the upper to the lower side of the wing. There is still an appreciable

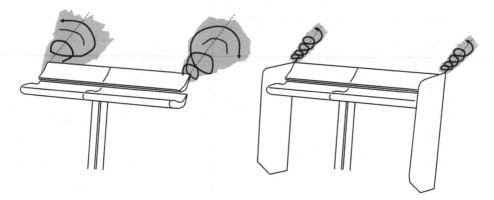

Fig. 6.63 Reduction of wake vortex size by end plates

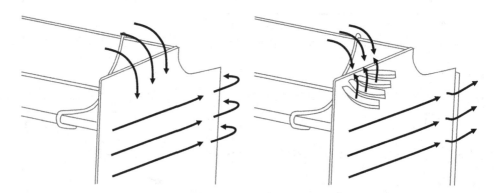

Fig. 6.64 Effect of a slotted end plate

underflow on the suction side, as shown in Fig. 6.65. This underflow is reduced by the horizontal footplate [23]. The air flowing from the upper side onto the footplate creates an highpressure region on the upper side of the footplate, which also leads to a moderate increase in the downforce on the front wing. As a result, the percentage downforce gain due to the footplate is higher than the downforce loss due to the reduction in span.

Another characteristic feature on the front wing of a Formula 1 car is the curvature of the end plate, which can be seen in Fig. 6.66a, b. For the 1998 season, the permissible vehicle width was reduced from 2000 mm to 1800 mm. The outer ends of the front wing were thus in front of the inner edges of the front tires. In order to prevent a drag-increasing frontal inflow of the front wheels, the end panels were retracted to the rear. For the 2008 season, the permissible front wing width was increased significantly so that the front wing ran across the entire width of the car. To redirect the air, the end panels are flared outwards for this case, which is illustrated for the front wing of the 2013 Red Bull RB9. By flowing around the tire, the pressure in the stagnation zone in front of the tire can be reduced, which has a positive effect on the drag of the vehicle.

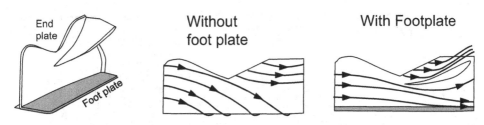

Fig. 6.65 Streamlines on the outside of the end plate

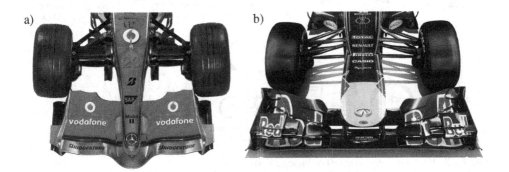

Fig. 6.66 End plate design as a function of front wing width. (**a**) McLaren-Mercedes MP4–23, 2008. (Courtesy © McLaren Technology Group 2018. All Rights Reserved); (**b**) Red Bull RB9, 2013. (Courtesy © David Clerihew and Red Bull Content Pool 2018. All Rights Reserved)

The front wing with its additional elements creates a complex vortex system, which is sketched in Fig. 6.67. The vortex system of the front wing essentially consists of end plate, foot plate, airfoil main and Y250 vortices. The vortices occur wherever there are strong pressure differences between highpressure and lowpressure regions. The Y250 vortex occurs at the transition between the mandatory standard profile introduced in 2009, which extends 250 mm from the longitudinal axis in the Y-direction, and the actual front wing profile. Since no significant positive and negative pressures are generated at the neutral standard profile, strong pressure differences occur in relation to the main profile, which then lead to a vortex-shaped compensating movement. A major goal of aerodynamic development is either to generate such vortices by vortex generators in order to increase downforce (Sect. 6.6.3) or to redirect them by vortex deflectors or baffles or "turning vanes" in such a way that they do not disturb the flow in sensitive areas. For example, the Y250 vortex is used to keep the wheel vortices away from the underbody inlet. The front wing and wheel vortices interact strongly. Smaller vortices are also created at the winglets or canard wings of the front wing. The front "turning vanes" and "bargeboards" generate vortices that are directed specifically to the underbody to use their low pressure core to generate downforce. How these vortices are dealt with depends heavily on the regulations.

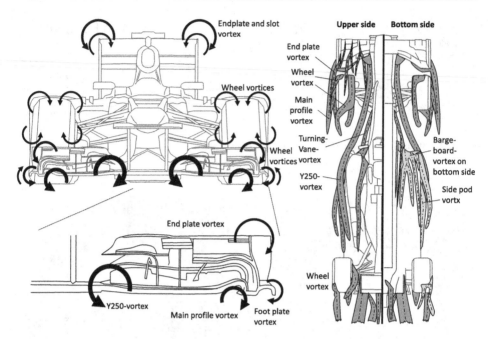

Fig. 6.67 Vortex system on an F1 vehicle (2013)

In particular, the front wing width, which has been changed several times from 2008 to the present day, has a major influence here. For the 2009 season, the use of flow aids in Formula 1 was severely restricted, which will be discussed again in Sect. 6.12.

6.5.7 Gurney Flap

The Gurney flap (Fig. 6.68) is a strip standing vertically on the trailing edge of the wing profile, the maximum height of which is about 20 mm or about 1–2% of the chord length. The component is named after former Formula One driver Dan Gurney, who first used this component on his racing cars in 1971. A Gurney flap generates two counter-rotating transverse vortices behind the trailing edge of the wing (cf. Figure 6.30) with the result that a negative pressure area is created behind the wing. This reduces the pressure gradient which the flow on the underside must overcome on its way to the trailing edge. This effect shifts the separation of the boundary layer to higher angles of attack, or more downforce can be generated at the same angle of attack. In addition, a stagnation zone with excess pressure forms in front of the Gurney flap, which also contributes to the gain in downforce. However, the increase in downforce also induces higher drag. However, at optimum Gurney flap height, the downforce gain is usually significantly higher than the increase in drag. In the case of wings on which signs of separation occur at an early stage, the

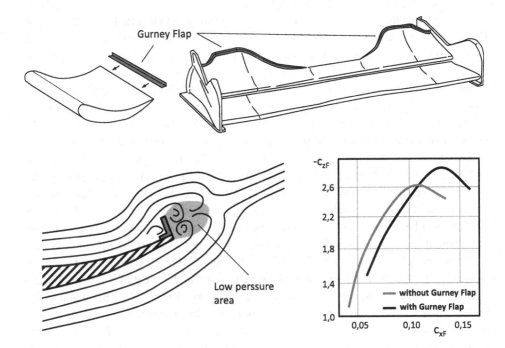

Fig. 6.68 Functionality of a Gurney Flap

installation of a Gurney flap can even lead to a decrease in aerodynamic drag. For these reasons, Gurney flaps provide numerous options for optimizing a wing profile. Gurney flaps are relatively easy to install, replace or remove from wing profiles, which is why they are often used to correct the aerodynamic balance of a racing vehicle. They have also been used in aviation, for example on the MD-11 and 787–8, following their application in motorsport [24].

6.5.8 DRS and F-Duct Systems

The sophisticated aerodynamics as well as the constantly improved drivability of modern racing vehicles due to modern development methods make overtaking the vehicle in front increasingly difficult. The following vehicle loses significant downforce in the wake of the vehicle in front, which makes it almost impossible to drive a tighter circle radius, which is necessary for out braking. This problem became increasingly apparent in the mid-2000s, as the crucial overtaking maneuvers were no longer performed on the track, but races were increasingly decided at the command posts and by pre-determined pit strategies. However, the lack of "real" overtaking maneuvers quickly makes a racing series less attractive to spectators.

To counteract this trend, Formula 1 introduced the so-called Drag Reduction System, or DRS for short, for the 2009 season, which initially allowed the flap on the front wing to be adjusted by up to 6°. From the 2011 season it was moved to its current position on the rear wing. This configuration is shown in Fig. 6.69. The upper part of the rear wing consists of an immovable main element and a rotating flap. The axis of rotation is located at the rear end of the flap. The flap may only take up two fixed positions, which are between a gap width of 10–50 mm. A change of position may only be made by the driver and is only permitted in certain areas of the track and only if the lead of the driver in front is a maximum of 1 s at defined sections of the track. The flap is usually actuated by a hydraulically operated linkage. Opening the flap reduces the downforce and aerodynamic drag of the vehicle and increases the top speed by approx. 4–5 km/h. Actuating the flap increases the aerodynamic balance and thus the oversteer tendency of the vehicle, so that the flap must be closed again in good time before braking for the next bend. It should be noted here that stable flow conditions only re-establish themselves at the wing profile after a certain time delay. To ensure driving stability, the actuating mechanism must be designed in such a way that the flap remains in the lower position in the event of a system failure.

The DRS was subsequently also introduced in Formula Renault 3.5 (2011) and in the DTM (2013). Figure 6.70 shows the system used in DTM. In contrast to Formula 1, only the use of a single-element wing profile is permitted on the rear wing in DTM. The angular position of this single-element can be changed by 18° via a pneumatically actuated linkage, although here too the conditions for actuation are limited by the sporting regulations. Rear wing and DRS are prescribed standard components. The avoidance of unstable driving

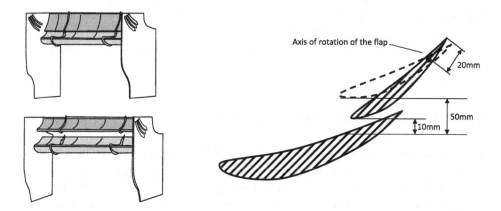

Fig. 6.69 DRS on the rear wing of an F1 car

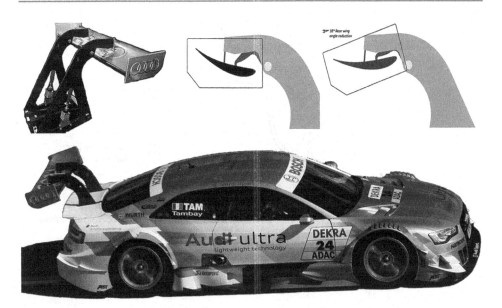

Fig. 6.70 DRS on the rear wing of an Audi RS5 DTM. (Courtesy of © AUDI AG 2018. All Rights Reserved)

conditions caused by unintentionally flat rear wings is achieved by sensor-based monitoring of the driving condition. In the event of a defined undercut or overshoot of limit values for the accelerator pedal position, brake pressure or lateral acceleration, the DRS is automatically brought into its steep normal position. Furthermore, the system is designed in such a way that the wing profile returns to the safe driving starting position even if the compressed air required for actuation is lost.

In the 2010 season, when the DRS was still on the front wing, McLaren introduced another aerodynamic system on the rear wing of the MP4/25 which reduced drag at high speed. This became known as the "F-Duct" system (Fig. 6.71). It got its name from the fact that a duct opening of this system was installed at the letter F of the corresponding sponsor lettering. As Fig. 6.72 shows, the F-duct system essentially consists of two intersecting flow ducts. The air inlet of the first duct is located at the airbox, and its regular outlet is located below the rear wing. The inlet of the second duct is located in the nose area of the monocoque. This duct has two exit openings. The first is installed in the cockpit of the vehicle and can be closed by the driver with his elbow, hand or knee. The second outlet opening of this channel is installed on the lower or suction side of the rear wing. The intersection of both channels represents a kind of pneumatic two-way valve, which is controlled by opening and closing the outlet opening in the cockpit.

Normally, the outlet opening in the cockpit is open, so that the airflow entering at the nose exits directly in the cockpit. In this state, the air flowing in at the airbox flows out below the rear wing. The flow around the rear wing is not affected. On the straights, where no downforce is required, the driver closes the cockpit-side outlet opening. Now the air

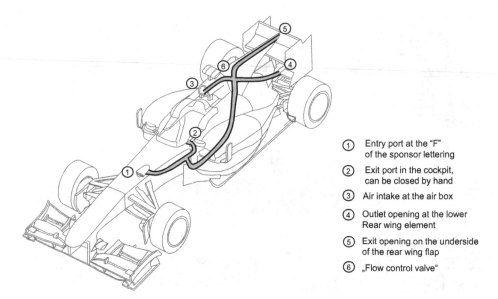

① Entry port at the "F"
   of the sponsor lettering

② Exit port in the cockpit,
   can be closed by hand

③ Air intake at the air box

④ Outlet opening at the lower
   Rear wing element

⑤ Exit opening on the underside
   of the rear wing flap

⑥ „Flow control valve"

Fig. 6.71 McLaren MP4–25 (2010) with F-duct system

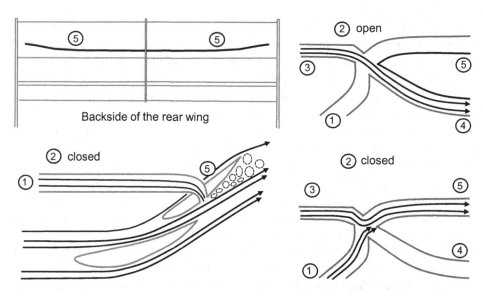

Fig. 6.72 Flow processes in the F-duct of the McLaren-Mercedes MP4–25 (2010)

flows from the nose towards the duct intersection. There it redirects the airflow entering the airbox so that it finds its way to the underside of the rear wing. The exit opening at the rear wing is arranged in such a way that the outflowing air meets the actual flow direction perpendicularly. This provokes a separation of the flow at the underside of the rear wing.

This reduces the negative pressure on the suction side. The force on the wing profile resulting from the negative pressure, which is composed of a downforce component and an air resistance component, decreases. In this case, the decrease in drag induced by downforce is greater than the increase in drag due to the detached flow. The total air resistance decreases, increasing the top speed of the vehicle by about 7 km/h. This effectively reverses the effect of a multi-element wing. Protests lodged against this system were rejected because the driver's body parts were not part of the vehicle and therefore could not be moveable aerodynamic devices.

For the 2011 season, F-duct systems were banned by declaring any driver-operated adjustment system that altered the aerodynamics of the car to be illegal. Explicitly exempted from this rule, however, was the drag reduction system now fitted to the rear wing. This inaccuracy in the regulations was exploited in the design of the Mercedes MGP W02. When the DRS system was activated, or when the upper wing section was raised, an opening was released that was hidden by the flap when the car was in normal operation. This opening was part of a duct system that had several exit ports on the front wing (Fig. 6.73). The highpressure at the top of the wing forced air into the duct system when the DRS was activated, which exited at the bottom of the front wing. Analogous to the drag reduction systems of the 2010 season, the release processes initiated by this caused a reduction in the overall drag. Protests against this system were also rejected. However, due to a clarification of the technical regulations, the F-duct systems were finally banned for the 2012 season. Corresponding articles were also adopted in the technical regulations of other racing series.

6.5.9   Flexible Wing Elements

Another aerodynamic tool that serves to resolve the conflict of goals between drag and downforce and regularly leads to discussions is the use of flexible aerodynamic elements. In this case, elastic deformations occurring under aerodynamic loads are utilized in order to specifically modify the aerodynamic properties of the vehicle, thereby circumventing the prohibition of moveable aerodynamic devices. The first application of such an elastic component known to the author was the flexible mounting of the rear wing on the transmission of the Lotus 72 [25]. The resulting aerodynamic force acting on the wing profile generated a torsional moment around the bearing. As speed increased, this torsional moment eventually became so large that the elasticity of the bearing led to the reduction in angle of attack visible in Fig. 6.74. The accompanying reduction in drag led to a higher top speed.

For various aerodynamic components, the FIA now prescribes load tests in which certain deformation limits may not be exceeded under a defined load. Nevertheless, designers occasionally succeed in designing aerodynamic components in such a way that they exhibit advantageous deformations even when the load tests are fulfilled in driving operation.

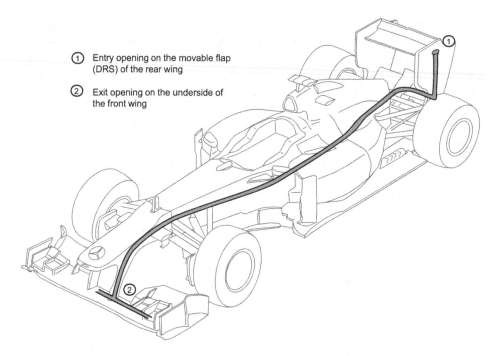

① Entry opening on the movable flap
(DRS) of the rear wing

② Exit opening on the underside of
the front wing

Fig. 6.73 F-Duct system on Mercedes MGP W02 (2011)

One example of this is the deformation of the flap on the rear wing of the Red Bull RB3 in 2007. The images from the TV cameras clearly showed how the gap between the main profile and the flap increasingly narrowed on long straights. This elastic deformation eventually led to a reinterpretation of the well-known advertising slogan: "Red Bull *bends your* wings". Figure 6.75 outlines the elastic and aerodynamic behavior of the wing profile. Closing the gap at least partially cancels the effect of the multi-element wing, and the downforce not required at high speed decreases, leading to a reduction in induced drag or increased top speed. The FIA eventually introduced a requirement for the use of shape-stabilizing templates during the season, which can be seen in Fig. 6.52b. In 2006, there was controversy over the front wing of the Ferrari 248 F1, which also visibly deformed. Although moving aerodynamic components are prohibited in principle, since practically all materials deform elastically (to a greater or lesser extent) under load, precise regulation at this point is impossible.

On the Red Bull RB6 from 2010, it was possible to bend the front wing so that the end plates approached the track, increasing the shielding effect of the end plate between the pressure and suction sides. This deformation was also clearly visible on TV and photographic footage. Then in 2011 it was the Red Bull RB7 that caused controversy due to an elastic deformation on the underbody. The FIA reacted again by prescribing new load tests. Deformations of the front wing were also observed on the Williams-Mercedes FW36 in 2014. Discussions also arose on the Toyota and Porsche LMP1 cars in 2013 and 2014 as the engine covers visibly deformed at high speed. A silencing of these discussions is not to be expected in the future.

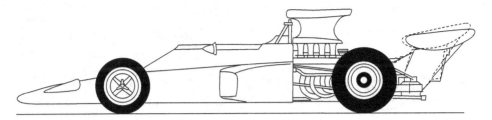

Fig. 6.74 Elastic rear wing attachment on the Lotus 72

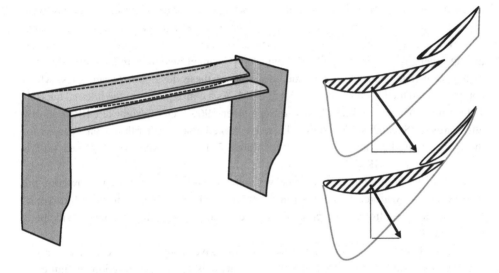

Fig. 6.75 Closing the gap by elastic deformation

6.6   Underbody and Diffusers

6.6.1   The Discovery of the Ground Effect

The Lotus 78 (Fig. 6.76), used in 1977 and 1978, is considered the forefather of modern racing car aerodynamics. It was the first racing vehicle whose underbody was consistently developed – taking into account the latest scientific findings – to exploit the ground effect. Historical explanations of the development of the Lotus 78 are summarized in [25–27], for example, where further interesting details on the Lotus 78 and its development as well as on the historical development of racing car aerodynamics in general can also be read.

The development of the Lotus 78 is closely associated with engineers Tony Rudd and Peter Wright, who first met at the BRM racing team in the late 1960s, and Colin Chapman, the famous founder of the Lotus sports car brand. BRM racing director at the time, Tony Rudd, commissioned his assistant Peter Wright to pre-develop an F1 racing car capable of generating downforce through a wing-profiled underbody. The aim was to find an

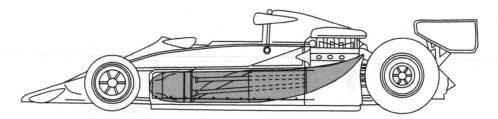

Fig. 6.76 Lotus 78, used in the F1 world championships 1977 and 1978

alternative to generating downforce through wing sections mounted on struts that were prone to failure. These investigations took place in secrecy away from everyday racing. However, problems with the regular racing cars meant that racing driver John Surtees, among others, had little sympathy for tying up personnel resources in such projects. The racing team management finally ordered a stop of these development activities, which abruptly ended the ground effect project at BRM. As a consequence of the internal quarrels Tony Rudd changed to Team Lotus in September 1969. Peter Wright left BRM shortly afterwards for Specialized Moldings. There he worked among other things on a project for the increase of the tank volume at the March 701 of the year 1970. Based on his investigations with BRM he designed a wing-shaped side pod, which had the advantage to take up the additional tank volume without increase of the air drag. However, the generation of downforce was not yet in the foreground, even though the March driver of the time, Mario Andretti, later reported a noticeable loss of downforce after the side pods were removed.

In 1975 Colin Chapman, team boss at Lotus, gave Tony Rudd the task to form a development team, which should show new potentials in the construction of future F1 cars parallel to the current racing business. Peter Wright, who had taken over the management of Technocraft (a subsidiary of Lotus) in 1974, was given responsibility for the subject area of aerodynamics at the express request of Rudd. The wind tunnel at Imperial College was hired to carry out the experimental tests, which had a moving flat belt to correctly represent the boundary layers between the vehicle and the road surface. The wind tunnel experiments were carried out using 1:4 scale wooden models. Cardboard strips, plasticine and tape were used extensively when swapping components to investigate different configurations. Investigations of the front wing configuration quickly showed that downforce increased as the aircraft approached the ground, corresponding to the ground effect considered in Sect. 6.5.4. This finding contradicted the view at the time that the wing had to be set as high as possible.

The concept for the later Lotus 78 (Fig. 6.76) was finally derived from the initial findings in late summer 1975. It was to have a tank at the height of the center of gravity. The oil cooler was accommodated in a narrow nose. The water cooler and petrol tanks were located in wing-profiled sidepods, again with the primary objective initially being to avoid additional drag. In the winter of 1975 the concept was to be optimized by further investigations in the wind tunnel. The intensive use of the wind tunnel led to a steady increase in the temperatures in the measuring section, so that plasticine and adhesive tape quickly lost their effect, which is why the side pods began to sag due to their already existing suction effect. As a result, the side boxes were fixed with wires and elongated cardboard strips. The cardboard strips closed the gap between the side pods and the road along the outside of the vehicle. After this measure, further tests showed a doubling of the downforce with only a minimal increase in drag. Sealing the underbody from the environment was the key to make efficient use of the ground effect through the underbody. However, it became apparent that for implementation in a real vehicle, the design of an efficient and wear-resistant seal was a challenge that should not be underestimated.

Initial tests of the new concept based on a Lotus 77 quickly revealed the sensitivity of the aerodynamic downforce to changes in ground clearance caused by body roll, which led to a noticeable loss of downforce on the inside of the vehicle. Riveted plastic skirts provided a remedy, but did not prove sufficiently resistant to wear. In further tests with the Lotus 77 and from August 1976 also with a Lotus 78, nylon brushes were used for sealing (Fig. 6.77a). These proved durable, but their sealing effect was far from adequate. Polypropylene skirts also deformed due to the enormous suction effect of the underbody. The development of form- and wear-resistant skirts was therefore the primary problem to be solved. One approach was to reduce the contact pressure of the skirt on the road surface by a "suck-up" design (Fig. 6.77b) to relieve the skirt. However, even this only achieved an inadequate compromise between initial performance and loss of performance after wear. The desired properties were finally achieved in conjunction with a ceramic tread, which was so durable that the aluminum skirt could even be designed as a "suck-down" skirt (Fig. 6.77c), increasing the sealing effect at increased vacuum. Test drives in the Lotus 78 confirmed the effectiveness of this solution.

In order to be able to use the advantage of the ground effect in a vehicle that was as mature as possible and to keep the concept secret from the competition for as long as possible, the Lotus 78 was scheduled for use in the 1977 season. The car won a total of five races that season (4 wins for Mario Andretti, 1 win for Gunnar Nilsson) and was regularly the fastest car in the field. Technical problems with the engine, however, did not allow the team to go beyond third place for Mario Andretti in the drivers' championship and second place in the constructors' championship. The Lotus 78 was also used for the first five races of the 1978 season – with one victory each for Andretti and Peterson – before its successor, the Lotus 79, was introduced (Fig. 6.78).

With the Lotus 79, some disadvantages of the Lotus 78 were eliminated, such as in particular the front-heavy aerodynamic balance resulting from the sidepods. This had to be compensated by a massive rear wing, whose high drag led to a too low top speed. In addition to an optimized aerodynamic balance, the Lotus 79 later had so-called board-in-a-box skirts (Fig. 6.77d), which were movably mounted in the sidepods. They were first introduced on the Wolf WR5 designed by Harvey Postlethwaite. Their mobility made it possible to compensate for pitch and roll. Surprisingly, despite the ban on movable aerodynamic devices, they were not deemed illegal. The Lotus 79 won six of the remaining 11 races, with Mario Andretti taking five victories and Ronnie Peterson one. Mario Andretti secured the drivers' title, and Lotus won the constructors' title. These were to

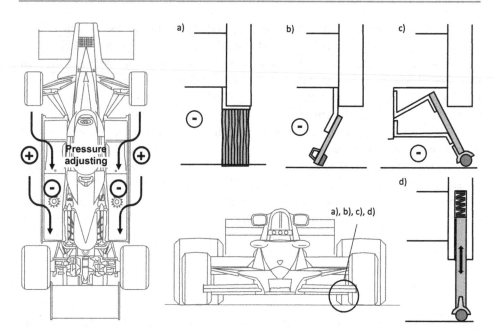

Fig. 6.77 Design forms of skirts to prevent pressure equalization in the Lotus 78 and 79 according to [27]: (**a**) Nylon brush skirts, (**b**) Suck-up skirst, (**c**) Suck-down skirts, (**d**) Board-in-a-Box skirts

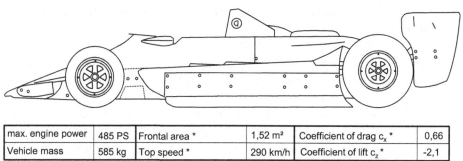

max. engine power	485 PS	Frontal area *	1,52 m²	Coefficient of drag c_x *	0,66
Vehicle mass	585 kg	Top speed *	290 km/h	Coefficient of lift c_z *	-2,1

* Author's estimates

Fig. 6.78 Lotus 79

be the last titles of Team Lotus, which to this day stands for groundbreaking innovations in racing like no other team. The dominance of the Lotus 79 finally made the competition realize what a powerful tool the effective use of the ground effect was, so that for the 1979 season practically every team started with a ground effect car or "wing car". Aerodynamics now finally gained at least the same importance as lightweight construction, engine, gearbox, brake and chassis development.

However, the enormous downforce and the lateral accelerations that could be achieved as a result also placed completely new demands on the design of racing vehicles. The vehicles had to be suspended ever more rigidly in order to be able to control the changes in the aerodynamic properties of the vehicles. By the 1981 season, cars were reaching spring rates of up to 700,000 N/m [28, 29], which was virtually equivalent to an unsprung car. Drivers increasingly complained about the lack of drivability of their cars, as they were almost impossible to control at the limit. Although a ban on movable skirts and an increase in the minimum ground clearance to 6 cm were introduced for the 1981 season, these changes were circumvented by the use of hydropneumatic chassis. The 1982 season saw an accumulation of serious accidents caused by ground effect vehicles. At the Belgian Grand Prix in Zolder, Ferrari driver Gilles Villeneuve died. A month later, Osella driver Ricardo Paletti was killed in a starting accident at the Canadian Grand Prix. These should remain the last two deaths until 1994, but another two months later Didier Pironi had such a serious accident with his Ferrari during practice for the German Grand Prix that he had to end his career. As a consequence of this, the use of a flat underbody between the trailing edge of the front tire and the leading edge of the rear tire was prescribed for the 1983 season. This ended the era of ground effect cars in Formula 1, but at the same time a new era began – that of the technically mature and successful turbo engines, which took Formula 1 into performance regions that it has never reached again to this day. The efficiency lost by restricting the underbody geometry was compensated for by the increased power of the turbo engines. The flattening of the underbody is still an integral part of the technical regulations of Formula 1 and most other racing series and limits the design of the underbody to the use of diffusers, which are located at the rear end of the underbody. Sealing the underbody with parts of the bodywork is no longer possible today. However, fins fitted in the diffuser reduce the influence of air flowing in from the side. Figure 6.79 outlines the nature of the Formula 1 underbody from 1979 to the present day.

After the fatal accidents of Roland Ratzenberger and Ayrton Senna during the 1994 Imola Grand Prix, the fitting of a 10 mm thick wooden plank to the underbody of F1 cars was prescribed at short notice. The aim of this regulation was to force the teams to increase the ground clearance, as the wear of the wooden plank could not exceed 1 mm. Thus, 5 hours after the Belgian Grand Prix, Michael Schumacher was disqualified because the thickness of the wooden plank was only 7.4 mm in places. It was the first disqualification of a race winner since 1985.

Finally, for the 1995 season, the stepped underbody was introduced, which is still mandatory today and also leads to an increase in ground clearance at the side diffuser channels. The prescribed height of the step was initially 25 mm. Over the years, however, the regulations on the design of the underbody have been repeatedly adapted and specified. Since 1997, the prescribed height of the step has been 50 mm. The introduction of the reference level shown in Fig. 3.6 took place in 2000.

6.6.2   Functioning of Diffusers

Figure 6.80 shows the geometric characteristics of a simple diffuser. Its characteristic feature is the upsweep, i.e. the rise starting at the end of the plane part from the underbody

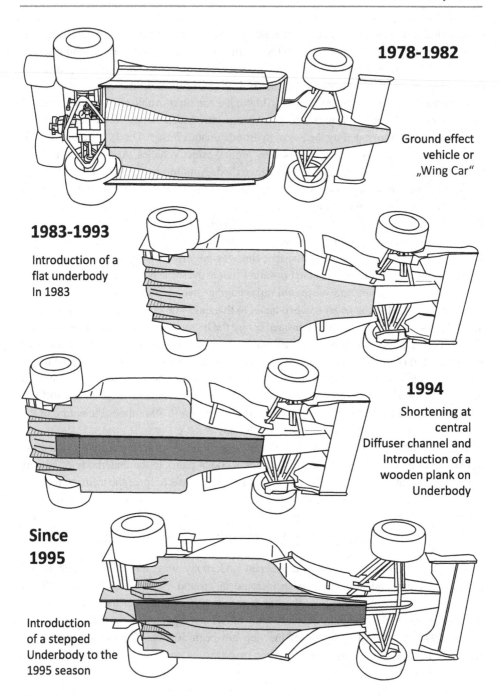

Fig. 6.79 Design of the underbody of F1 cars since 1978

up to the diffuser height h_d. The diffuser height h_d, in conjunction with the diffuser length n, determines the angle of rise θ. Of central importance to the operation of the diffuser is the pressure p_1 prevailing at the underbody, or the applied pressure difference $p_1 - p_0$, as it is

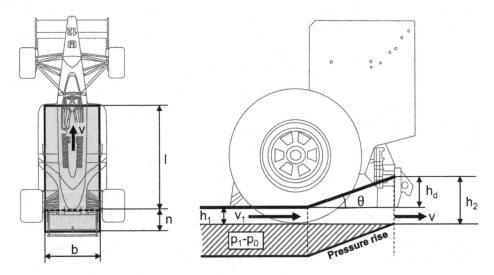

Fig. 6.80 Geometry of a simplified underbody diffuser

responsible for the downforce generated at the underbody (i.e. $p_1 - p_0 < 0$). Under the simplifying assumptions that the ambient pressure p_0 prevails directly behind the diffuser closure and that the exit velocity corresponds to the vehicle velocity v, the pressure difference is obtained according to Eq. (6.30) to:

$$p_1 - p_0 = \frac{\rho_L}{2} \cdot v^2 \cdot \left(\frac{h_2^2}{h_1^2} - 1\right) = \frac{\rho_L}{2} \cdot v^2 \left(\left(1 + \frac{n \cdot \tan(\theta)}{h_1}\right) - 1\right), \tag{6.39}$$

The following downforce acts on the flat surface of the underbody (without the part of the diffuser):

$$F_{z,Diff} = (p_1 - p_0) \cdot b \cdot l. \tag{6.40}$$

If we begin our consideration of the diffuser at the rear end, we can think of the diffuser as a pump that "sucks" or accelerates the air under the vehicle floor, thus lowering the pressure level under the underbody. The potential of the diffuser to generate downforce is largely determined by the ground clearance h_1 of the vehicle and the diffuser height h_d. The downforce is greater the lower the ground clearance and the higher the diffuser termination or the steeper the upsweep angle. The influence of these two parameters can be seen in Fig. 6.81. It shows the course of the downforce and drag coefficients for different upsweep angles with variation of the ground clearance on a vehicle-like basic body.

With an upsweep angle $\theta = 5°$, the downforce coefficient increases as ground clearance decreases, so that its maximum value exceeds the initial value from high ground clearance by a factor of four. At the same time, drag increases by only a moderate 20%. This explains

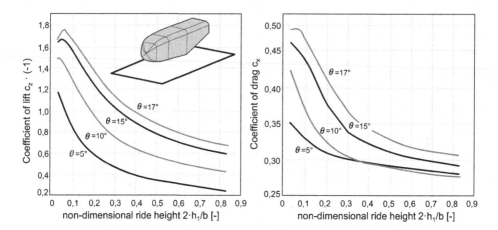

Fig. 6.81 Coefficients at the underbody (according to [30]). (Courtesy of © Prof. Xin Zhang 2018. All Rights Reserved)

Colin Chapman's saying, "Ground effect is getting something for nothing." So, like airfoils, the underbody exhibits a ground effect, which is already accounted for in Eq. (6.39). Increasing the upsweep angle leads to an increase in downforce coefficients, as also predicted by Eq. (6.39). The impending drop behind the maximum downweep coefficient indicates an onset of flow separation, which can be explained by the now steeper pressure rise on the way to the car's rear end.

The consideration of the drag coefficients has a special feature in that with an upsweep angle $\theta = 10°$ and sufficient ground clearance, the drag even partially decreases. Accordingly, diffusers can be designed for maximum downforce or a reduction in drag. The second case is mainly applied to production passenger cars. The further increase of the angle of attack to 15° leads again to an increase of the downforce coefficients, but also the effect of the incipient separation becomes stronger. In this range, the diffuser reaches its maximum downforce, and a further increase of the angle would reduce the efficiency of the diffuser.

The flow process at the underbody of a racing vehicle is more complex than expressed by Eq. (6.39), which is why it can only be used as an approximate equation, but not to determine real pressure distributions. It must also be taken into account that the flow process at the underbody is disturbed by air flowing in from the side, that the flow process is not frictionless due to roughness at the underbody, that the underbody interacts with the rotating rear wheels and that detachment phenomena occur. Determining the performance of a diffuser or its design therefore requires extensive experimental investigations in the wind tunnel or complex CFD simulations.

Figure 6.82 sketches a realistic pressure distribution on the underbody in the form of the pressure coefficient. According to this, a pronounced negative pressure peak occurs near the diffuser upsweep, which shifts the aerodynamic balance in favor of the rear axle. As the diffuser rises, this negative pressure is returned to the ambient level. Here, the flow in the

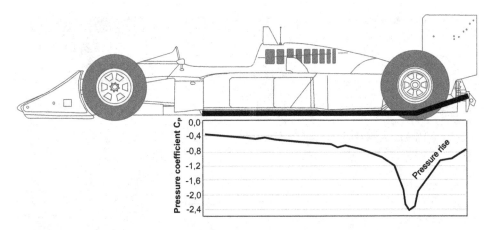

Fig. 6.82 Realistic pressure distribution on the underbody

diffuser must overcome a significant pressure rise, which favors the onset of flow separation. The steeper the upsweep angle, the greater the pressure rise that must be overcome and thus the greater the risk of detachment setting in. The phenomenon of detachment places a natural limit on the generation of downforce in the diffuser. In addition, the flow in a simple diffuser is disturbed by air flowing in from the side. The effects of detachment and side inflowing air are shown in Fig. 6.83 for various upsweep angles.

For small upsweep angles, the flow in the diffuser is still largely laminar except in the area of the air flowing in laterally. At an upsweep angle $\theta = 10°$, a separation bubble already forms. At $\theta = 15°$, the flow is then detached over a wide area. The main design goals of diffusers for racing vehicles are therefore to minimize the disturbing effects of air flowing in from the side and to avoid premature separation phenomena.

In the case of racing cars with closed wheels, such as today's DTM cars or the Le Mans prototypes, where practically no front wing profiles can be used, a diffuser is used on the underbody at the front and rear axle. Figure 6.84 shows the typical arrangement of front and rear diffuser of such cars using the example of the underbody of the Peugeot 908 HdI FAP. The underbody is divided into two sections. The rear section of the underbody starts from the wheel center line of the front axle. This section is initially flat. The rise of the rear diffuser begins well forward of the leading edge of the rear wheels. The rear diffuser is of the classic twin-tunnel design with the gearbox installed between the two ducts. This arrangement results in a low-slung gearbox and reduces the height of the vehicle's center of gravity. Also typical are the fins installed in the diffuser ducts, which stabilize the flow and at least partially isolate it from disturbing influences.

The front part represents the front diffuser, which extends over the entire distance between the two front wheels. The front diffuser is vented and the pressure potential is created through lateral ducts leading to the outside, which are located behind the wheel arches (cf. Figure 6.85). If we assume for the sake of simplicity that ambient pressure again prevails to the side of the vehicle, then the mode of operation of the front diffuser is

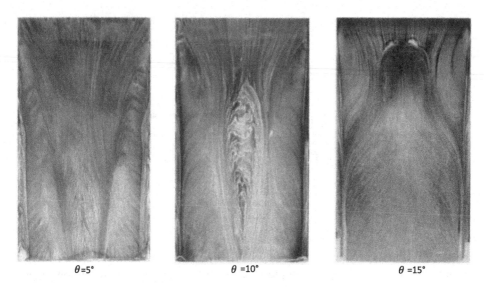

θ=5° θ =10° θ =15°

Fig. 6.83 Underbody detachment phenomena. (Adapted from [30, 31]; courtesy © The American Society of Mechanical Engineers 2018. All Rights Reserved)

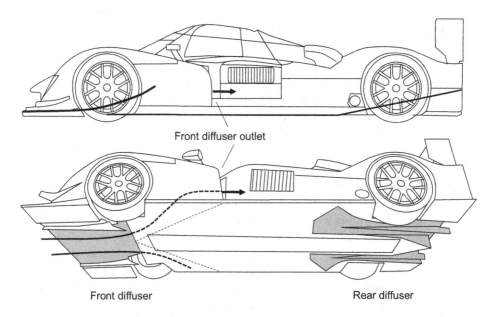

Fig. 6.84 Front and rear diffuser on the underbody of an LMP1

identical to that of the rear diffuser. In the case of racing cars with closed wheels, which generally do not allow efficient use of front wing profiles, it is much more difficult to achieve an aerodynamic balance than in the case of formula cars, which is why additional

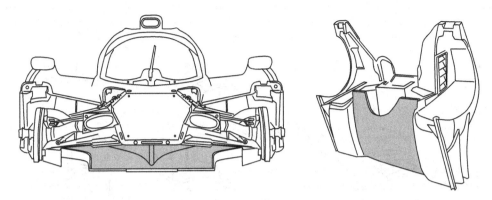

Fig. 6.85 Front diffuser and air duct on a Ligier LMP1

measures must be taken to increase downforce at the front axle, especially in the case of these cars. Some of these measures are explained in Sects. 6.7 and 6.8.

6.6.3   Performance Optimization on Diffusers

From Sect. 6.6.2 it can be summarized that the performance potential of a diffuser is largely determined by its geometric properties, its sealing against the environment and its tendency to flow separation. Due to the restrictive regulations, there is little scope for the geometric design of the diffuser. Even a classic sealing of the diffuser is no longer possible today due to various regulations. Due to these strict limitations, numerous aerodynamic tricks have been developed to optimize the performance of the diffuser and achieve the decisive advantage over the competition (Fig. 6.86).

Figure 6.87 shows various designs of so-called vortex generators, which are arranged in front of the side pod. These are used for the targeted generation of drag vortices which run along the side pod. As there is negative pressure in the vortex core, the pressure equalization between the underbody and the environment is prevented or significantly reduced. The drag vortex thus has a sealing effect.

The generation of negative pressure by vortices explained in Sect. 6.4 and illustrated in Fig. 6.32 can be used directly to generate downforce. An example of this is the front diffuser of the Audi R15 plus shown in Fig. 6.87. Small triangular vortex generators are arranged across the entire width of the front diffuser. The generated vortices graze the rear diffuser surface and induce strong negative pressure areas there, which pull the body downwards and thus increase the downforce generation of the diffuser.

One solution to increase diffuser performance that became trendy during the 1980s was to locate the exhaust exit in the side or central diffuser channel. Figure 6.88 shows this

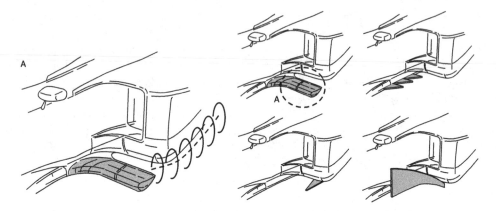

Fig. 6.86 Vortex generators for generating side-sealing vortices (see also [16])

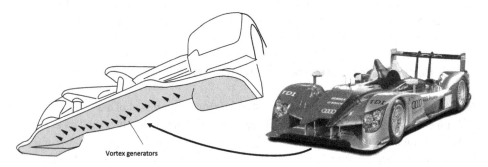

Fig. 6.87 Vortex generators on the front diffuser of the Audi R15 plus. (Photo on the right courtesy of © Audi AG 2018. All Rights Reserved)

Fig. 6.88 Arrangement of the exhaust exit in the lateral diffuser channels

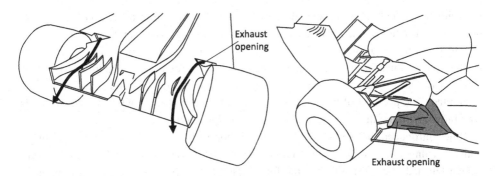

Fig. 6.89 Blowing the diffuser on the Red Bull RB7

exemplified by the underbody of the 1988 McLaren MP4/4. The energy of the outflowing exhaust gases accelerates the flow at the underbody and inhibits the phenomenon of detachment, resulting in an increase in the downforce generated. A disadvantage of this principle is that the downforce generated is dependent on the throttle position, and there is a risk of abrupt loss of downforce at the rear axle during a power-off load change. This dependence on the throttle position requires an adaptation of the driving style.

In 1999, Ferrari began to move the exhaust exits to the top of the sidepods and use the energy of the exhaust gases to assist the flow onto the lower rear wing profile. This trend was gradually followed by the other Formula One racing teams and, aided by a change in the regulations introduced for the season, the blown diffuser principle returned in 2011. The aforementioned change in the regulations moved the beginning of the diffuser 330 mm backwards, so that it now had to start directly at the level of the axle center line. At the same time, the permissible diffuser height of the side diffuser ducts was raised to 175 mm. The lateral diffuser channels and the central channel thus had the same height. These geometric changes caused a significantly steeper diffuser upsweep, which favored the tendency to detach.

To avoid detachment, various techniques have been developed for blowing on the diffuser. Figure 6.89 shows the solution used on the Red Bull RB7 from 2011. Due to regulations, it is no longer possible to locate the exhaust ports directly in the diffuser channels. On the Red Bull RB7, the exhaust port is positioned on the top in the "coke-bottle" section so that the exhaust gases are blown through the opening between the underbody and rear wheels into the side diffuser channels. Comparable concepts can also be found on other cars entered from the 2010 season onwards. The electronic engine control system also made it possible to maintain the blowing into the diffuser even when the accelerator pedal was pulled back. A distinction is made here between so-called cold and hot blowing. With both techniques, the throttle valves remain open when the accelerator pedal is released, so that air continues to flow into the combustion chambers. With *cold blowing,* fuel injection is suspended. That is, only the incoming air is compressed and blown into the exhaust during the exhaust stroke. The energy of the compressed air is fed to the diffuser. *Hot blowing* only suspends ignition, not fuel injection. The compressed mixture then ignites in the hot exhaust pipe. This allows significantly more energy to be supplied to the diffuser. Driver-independent control of the throttle and injection is now prohibited. The position of the exhaust port has also been more strictly limited, which is why blowing on the diffuser no longer seems possible today.

Rules introduced in 2009 allowed for the development of the infamous double diffuser, which was instrumental in Jenson Button winning the drivers' title in a Brawn GP. Team Brawn GP had been formed from the former Honda works team, which had previously made a surprise withdrawal from Formula One. Former technical director Ross Brawn took over the team, and Mercedes stepped in as engine supplier at short notice. A loophole in the regulations allowed holes to be made in the step of the underbody, as shown in Fig. 6.90. These holes were in front of the actual diffuser rise and formed the openings for additional

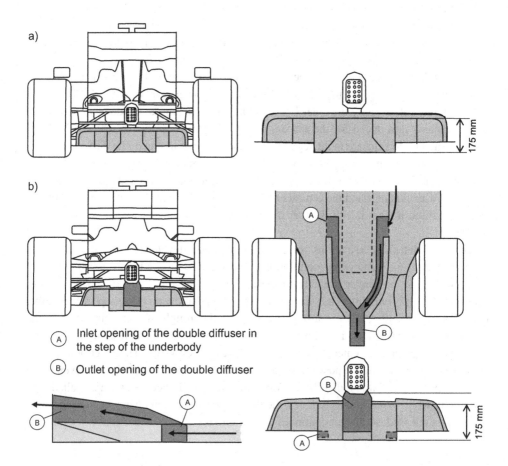

Fig. 6.90 Vehicle (**a**) with standard diffuser, (**b**) with double diffuser (2009)

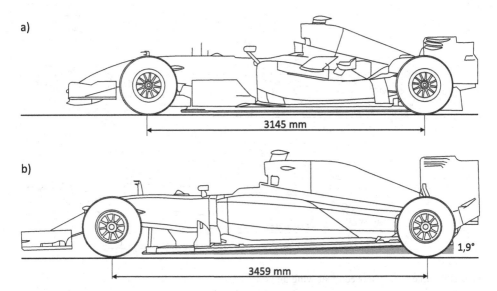

Fig. 6.91 Increase of the diffuser height by vehicle adjustment: (**a**) Red Bull RB3 (2007) without adjustment, (**b**) Red Bull RB11 (2015) with approx. 1.9° adjustment

diffuser ducts which ran above the actual diffuser roof. This met the requirements for the prescribed diffuser height. However, with the duct above, the effective diffuser height could be increased. According to Eq. (6.39), this allows the pressure difference between the underbody and the environment to be increased, resulting in a corresponding increase in downforce. The aforementioned gap in the regulations was closed for the 2011 season.

One way of increasing the effective diffuser height that is still permissible today is to raise the entire vehicle. Figure 6.91 illustrates this measure using the example of the Red Bull RB3 from 2007 and the Red Bull RB11 from 2015. It can be clearly seen that the underbody of the RB11 is no longer parallel to the road, but rises towards the rear. This is possible because the reference plane for all measurements is the underbody itself, or the so-called reference plane, and not the road. So the ground clearance is not directly prescribed. In 2016, the angle of attack on Formula 1 cars was between 1° and 2°. With an average wheelbase of these cars of about 3500 mm, the difference in height between the front and rear axles is 6–12 cm. Such an angle of attack of the vehicle can now also be observed in sports car prototypes.

According to Eq. (6.39), lowering the pressure level behind the diffuser is another way to increase the vacuum level at the underbody. This can be achieved by bringing the rear wing and diffuser into interaction. Figure 6.92 illustrates the principle using a Formula 1 car and sports car prototype as examples. In both vehicles, the rear wing or its lower plane is arranged in such a way that the negative pressure area created on the underside of the wing projects into the area behind the diffuser. This measure reduces the pressure acting on the underbody p_1, which increases the differential pressure with respect to the surroundings. As a result, the downforce generated at the underbody increases. However, the air

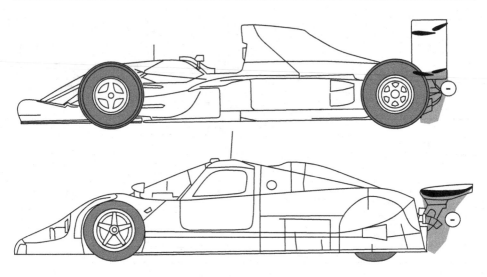

Fig. 6.92 Interaction between rear wing and diffuser

flowing out from the diffuser possibly interferes with the inflow to the rear wing. The exact determination of the relative position of the wing elements must therefore be made taking into account the interactions of all aerodynamic components with each other. This is another example of the complexity of the flow processes occurring on high-performance racing vehicles.

6.7   Spoilers, Splitters and Louvers

In addition to wing profiles and underbody, there are numerous other ways to increase the downforce of a vehicle or at least to reduce the lift forces caused by the body. Two relatively simple elements are a so-called airdam spoiler at the front of the vehicle and a rear spoiler, such as those used on the 2005 Dodge Charger (Fig. 6.93) for the NASCAR Nextel Cup. These elements are particularly common on cars with near-production silhouettes or prototypes with enclosed wheels. An Airdam spoiler significantly reduces the free area between the front end of the car and the road. This measure reduces lift at the front axle and increases the air volume flow into the frontal inlets for engine cooling. At the same time, under given boundary conditions, even the aerodynamic drag can be reduced. As an unintended side effect, however, the shielding of the underbody worsens the cooling of the front brakes and, in the case of vehicles with front-mounted engines, the cooling of the oil sump (cf. [9]). This may have to be counteracted by introducing additional supply openings.

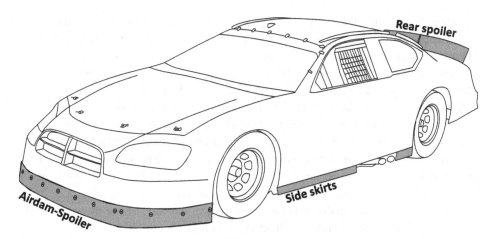

Fig. 6.93 Dodge Charger for participation in the NASCAR Nextel Cup 2005

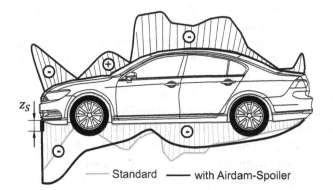

z_S	Δc_{zV}	Δc_{zH}	Δc_x
0	0	0	0
50	-0,18	0,01	-0,03
100	-0,24	0,03	-0,04
150	-0,26	0,08	-0,03
200	-0,27	0,09	0,00
250	-0,28	0,08	0,04
300	-0,28	0,07	0,08

Standard —— with Airdam-Spoiler

Fig. 6.94 Effect of an Airdam spoiler [32]

Figure 6.94 illustrates the effect of an Airdam spoiler on lift. The principle distribution of the pressure coefficient C_p on the upper side of the body and on the underbody of the vehicle is shown. On the convex surfaces of the body, i.e. in particular on the bonnet and roof, there are areas of significant negative pressure which are responsible for the axle lift of production vehicles. This is in contrast to the negative pressure on the vehicle floor, which cannot compensate for the lift on the upper side. The use of an airdam spoiler primarily increases the negative pressure on the underside of the front end, which is due to the flow behind the airdam spoiler detaching and creating a wake area. The increased negative pressure at the underbody reduces lift or increases downforce at the front axle. This effect can be amplified by side-sealing skirts that impede the pressure equalization between the environment and the underbody. The effect of the Airdam spoiler on axle downforce coefficients and drag as a function of spoiler depth z_S can be seen in the attached diagram. The main effect is the continuous increase in the front axle downforce coefficient Δc_{zV}. However, at spoiler depths of up to 200 mm, aerodynamic drag is also moderately reduced because the reduced air masses flowing off via the underbody can no longer swirl on the underside of the vehicle between engine, transmission and chassis components. However, the effect of drag reduction is not observed on vehicles with a shrouded underbody. As the spoiler depth increases, the frontal area of the vehicle also increases and the stagnation zone increases. The increasing highpressure at the front end compensates for the above-mentioned effect, whereby the air resistance finally increases from 200 mm spoiler depth.

As a further side effect, a slight increase in rear axle lift can be observed, which essentially results from a reduced negative pressure in the rear area of the vehicle. The optimum spoiler depth is therefore dependent on various parameters.

The use of massive rear spoilers is still characteristic of stock cars used in the various NASCAR racing series. Figure 6.95 shows the basic effect of a rear spoiler. The schematic diagram shows the flow over the body of a production car (top) and a stock car (bottom). In the case of the standard vehicle, the airflow is in contact with the roof well into the sloping area, which is generally favorable for reducing air resistance. However, the fast flow and the resulting negative pressure area lead to lift at the rear axle. The installation of a rear spoiler has two main effects. Firstly, the air is decelerated as it flows towards the spoiler, creating a stagnation zone with highpressure in front of the spoiler. The excess pressure on the top of the body results in a downforce. The change in the downforce coefficient due to the rear spoiler is mainly influenced by its height h and its angle of incidence θ. The change in rear axle downforce coefficient Δc_{zH} is proportional to its height and angle of attack. Remarkably, the addition of rear spoilers according to [32] is even accompanied by a slight drag reduction up to 400 mm in height. The second effect of a tail spoiler is the forward displacement of the flow separation. This reduces the negative pressure and thus the suction effect in the rear window area. The air resistance decreases moderately. From a height of 400 mm, however, the influence of the stagnation zone and the separation behind the spoiler are dominant, whereby the air drag begins to increase.

As a side effect of the rear spoiler, the front axle lift increases moderately, which in turn illustrates that all aerodynamic components on a racing vehicle interact with each other. In general, the effect of the rear spoiler depends on the body shape. Airdam and rear spoiler must be matched to each other. The total downforce coefficient c_z of a stock car, as shown in Fig. 6.93, is about -0.42 with a drag coefficient c_x of 0.39 [33]. Airdam spoiler and rear spoiler are the only significant aerodynamic aids, so the total downforce achieved can be explained by Figs. 6.94 and 6.95.

Another relatively simple measure to increase downforce is the use of splitters. The principle of a splitter is to use the highpressure in front of a stagnation zone to generate downforce. Figure 6.96 shows the arrangement of a front splitter on the Audi RS5 DTM. The front splitter is an extension of the Airdam spoiler. Its principle of operation is shown in Fig. 6.97. As explained in Sect. 6.4, a stagnation zone with excess pressure is formed in

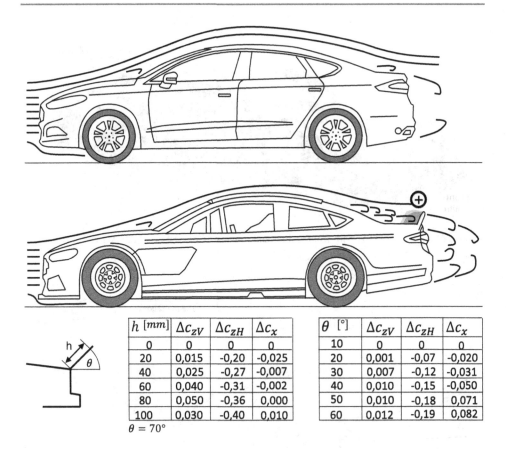

h [mm]	Δc_{zV}	Δc_{zH}	Δc_x
0	0	0	0
20	0,015	-0,20	-0,025
40	0,025	-0,27	-0,007
60	0,040	-0,31	-0,002
80	0,050	-0,36	0,000
100	0,030	-0,40	0,010

$\theta = 70°$

θ [°]	Δc_{zV}	Δc_{zH}	Δc_x
10	0	0	0
20	0,001	-0,07	-0,020
30	0,007	-0,12	-0,031
40	0,010	-0,15	-0,050
50	0,010	-0,18	0,071
60	0,012	-0,19	0,082

Fig. 6.95 Effect of a rear spoiler [32]

front of the vehicle, which primarily generates air resistance. By fitting a front splitter, a flat horizontal surface is created across the entire width of the front of the vehicle, on which the highpressure of the stagnation zone acts. This results in a downforce $F_{z,\ splitter}$. The gain in downforce is approximately proportional to the surface area of the splitter and thus to the length of its overhang. As a rule, the maximum downforce is reached at about 100 mm overhang. Up to this value, the downforce increases approximately linearly with the length. After that, it remains at a constant level, as there is no longer any overpressure at a great distance from the front end of the vehicle. The change in the front-axle downforce coefficient Δc_{zV} at maximum downforce is about 0.2. The changes in the drag coefficient are given as 0.002–0.02. Although front splitters are mainly used on vehicles with closed wheels, as they do not have front wings, the principle of a splitter is also used on monoposti.

Louvers

Front-
Splitter

Fig. 6.96 Audi RS5 DTM with front splitter and louvers. (Courtesy © Audi AG 2018. All Rights Reserved)

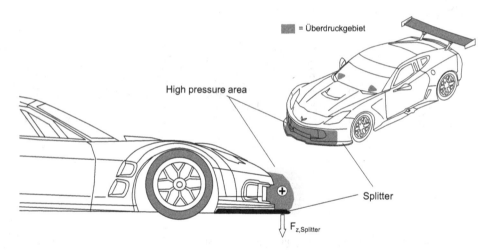

= Überdruckgebiet

High pressure area

Splitter

$F_{z,Splitter}$

Fig. 6.97 Functionality of a front splitter

Figure 6.98 shows the use of splitters using the example of a Formula 1 vehicle. On an F1 vehicle, stagnation zones with excess pressure form in front of the rear wheels and in the transition area between the raised monocoque and the underbody. In Formula 1 cars, the entire permissible vehicle width is used to achieve the largest possible flat surface on the underbody. This maximizes the downforce generation of the underbody. In the "coke-bottle" section, this results in a free-standing area on the upper side in front of the rear wheels. The highpressure that builds up in the stagnation zone in front of the rear wheels acts on it. To make efficient use of this effect, the gap between the base plate and the wheel must be as small as possible in order to achieve a good seal. For the 2005 season, however,

Fig. 6.98 Splitter on an F1 car: as an underbody attachment (A) and in front of the rear wheel of an F1 car (B)

this area was more heavily regulated by the FIA, resulting in a significantly larger gap and a reduction in the resulting downforce. Investigations in the laboratory showed a loss of downforce of about 5% in relation to the total downforce [34]. Thus, it can be assumed that the downforce coefficient of this splitter was about 0.15. As already described in Sect. 6.5.5, the trend towards high noses began in the early 1990s. As a result, the underbody no longer represented a continuous plane, but rather a height difference had to be overcome between the nose and the underbody. As Fig. 6.98b shows, a stagnation zone forms in this transition area below the monocoque.

The resulting highpressure is utilized by continuing the underbody below the monocoque in the form of a splitter. The highpressure acts on the resulting flat surface and generates a downforce. Lateral elevations impede the flow of air over the splitter and improve the downforce effect.

Another aerodynamic aid on vehicles with enclosed wheels are louvers. Louvers are openings such as those found mainly on wheel arches. With enclosed wheels, a rotational flow is created that generates positive pressure in the wheel arch. This positive pressure, together with the negative pressure on the convex body surfaces, contributes significantly to axle lift. The Louvers create a pressure balance between the wheel arch and the top of the

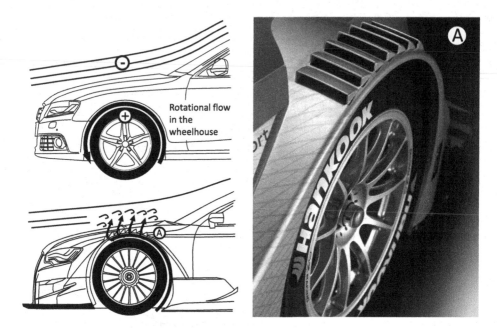

Fig. 6.99 Wheelhouse venting through Louvers. (Photo at right courtesy of © Audi AG 2018. All Rights Reserved)

body (Fig. 6.99). The overpressure in the wheel arch decreases. The air flowing out of the wheel arch disturbs the flow over the top of the body, which reduces the negative pressure prevailing there. Both of these effects reduce axle lift or create a downforce. However, this comes at the cost of an increase in aerodynamic drag, as the resulting turbulence increases frictional resistance.

Louvers can be used on the front and rear axle, as Fig. 6.100 shows using the example of an Audi R10. Depending on the track characteristics, different louvers variants may be used. In the high-downforce set-up of the Audi R10, the louvers on the front axle extend over the entire wheel arch length, while in the low-downforce set-up they only take up about a quarter of this length. On the rear axle, the low-downforce configuration even dispenses with louvers completely. Louvers are also often found, as on the Ferrari 458 GTE, on the engine or front hood or in otherlowpressure areas of the top of the body. In such cases, they are used as cooling air outlets. Placement in a negative pressure area increases the pressure differential between the inlet and outlet sides, increasing the airflow through the radiator and thus its cooling capacity. The outgoing airflow in turn disturbs the airflow over the surface, helping to reduce lift.

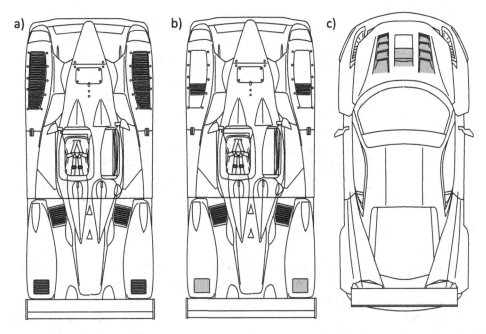

Fig. 6.100 Audi R10 (LMP1) (**a**) in high-downforce configuration, (**b**) in low-downforce configuration; (**c**) Ferrari 458 GTE with front hood venting

6.8 Influence of the External Contour of the Vehicle

Downforce can also be increased within certain limits by targeted measures on the outer contour of a vehicle. A significant influencing factor is, for example, the shape of the nose. Test drives with suspension travel transducers were carried out with the Lotus 38 (Fig. 6.101a) at the Indianapolis Motorspeedway in August 1967. The evaluations of the measurements showed that the suspension travel was greater on the stationary vehicle than in the driving conditions on the oval [25]. Thus, the vehicle had to generate such high lift forces through its body that the vehicle moved upwards at higher speeds. One reason for this was the relatively high nose cone. This favors the formation of a stagnation point on the underside of the body. The resulting highpressure on the underside of the vehicle generates a lift force.

On its successor, the Lotus 56 (Fig. 6.101b), a low-slung nose tip was used in conjunction with a wedge-shaped body. Due to a low nose, the stagnation point moves to the upper side of the body. With such a body shape, moderate downforce coefficients can already be achieved even without additional aerodynamic components. This wedge shape was also characteristic of the later Lotus 72, with which Lotus won a total of three drivers' titles. Until the mid-1990s, the noses of Formula 1 cars were characterized by a low-flying nose. For maximum use of the permissible width of the wing profile and more efficient flow

a)

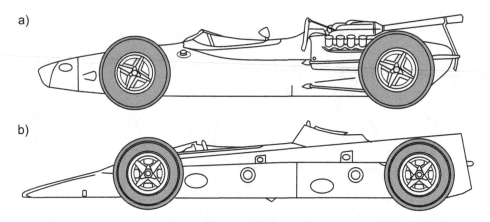

b)

Fig. 6.101 (**a**) Lotus 38 with raised nose, (**b**) Lotus 56 with wedge shape

to the underbody, the nose tips are now much higher, and the wing is suspended above two pylons ("high-nose" concept; Fig. 6.57b). However, an attempt is still made to pull the nose down as far as possible in order to move the stagnation point to the upper side.

The Porsche 917 K, with which Porsche won the 24 Hours of Le Mans in 1970 and 1971, also had a wedge-like shape (Fig. 6.102b). "K" denotes the short-tail variant, for which the term "high tailer" is used in English. The Porsche 917 was originally designed as a drag-optimized long-tail version or "long tailer" (Fig. 6.102a). Due to the lower drag, this variant achieved a significantly higher top speed than the short-tail version, which was initially considered crucial for successful participation in Le Mans. However, a lack of high-speed stability and insufficient cornering speeds meant that this calculation did not work out. It was only the short-tail version that brought Porsche its long-awaited first overall victory at the 24 Hours of Le Mans in 1970. In 1971, the 917 K, without wing profiles or a special design of the underbody, achieved a total downforce coefficient of $c_z = -0.31$ [35] with a drag coefficient of $c_W = 0.42$. The long-tail version, on the other hand, generated noticeable lift, especially at high speeds.

Another typical measure on the body of sports car prototypes is the shaping of a concave vehicle front. By using a concave instead of a strongly convex shaped vehicle front, the front axle downforce can be increased in vehicles with enclosed wheels. A recent example of such a measure is the front wheel arches of the Lola Mazda LMP2 (Fig. 6.103). Such shaping can also be observed on older sports car prototypes such as the Porsche 917/30 or the Porsche 956. However, on the other hand, the area of a convex shaped wheel arch following behind the concave section favors the occurrence of undercuts on this surface, which in turn leads to the generation of a lift force. This tendency can be counteracted by fitting louvers or cooling air outlets in these areas.

a)

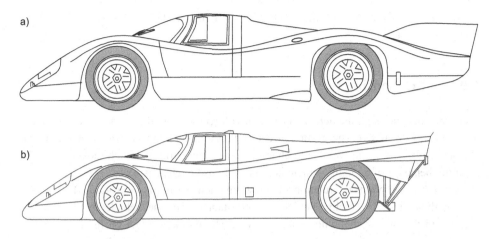

b)

Fig. 6.102 (**a**) Porsche 917 L as "Long-Tailer", (**b**) Porsche 917 K as "High-Tailer"

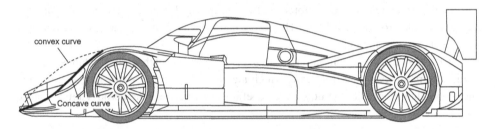

Fig. 6.103 Lola Mazda LMP2 with concave wheel arches

Fig. 6.104 Audi R15 (2009). (Courtesy © Audi AG 2018. All Rights Reserved)

Figure 6.104 shows the Audi R15, which has conspicuously curved wheel arch linings on the rear axle. Here, too, it can be assumed that a stagnation zone is formed in front of the wheel arches, which causes a downforce-generating overpressure area on the upper side of the body. Since the permissible rear wing width for LMP1 category prototypes was reduced in 2009, this measure has presumably been able to compensate for some of the downforce lost at the rear axle.

6.9   Pitch Sensitivity

In Sects. 6.5.1 and 6.6.2 the dependence of downforce on ground clearance for airfoils and diffusers was discussed. The ground effect on these components causes the aerodynamic characteristics of a racing vehicle to change due to heave, roll and pitch movements. The resulting behavior of a racing vehicle is referred to as "pitch sensitivity".

Figure 6.105 explains the relationship between pitch sensitivity or ground effect and the driving dynamics of a racing vehicle. At constant speed, the vehicle in the example considered has an aerodynamic balance of 0.4. The moment effect on the body caused by the inertia force during acceleration is supported by the chassis. With this support, the vehicle's spring travel increases on the rear axle and decreases on the front axle, resulting in a negative pitch angle. The compression on the rear axle reduces the ground clearance and thus increases the rear axle downforce coefficient. In parallel, downforce on the front axle decreases as the front wing is lifted. The aerodynamic balance shifts in favor of the rear axle and favors power-on understeer.

During braking, the conditions are reversed and a positive pitch angle is established. The vehicle compresses on the front axle, which increases the front axle downforce value. On the rebounding rear axle, the downforce coefficient decreases. The aerodynamic balance

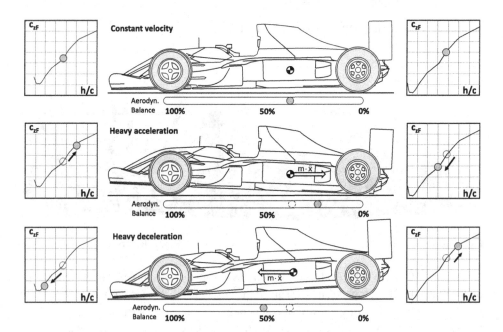

Fig. 6.105 Change in aerodynamic balance due to braking and acceleration

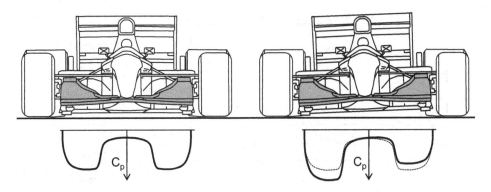

Fig. 6.106 Influence of roll on downforce at the front wing

increases and favors a tendency to oversteer. It can also be seen that the front wing approaches its maximum downforce, which can lead to the "pumping" or "aero-bouncing" described in Sect. 6.5.4. These aerodynamic effects on the driving dynamics overlap with the effects discussed in Sect. 5.5 and must be taken into account when tuning the chassis.

The ground clearances on the left and right sides also change when the vehicle rolls. Figure 6.106 shows the influence of the rolling motion on the pressure distribution at the front wing. The pressure difference increases on the outer side of the curve, while it decreases on the inner side of the curve. Depending on whether an aerodynamic component reacts more sensitively to an increase or reduction in ground clearance and whether the vehicle reacts more sensitively at the front or rear, roll thus also influences the overall downforce and aerodynamic balance of the vehicle.

A detailed knowledge of the downforce values at the individual wheels as a function of the spatial position of the body is therefore required for an exact aero- and driving-dynamic description of the vehicle behavior. The position of the body is completely determined by the lift at the center of gravity as well as by the roll and pitch angles. Alternatively, the description of the body position can also be uniquely determined by three wheel spring travels. By determining the downforce coefficients as a function of the body position, an aerodynamic "map" of the body is created, which is also referred to as the "aeromap" of a racing vehicle. "Aeromapping" describes the specific design of aerodynamics in conjunction with chassis tuning. Such an aeromap is also required, for example, for the use of simulation models in the driving simulator and is usually the result of extensive CFD or wind tunnel studies.

Figure 6.107 shows such an aeromap for the same directional compression and rebound of a simulated racing vehicle. The roll angle is zero, so that for this special case the position of the body is clearly defined by the ground clearances at the front and rear axles. The diagrams show the downforce coefficients of the front and rear axles as a function of the front ground clearance h_V and the rear ground clearance h_H. The lines of constant downforce coefficients are plotted, resulting in a kind of map. With known spring deflections, the

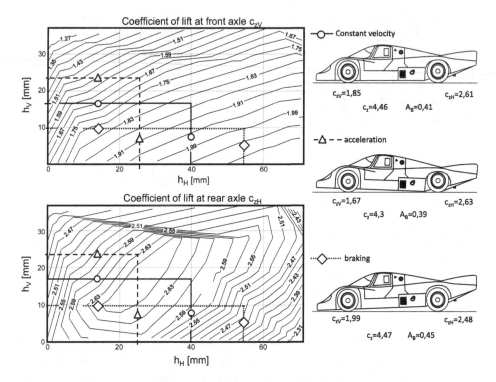

Fig. 6.107 Aeromapping. (Adapted from [21]; courtesy of © The American Society of Mechanical Engineers 2018. All Rights Reserved)

downforce values can be determined. This is illustrated by way of example for a stationary straight-ahead drive as well as an acceleration and a braking process.

In the stationary case, which corresponds to driving at constant speed, the aerodynamic balance is 0.41 with a total downforce coefficient of 4.46. When accelerating, the suspension travel changes due to the pitching motion. The ground clearance of the front wing increases and that at the rear decreases. For this condition, the axle downforce coefficients are read again. The aerodynamic balance decreases by about 5%, resulting in a greater tendency for the vehicle to understeer. During braking, the front axle dips and the rear axle springs out. The aerodynamic balance increases by about 10%, the vehicle tends to oversteer more. The dynamic behavior of a racing vehicle is thus significantly influenced by its aerodynamic properties. The extent to which the suspension travel changes as a function of the driving condition depends on the suspension geometry and tuning.

Via the ground effect, the aerodynamic properties of a racing vehicle are closely linked to chassis design and tuning. The chassis geometry as well as the adjustment of body springs, dampers and stabilizers must be carried out in such a way that the position of the body is always kept within its aerodynamically optimal operating window. In particular, it must be taken into account that the effects of pitch and wheel load sensitivity may overlap in opposite directions, so that the conventional approach to tuning racing vehicles is only valid to a limited extent.

An ingenious idea for the avoidance of negative effects by the pitch sensitivity had again the engineers of Lotus in 1981. They entered the 1981 US Grand Prix with the Lotus 88, which had so called twin chassis (Fig. 6.108). This consisted of the rolling chassis and an aerodynamic chassis, each connected to the wheel carrier by separate suspension. The rolling chassis consisted of the cockpit and all the components required to propel the vehicle. Since this part of the vehicle hardly affected the aerodynamics, a relatively soft suspension could be used, which had a positive effect on the traction of the vehicle. On the other hand, very stiff suspensions were used on the aerodynamic chassis, which practically only consisted of the components for generating downforce, resulting in only minor changes to the aerodynamic properties. The requirement of the technical regulations for the aerodynamic components to be attached to the sprung masses was thus basically fulfilled. Nevertheless, the car was deemed illegal before it even started, and the Lotus 88 did not contest a single race. This decision is still controversially discussed today. But the Lotus 88 impressively demonstrates how closely aerodynamics is linked to the characteristics of the chassis.

In addition to influencing the driving dynamics, the aerodynamic properties of a racing vehicle also have an influence on the driving condition that should not be underestimated from a safety point of view. Excessive deviations from the roll, pitch and side slip angles intended for the vehicle can lead to aerodynamically critical conditions. In the 1980s and 1990s, for example, a so-called skid block made of magnesium attached to the underbody of F1 cars prevented the rear underbody from bouncing. This prevented a sudden loss of control due to the rear of the car touching down. The function of the skid block is virtually

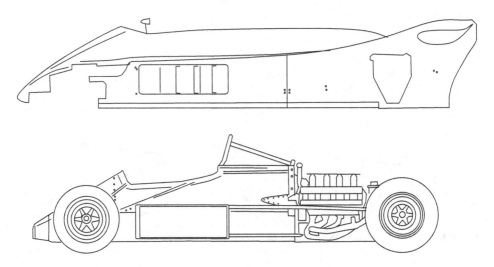

Fig. 6.108 Lotus 88 with "Twin-Chassis" (1981)

equivalent to an infinitely stiff compression stop spring. A side effect was the spectacular flying sparks typical of the vehicles of the time.

6.10 Crosswind and Off-Design Conditions

In this section, the aerodynamic properties are examined in more detail as a function of the angle of attack τ_L, which has so far only been briefly mentioned in Sect. 6.1. The consideration of the angle of attack is important both for relatively small angles, i.e. for the normal range of driving dynamics, and for very large angles of attack, as they occur when control of the vehicle is lost. In the latter case, one speaks of "off-design" conditions, because the aerodynamic behavior of the vehicle is not specifically designed for these cases. However, due to numerous incidents, especially in the field of sports car prototypes, these conditions have increasingly come into the focus of the rule-making sports authorities. As Fig. 6.109 shows, the vector of the resulting incident flow velocity $\vec{v}_{res}$ is determined by vehicle speed, side slip angle and the wind speed and its angle to the vehicle.

The vectorial addition of the vehicle and wind velocity vector yields the following relationship for the vector of the resulting incident flow velocity.

$$\vec{v}_{res} = -\vec{v} + \vec{v}_w = -|v| \cdot (\cos(\beta) - \sin(\beta)) + |v_W| \cdot \begin{pmatrix} -\cos(\delta) \\ -\sin(\delta) \end{pmatrix} \qquad (6.41)$$

From the components of the vector of the resultant incident flow velocity $\vec{v}_{res}$, the magnitude of the incident flow velocity and the incident flow angle τ_L can be determined as follows:

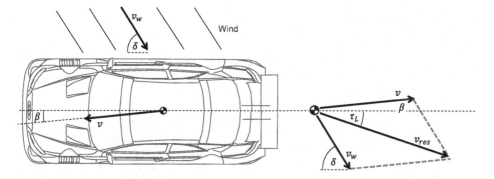

Fig. 6.109 Determination of the resulting face velocity

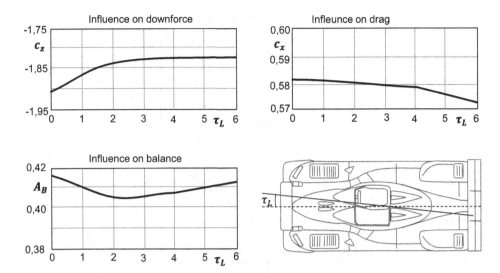

Fig. 6.110 Influence of the angle of attack on the aerodynamics of the Zytek Z11 SN. (Adapted from [36]; courtesy of © Simon McBeath 2018. All Rights Reserved)

$$|v_{res}| = \sqrt{(-|v| \cdot \cos(\beta) - |v_w| \cdot \cos(\delta))^2 + (|v| \cdot \sin(\beta) - |v_w| \cdot \sin(\delta))^2}, \quad (6.42)$$

$$\tau_L = \tan^{-1}\left(\frac{|v| \cdot \sin(\beta) - |v_w| \cdot \sin(\delta)}{-|v| \cdot \cos(\beta) - |v_w| \cdot \cos(\delta)}\right). \quad (6.43)$$

The angles of attack resulting from the side slip angle and even from strong crosswinds remain relatively small. Nevertheless, they can significantly influence the handling characteristics of a racing vehicle. The change in aerodynamic properties due to the angle of attack is shown in Fig. 6.110 using the example of the Zytek Z11 SN, an open sports car prototype in the LMP2 category (cf. [5]). The increase of the angle of attack up to 6° leads primarily to a continuous decrease of the downforce coefficient. Due to the reduction of the induced drag, the total air resistance of the vehicle also decreases in this range. The loss of downforce at the front axle is greater than at the rear axle, which is reflected in a decrease in the aerodynamic balance. The increase in the angle of attack thus increases the understeer tendency of the vehicle.

In addition to the change in downforce and drag coefficient, a non-zero angle of attack also leads to the build-up of an airside force F_{Ly}, which can be described by the airside force coefficient c_{Ly}. This airside force generates a yaw moment about the vehicle's center of gravity. Unlike production passenger cars, high-performance race cars are designed so that the pressure point of the airside force is behind the center of gravity, not in front of it. The rearward position of the pressure point is primarily a result of the relatively large end plates of the rear wing. The effect can be further enhanced by the addition of fins. Figure 6.111 illustrates the principle conditions for a production car and a racing car.

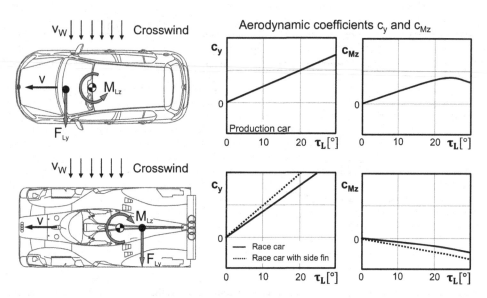

Fig. 6.111 Influence of the pressure center position on the crosswind moment build-up of series and racing vehicles

In a standard passenger car, a positive air moment M_{Ly}, which further increases the angle of attack when the steering wheel is released, arises due to the air lateral force and the submission of the pressure point. This is therefore referred to as an aerodynamically unstable design, as the disturbance is further amplified without correction. The car driver must steer against the wind to correct the disturbance. This corresponds to the intuitive reaction of a normal driver. A backward position of the pressure point, on the other hand, leads to a negative air moment, which turns the vehicle against the angle of attack. This is referred to as an aerodynamically stable design, since the vehicle attempts to return to an equilibrium position automatically, i.e. without any corrective intervention. Such a design ensures that disturbances subside more quickly and can, if necessary, provide the driver with the reaction time required for a correction in the event of an impending loss of control.

The loss of control over a racing vehicle or the occurrence of critical incident flow conditions can have dramatic consequences. Essentially, three scenarios can be distinguished here in which the resulting incident flow conditions deviate greatly from the conditions in the normal operating range (Fig. 6.112). As already mentioned, such conditions are referred to as "off-design" conditions.

The first scenario describes the "backflip", which was frequently observed on closed sports car prototypes in the past and still occurs regularly today. A major reason for the critical behavior is the high lift forces generated above the body, especially in sports car prototypes. In normal operation, these are overcompensated by the downforce forces. However, if there is a serious disturbance of the incident flow in the area of the front end, as caused mainly by the wing and diffuser upwinds of a vehicle in front or by ground

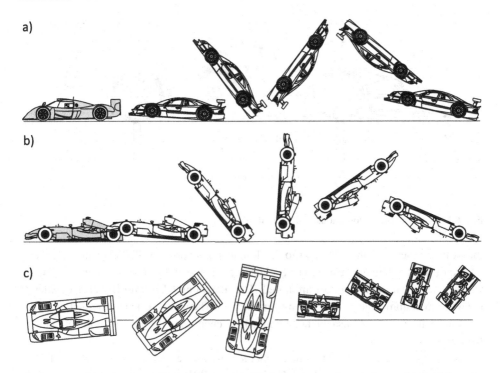

Fig. 6.112 Critical inflow conditions: (**a**) Backflip of a sports car prototype, (**b**) Backflip of a monoposti and (**c**) "Blow-over" due to crossflow after loss of control

unevenness, this can lead to a strong reduction of the front axle downforce. The body lift forces on the front axle and the drag force on the rear wing then cause a pitching moment that sucks the front end into the air. These conditions are shown in Fig. 6.113. In the worst case, the lift is so great that the front end lifts off the ground and becomes banked. The lean causes a stagnation zone with excess pressure to form in the front area of the underbody, further levering the front end of the car. In extreme cases, the vehicle rises and turns backwards around its y-axis.

At Le Mans in 1999, such behavior could be observed several times on the Mercedes CLR. During qualifying on Thursday evening, Mark Webber's #4 CLR took off before the Indianapolis corner and rolled over several times. Again with Mark Webber at the wheel, the slightly aerodynamically modified car took part in Saturday's warm-up, but this time the No. 4 CLR only made it as far as the then-present hill before the Mulsanne corner, where the car again lifted off the track, rolled over and came to rest on its roof. Despite this second accident, Mercedes-Benz decided to let the other two cars start the race with the numbers 5 and 6. This involved making further modifications to them and instructing the drivers not to follow other cars too closely over major bumps. Despite all the precautions, another accident with a CLR occurred in the race after a good 4 hours. This time it involved the #5 car of driver Peter Dumbreck. Dumbreck was following a Toyota GT-One, his CLR

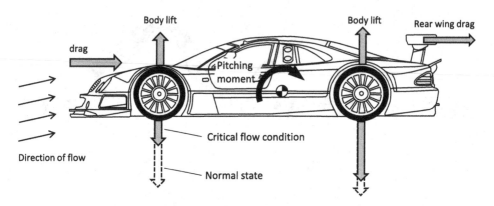

Fig. 6.113 Critical flow conditions with frontal inflow

took off in a bend on the approach to the Indianapolis corner, rolled over in the air and landed off the track in the bushes. The accident could be seen live on television. Dumbreck was rescued from the car almost uninjured. The remaining CLR with Bernd Schneider at the wheel was immediately withdrawn from the race. The incidents led to a reduction of the altitude changes on the Mulsanne and to numerous changes in the aerodynamic regulations for Le Mans prototypes.

As early as 1998, an almost identical incident was observed on the Porsche 911 GT1 during the race at Road Atlanta. At the Nürburgring in 2015, a tragic incident occurred with a Nissan GT-R GT3 during a VLN race. On the crest of the Quiddelbacher Höhe, the front of the car was sucked into the air and the vehicle, which was standing vertically in the air, was catapulted into the safety fence. A spectator standing behind the safety fence was fatally injured. The aerodynamic behavior in off-design conditions and the avoidance of uncontrollable flight situations are therefore increasingly in the focus of the regulatory organizations today.

Critical angles of attack can also occur with monoposti, despite the low body lift forces compared to prototypes, as shown in the second scenario in Fig. 6.112b. A typical cause for monoposti in the occurrence of such conditions is the contact of a following vehicle with the rear tires of the vehicle in front. The rear tires pull the oncoming vehicle into an inclined position, causing a stagnation zone with highpressure to form on the underbody. Loss of downforce and excess pressure at the underbody push the vehicle further upwards and aggravate the situation. Again, in extreme cases, the vehicle rises and rolls over backwards. An example of this was the collision between Mark Webber and Heikki Kovalainen in Valencia at the European Grand Prix in the 2010 season.

The third scenario results from a loss of control of the vehicle, in which such high float angles occur that the vehicle experiences a transverse flow of about 90°. As Fig. 6.114 illustrates, in this case a stagnation zone with excess pressure forms along the side of the vehicle. As the air flows down the wheel arches and roof, it is strongly accelerated, creating areas of strong negative pressure on the wheel arches facing into the wind and on the

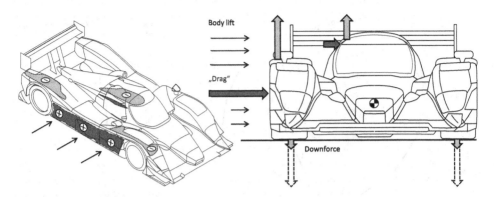

Fig. 6.114 Critical flow conditions in case of crossflow

cockpit. Diffuser and wings are practically ineffective in this situation. The lift forces created in the negative pressure areas and the pressure force acting on the lateral surface of the vehicle create a moment which causes the vehicle to roll to the opposite side. Again, this can result in a skew that leads to the formation of a stagnation zone under the underbody. The initial lifting of the vehicle can also be caused by bumps in the run-off zones. In extreme cases, the vehicle lifts off to the side and rolls over, which is then referred to as a "blow-over" (Fig. 6.112c).

The causes of such accidents were systematically investigated in [37], among others. As explained above, the decisive influencing factors are the vehicle's inclination and speed.

Figure 6.115 shows an example of how lift and downforce on an Audi R8 behave as a function of the angle of attack and vehicle speed. In the circular diagram, the critical speed at which the front and rear axles or the total downforce is compensated by the lift is entered for each angle of attack τ_L. In normal driving operation, which takes place with angles of attack between $-20 - +20°$, the aerodynamic behavior is not critical. There are no conditions that generate real lift. Only at larger angles of attack can a complete unloading of the axle occur, whereby especially at the front axle the critical speed decreases with increasing angle of attack. Thus at point A at a speed of approx. 230 km/h the front axle load is compensated by the front axle lift at a 90° angle of attack (the vehicle is therefore standing transversely), as the wings and underbody hardly generate any downforce at such high angles of attack. Note: The critical speeds can be significantly reduced by external circumstances such as vehicles in front, bumps or strong winds. Initial lean angles can also be caused by slipping over curbs or other causes that are not taken into account in the diagram above.

These aerodynamic phenomena are still a safety problem for sports car prototypes today, as demonstrated, for example, by the accident involving Stéphane Ortelli in a Courage Oreca LMP1 at Monza in 2008. A Peugeot 908 HDI FAP also lifted off the side during a test drive in 2008. The particular danger posed by vehicles taking off is that the safety measures on the track can no longer perform their function of mitigating the consequences of accidents. This results in an increased risk for drivers and persons who are close to the

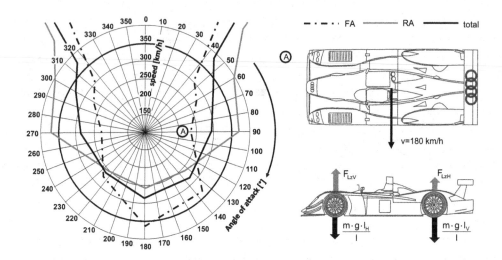

Fig. 6.115 Critical vehicle speeds as a function of angle of attack. (Adapted from [9]; courtesy of © Springer Fachmedien Wiesbaden 2018, All Rights Reserved)

track. For these reasons, the FIA has prescribed two design measures for LMP vehicles, which can be clearly seen in Fig. 6.116 using the example of an Audi R18.

The first measure is the installation of a stabilizing fin (since 2011) in the middle of the vehicle between the cockpit and the rear wing. This has two functions: Firstly, it generates a stabilizing air moment under cross-flow conditions by shifting the pressure point to the rear, which counteracts a rotation of the vehicle around its vertical axis. Under certain circumstances, this can prevent the occurrence of a cross-flow. If a cross-flow occurs despite the guide fin, a dynamic pressure forms in front of the guide fin, which reduces the negative pressure over the wheel arches and thus the tendency to tilt.

The second measure reinforces this effect by prescribing an area at the front and rear axles in which the wheel arches must have a completely open surface (since 2012). The negative pressure created above the wheel arches cannot generate any lift force without a corresponding surface. This further reduces the vehicle's tendency to tip over and thus the likelihood of an uncontrolled condition.

Due to the high density of vehicles, the ovals of the US racing series are repeatedly the scene of a large number of accidents involving spinning and rear-ended vehicles. Especially in stock cars, body lift was a massive problem for years. Apart from blow-over scenarios, stock cars turned by about 180° in particular posed a high risk. The rotation of the vehicle regularly caused the rear of the vehicle to lift off. As the braking effect in these conditions was very low, the vehicles sometimes hit the track boundaries at high speeds. In the mid-1990s, NASCAR finally introduced roof-mounted deflector flaps ("roof-flaps"), which are shown in Fig. 6.117. As soon as the car is rearward in the wind and a high negative pressure is applied to the roof, the flaps open automatically and position themselves vertically into the wind. The body lift is reduced and the vehicle does not get into a

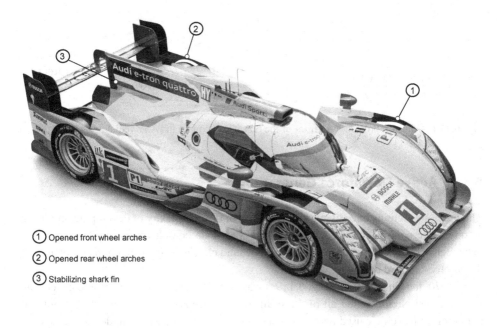

Fig. 6.116 Measures to reduce aerodynamic overturning moments on the Audi R18 (2012). (Courtesy © Audi AG 2018. All Rights Reserved)

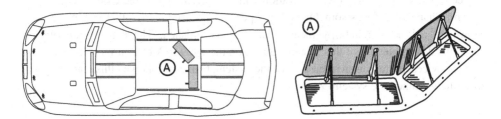

Fig. 6.117 Flow disturbing deflector flaps on a NASCAR stock car

flying attitude. The increased air resistance, which slows the vehicle down, also contributes to the reduction in lift.

The catapult effect of the rear wheels of monoposti has a general hazard potential, as described above. For this reason, between 2012 and 2017 IndyCar prescribed a collision guard for its vehicles that was located directly behind the rear wheels (Fig. 6.118). This measure was one of the consequences of the fatal accident involving Dan Wheldon at Las Vegas Motor Speedway (cf. Sect. 2.1.12).

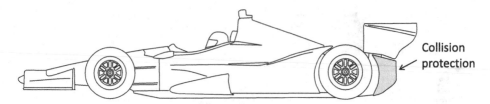

Fig. 6.118 Collision protection in front of the rear wheels on an IndyCar vehicle

6.11   Active Downforce Generation

The previous sections have given an impression of the complexity of developing and using aerodynamic components to generate downforce. This complexity is largely due to the dependence of downforce on vehicle speed, ground clearance, detachment phenomena and downforce-induced drag.

In 1970, the Chaparral 2 J (Fig. 6.119a) was a vehicle that used a downforce generation principle that circumvented all of these difficulties. Instead of massively sized front and rear wings, this vehicle used two 70 hp. fans which extracted air from under the vehicle. In connection with an underbody sealed by skirts one could produce in such a way – according to the program booklet to the Monterey Castrol GTX Grand Prix – already at the standstill a down force of approx. 1000 Pounds. The main advantage of this principle is a speed-independent downforce generation with an extremely low increase in drag.

The Chaparall 2 J was sometimes 2–3 s per lap faster than any other car in the field. However, a lack of reliability meant that it never finished. The competition quickly recognized the potential of this system. Based on the 1969 FIA ban on moveable aerodynamic devices, protests were filed against the system. Can-Am followed this reasoning and eventually banned the system.

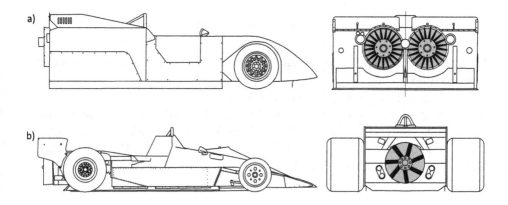

Fig. 6.119 (**a**) Chaparral 2 J-Can-Am vehicle (1970) and (**b**) Brabham BT46 F1 vehicle (1978)

In 1978 Gordon Murray copied this principle, despite the valid ban on movable aerodynamic devices, and used it in the Brabham BT46 (Fig. 6.119b). The regulations at that time only prohibited the use of moveable aerodynamic components which served "primarily" to generate downforce. For this reason, three radiators were connected upstream of the fan, so that the primary task of the system was to cool the engine and not to generate downforce. As long as a maximum of 49% of the fan power was used to generate downforce, the system was therefore legal. This view was also shared by the FIA, which classified the system as conforming to the rules. With this car Lauda won the Swedish Grand Prix in 1978. However, the exploitation of this grey area in the regulations led to considerable disgruntlement among the competition. Team owner Ecclestone, who was also chairman of the team association FOCA at the time, was threatened with withdrawal of confidence. To ensure FOCA's cohesion, Brabham withdrew the BT46 after just one race. The era of the "vacuum cleaners" was thus finally over, as the loopholes in the regulations were closed for the following season.

In racing vehicle technology, the drag reduction system (DRS) is therefore the only approved exception in the field of moveable aerodynamic components today. However, active aerodynamic elements are regularly used in sporty production vehicles in particular. Porsche's PAA system has already been mentioned in Sect. 6.1. Figure 6.120 shows two further examples of movable and active aerodynamic components on the Audi R8 and Mercedes-Benz SLR McLaren. These are movable spoilers which are automatically extended depending on the vehicle speed. When extended, they generate downforce at the rear axle, increasing high-speed and braking stability. At lower cruising speeds, the spoilers remain retracted to optimize drag.

Fig. 6.120 Active spoilers (**a**) on Audi R8 (2010). (Courtesy of © Audi AG 2018. All Rights Reserved), (**b**) on McLaren-Mercedes SLR. (Courtesy of © Daimler AG 2018. All Rights Reserved)

6.12   Aerodynamic Development Trends and Influence of the Regulations

The aerodynamic development of racing vehicles is a constant competition between the inventive spirit of the designers and the sporting authorities who set the rules. The technical regulations defined by the sports authorities have the task of guaranteeing the framework conditions for fair, exciting and, above all, safe competition. Within these limits, the engineers try to find solutions that give them the decisive advantage ("unfair advantage") on the track. The best example of this was the discovery of the ground effect and the stricter regulation of the underbody design from 1983 onwards. Such upheavals in the regulations have repeatedly changed the appearance of the vehicles or the design of the aerodynamic components in the long term. The following is a chronological list of the most important changes to the Formula 1 regulations relating to aerodynamics:

- 1969: Prohibition of moveable aerodynamic devices, attachment of aerodynamic components to the sprung masses, maximum height
- 1976: Prohibition of high air boxes
- 1981: ban on moveable skirts, minimum ground clearance
- 1983: Flattening of the underbody between the rear edges of the front tires and the front edges of the rear tires
- 1991: Maximum front wing width reduced from 1500 to 1400 mm
- 1993: Maximum vehicle width reduced from 2150 to 2000 mm, maximum rear wing height reduced from 1000 to 900 mm
- 1994: Prohibition of active chassis, shortening of the central diffuser channel, post-Imola changes: maximum rear wing height reduced by 100 mm, minimum ground clearance of the front wing increased, no protruding of front wing components over the front wheels, introduction of a 10 mm thick wooden panel on the underbody, rear wing elements must not protrude over the rear axle in front
- 1995: Adoption of the changes from 1994, introduction of a stepped underbody, outer areas are higher than the central area
- 1998: Maximum vehicle width reduced from 2000 to 1800 mm
- 2001: Increase of the ground clearance of the front wing by 50 mm, exception 500 mm in the middle of the vehicle, limitation of the number of rear wing elements to three in the upper plane and to one in the lower plane with precise height specifications
- 2004: Number of rear wing elements in the upper plane reduced to two, rear wing end plate extended
- 2005: outer sections of the front wing raised by a further 5 cm, rear wing shortened by 150 mm in the upper plane and end plates lengthened, rise height of the outer diffuser channels reduced
- 2007: Intermediate elements on the rear wing prescribed to prevent elastic deformation

- 2009: front wing width increased by 40 cm and lowered by 10 cm, DRS on front wing, rear wing width reduced by 25 cm and raised by 15 cm, profile height increased by 2 cm, use of flow aids severely restricted, start of diffuser rise shifted backwards
- 2011: DRS at the rear wing with main profile and adjustable flap (10–50 mm adjustable, only positions "open" and "closed" allowed, an additional lower profile at the rear wing located near the diffuser
- 2014: Lower rear wing element ("beam wing") prohibited, depth of rear wing main elements reduced
- 2015: Safety modifications to the nose
- 2017: Vehicle width increased from 1800 to 2000 mm, front and rear wing dimensions significantly changed (Sect. 3.2.1)

More stringent regulations of the underbody geometry usually have a direct effect on the wing design in order to compensate for the loss of downforce by optimizing it. Since the flow conditions at the wings interact strongly with the body, there are also feedback effects on the periphery of the vehicle. Figure 6.121 shows the front view of Formula 1 cars from different eras. The changes in vehicle design are a constant interplay between the desire of the designers to gain a competitive advantage through increased downforce and the efforts of the regulatory authorities to keep the downforce level within a reasonable range, whereby this is determined in particular by safety and financial aspects. Figure 6.121 clearly shows the trend to continuously make the nose and safety cell narrower, thus increasing the usable area of the front wing. Beginning in the 1990s, the "high nose" concept is gaining acceptance. At the same time, the use of all kinds of flow aids increases dramatically. This development reaches its peak in 2008, as Fig. 6.122a shows using the McLaren-Mercedes MP4–22 as an example. The main purpose of the flow aids on the MP4–22 is to control the vortices created on the car and the airflow to the underbody and rear wing. For the 2009 season, the use of flow aids has been rigorously limited, as shown in Figure 6.122b using the BMW Sauber F1.09 as an example. Today's Formula 1 cars are therefore returning to a more classical shape, at least to some extent.

The position of the front wing relative to the front wheels has changed significantly due to reductions in vehicle width in 1993 and 1998 and the widening of the front wing in 2009. The widening of the front wing was intended to reduce the understeer tendency of the vehicles when overtaking. The configuration of the front wing then also changed significantly in the following years. Until well into the 2000s, relatively simple two-element front wings were used. The front wings of today's Formula One cars have up to six airfoil elements, which is illustrated in Fig. 6.123 by comparing a Leyton House CG901 with a Red Bull RB14. The wing of the Leyton House shows an endplate spur to affect the endplate and wheel vortices, which were also banned in the 1990s. One reason for the now high number of airfoil elements is the location of the airfoil in front of the front wheels. The highpressure area there increases the pressure rise that the flow on the suction side of the front wing has to overcome. Multi-element wings counteract the resulting tendency to detach.

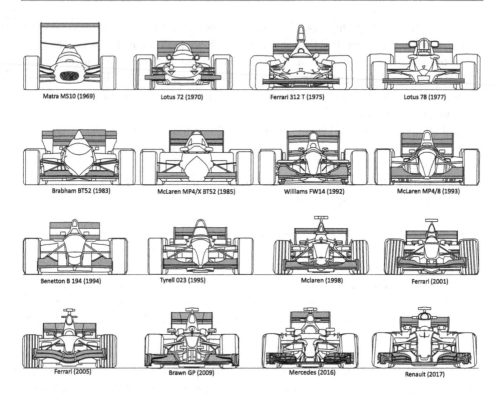

Fig. 6.121 Influence of changes to the regulations on aerodynamic vehicle design

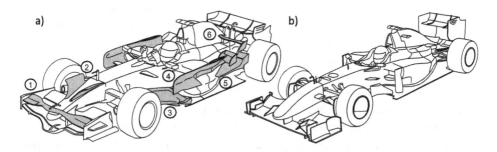

Fig. 6.122 Restriction of flow aids (1) to (6) for the 2009 Formula 1 World Championship: (**a**) McLaren-Mercedes MP4–22, (**b**) BMW-Sauber F1.09

Another characteristic of modern F1 cars is the extremely narrow rear end, also known as the "coke-bottle" section due to its resemblance to a bottle neck. The change in the rear end is illustrated in Fig. 6.124 by contrasting three McLaren, MP4/1, MP4/2 and MP4–31. McLaren was the first team to introduce a retracted rear end after the ban on ground effect vehicles [38]. The retraction had become possible because there was no longer allowed to

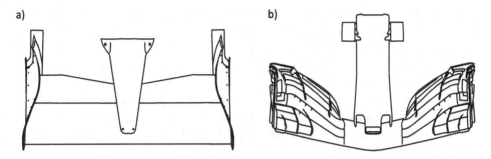

Fig. 6.123 Front wing (**a**) on Leyton House CG901 (1990), (**b**) on Red Bull RB14 (2018)

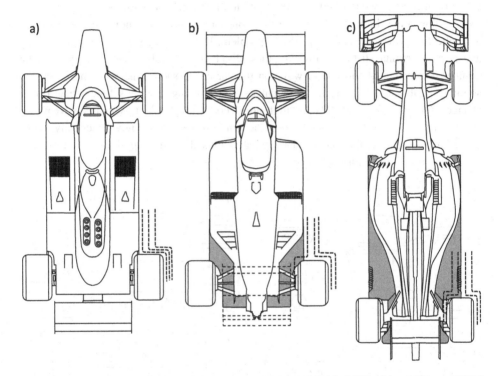

Fig. 6.124 Rear end or "coke-bottle" section (**a**) on McLaren MP4/1 (1982), (**b**) on McLaren MP4/2 (1984), McLaren MP4–31 (2016)

be a diffuser rise in front of the rear wheels due to the straightening of the underbody. On the one hand, the retracted rear end reduces the highpressure in front of the rear wheels, since the air can now flow off both the inside and the outside. On the other hand, the airflow to the lower rear wing profiles is optimized. In this way, some of the rear downforce that had been lost at the underbody could be compensated for. In the 2010s, Red Bull in particular began to make the rear of the car more and more compact.

The technical regulations for sports car prototypes basically contain comparable measures that are intended to regulate the downforce level of the vehicles. Over the last 20 years, it has been observed that the aerodynamics of sports car prototypes have increasingly been oriented towards formula cars. Figure 6.125 shows the front views of various sports car prototypes.

Since the late 1990s and early 2000s, sports car prototypes have featured a Formula 1-style nose, with today's prototypes even featuring a "high nose" configuration. Also clearly visible is the consistent effort to reduce the frontal area. Front radiators, such as those still fitted to the Sauber C9, were replaced by side radiators as early as the beginning of the 1990s. Between 2000 and 2008, an open prototype won the 24 Hours of Le Mans a total of seven times. The main advantages of this concept are lower weight and a lower center of gravity, as well as better accessibility to the driver's position, which significantly reduces downtime during pit stops. Furthermore, no interior air conditioning is required and reflective windscreens do not pose a problem. Disadvantages are the increased drag and a lower torsional stiffness. However, these disadvantages only came into play when the number of mechanics allowed to work on the car at the same time was limited by the sporting regulations. As a result, tire changes and refueling now took longer than driver changes. On the 2016 Audi R18, a "high-nose" concept is evident, which is extreme for sports car prototypes. The pylons support the front diffuser, which now practically works like a front wing due to the recesses. Due to regulations, the rear wing of the Audi R18 does not extend across the entire width of the car.

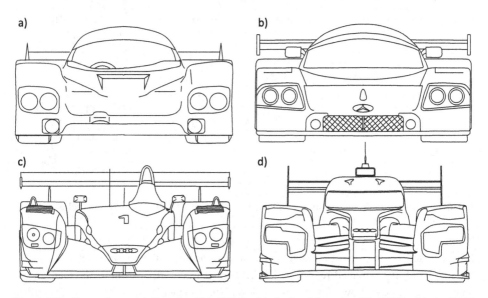

Fig. 6.125 Front views: (**a**) Porsche 956 (1982), (**b**) Sauber-Mercedes C9 (1989), (**c**) Audi R8 (2000) (**d**) Audi R18 e-tron quattro (2016)

Fig. 6.126 Splitter on the underbody of the Audi R18 e-tron quattro (2016). (Courtesy © Audi AG 2018. All Rights Reserved)

In Fig. 6.126 it can be seen that the Audi R18 uses another element, the forward-facing splitter, which was previously found mainly on F1 cars. The dismantled front fairing reveals that the high-lying monocoque design and the entire chassis connection hardly show any difference to modern F1 cars at first glance. As with Formula 1 cars, only a minimal number of suspension components are in the direct airflow of the front diffuser, which already defines a key design objective for the chassis of high performance racing cars considered in Chap. 7.

References

1. Zörner, C., Islam, M., Lindener, N.: Aerodynamik und Aeroakustik des neuen Audi A8. ATZ Automob. Z. **112**, 420–429 (2010). https://doi.org/10.1007/BF03222175
2. Meder, J., Wiegand, T., Pfadenhauer, M.: Adaptive Aerodynamik des neuen Porsche 911. ATZ Automob. Z. **116**, 58–63 (2014). https://doi.org/10.1007/s35148-014-0045-7
3. McBeath, S.: Aerodynamic freedom. Racecar Eng. **20**(12), 21–22 (2010)
4. McBeath, S.: All change please .... Racecar Eng. **22**(3), 47–48 (2012)
5. McBeath, S.: Balancing act. Racecar Eng. **22**(11), 45–46 (2012)
6. Wright, P.: Ferrari Formula 1 – under the Skin of the Championship-Winning F1-2000. Society of Automotive Engineers, Warrendale (2003)

7. Rendle, S.: Williams FW14B – Owners' Workshop Manual. Haynes Publishing, Sparkford (2016)
8. Anderson, J.D.: Fundamentals of Aerodynamics. McGraw-Hill, New York (2011)
9. Hucho, W.H.: Aerodynamik des Automobils, pp. 157–284. Vieweg, Wiesbaden (2005)
10. Ludvigsen, K.: Mercedes-Benz Renn- und Sportwagen. Bleicher, Gerlingen (1993)
11. Bamsey, I.: The Anatomy & Development of the Sports Prototype Racing Car. Motorbooks International, Osceola (1991)
12. Cimarosti, A.: The Complete History of Motor Racing. Aurum Press Limited, London (1997)
13. Abbott, I.H., von Doenhoff, A.E.: Theory of Wing Sections. Dover Publications, New York (1959)
14. Zickuhr, T.: How wings work. Racecar Eng. 21(3), 86–88 (2011)
15. Katz, J.: Race Car Aerodynamics – Designing for Speed. Bentley Publishers, Cambridge (2006)
16. Lutz, T.: Profilentwurf. Vorlesungsumdruck. Institut für Aerodynamik und Gasdynamik, Universität Stuttgart (2008)
17. Demuth, R.: Aerodynamik von Hochleistungsfahrzeugen. Vorlesungsumdruck, TU München (2011)
18. Hünecke, K.: Die Technik des modernen Verkehrsflugzeugs. Motorbuch, Stuttgart (2008)
19. Benzing, E.: Dall' aerodinamica alla potenza in Formula 1. Giorgio Nada Editore, Vimodrone (2004)
20. Benzing, E.: Wings – Their Design and Application to Race Cars. Giorgio Nada Editore, Vimodrone (2012)
21. Zhang, X., Toet, W., Zerihan, J.: Ground effect aerodynamics of race cars. Appl. Mech. Rev. 59(1), 33–49 (2006). https://doi.org/10.1115/1.2110263
22. Zhang, X., Zerihan, J.: Aerodynamics of a double-element wing in ground effect. AIAA J. 41(6), 1007–1016 (2003). https://doi.org/10.2514/2.2057
23. McBeath, S.: Competition Car Aerodynamics. Haynes Publishing, Sparkford (2006)
24. Binks, D., DeWitt, N.: Making it Faster – Tales from the Endless Search for Speed. CreateSpace, North Charleston (2013)
25. Ludvigsen, K.: Colin Chapman – Inside the Innovator. Haynes Publishing, Sparkford (2010)
26. Tipler, J.: Lotus 78 und 79 – The Ground Effect Cars. The Crosswood Press Ltd, Ramsbury (2003)
27. Wright, P.: Formula 1 Technology. Society of Automotive Engineers, Warrendale (2001)
28. Henry, A.: Grand Prix Car Design & Technology in the 1980s. Hazleton Publishing, Richmond (1988)
29. Incandela, S.: The Anatomy & Development of the Formula One Racing Car from 1975. Haynes Publishing, Sparkford (1990)
30. Mahon, S., Zhang, X., Gage, C.: The evolution of edge vortices underneath a diffuser equipped bluff body. In: 12th International Symposium on Applications of Laser Techniques to Fluid Mechanics (2004)
31. Ruhrmann, A., Zhang, X.: Influence of diffuser angle on a bluff body in ground effect. J. Fluids Eng. 125(2), 332–338 (2003). https://doi.org/10.1115/1.1537252
32. Schenkel, F.: The origins of drag and lift reductions on automobiles with front and rear spoilers. SAE-Technical Paper. 770389 (1977). https://doi.org/10.4271/770389
33. Brzustowicz, J., Lounsberry, T., de La Rode, J.: Experimental & computational simulations utilized during the aerodynamic development of the dodge intrepid R/T race car. SAE-Technical Paper. 2002-01-3334 (2002). https://doi.org/10.4271/2002-01-3334
34. Piola, G.: Formula 1 Technical Analysis 2005/2006. Giorgio Nada Editore, Vimodrone (2006)
35. Frère, P.: Porsche Racing Cars of the Seventies. Patrick Stephens Ltd, Somerset (1980)
36. McBeath, S.: Mirror image. Racecar Eng. 22(6), 43–44 (2012)

37. Dominy, R.G., Ryan, A., Sims-Williams, D.B.: The aerodynamic stability of a Le Mans prototype race car under off-design pitch conditions. SAE-Technical Paper. **2000-01-0872** (2000). https://doi.org/10.4271/2000-01-0872
38. Newey, A.: How to Build a Car. HarperCollinsPublisher, London (2017)

Suspension, Steering and Brake System

This chapter provides an overview of the contribution made by the suspension, steering and brake system to the dynamic performance potential of a racing vehicle. Design details of individual chassis components and parts for use in motorsport are described comprehensively in [1].

7.1 Tasks of the Suspension, the Steering and Brake System

The main task of the suspension, the steering and brake system as a part of the chassis is to transmit the longitudinal and lateral forces required for propulsion, braking and steering to the road surface. Accordingly, the following modules are assigned to chassis:

- Suspension
- Steering system
- Brake system
- Tire and rim

For organizational reasons, some of the following modules are also included in the scope of the chassis:

- Drive shafts
- Foot pedal or pedal set
- Fuel system and exhaust system

Figure 7.1 shows the assignment of individual components to the modules using the example of the front and rear axles of the Audi R8. The characteristics of the tire were

L. Frömmig, *Basic Course in Race Car Technology*, https://doi.org/10.1007/978-3-658-38470-8_7

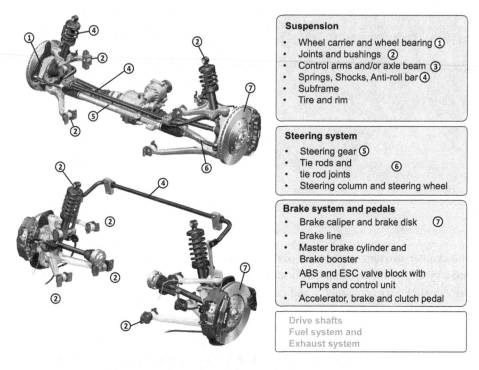

Suspension
- Wheel carrier and wheel bearing ①
- Joints and bushings ②
- Control arms and/or axle beam ③
- Springs, Shocks, Anti-roll bar ④
- Subframe
- Tire and rim

Steering system
- Steering gear ⑤
- Tie rods and ⑥
- tie rod joints
- Steering column and steering wheel

Brake system and pedals
- Brake caliper and brake disk ⑦
- Brake line
- Master brake cylinder and Brake booster
- ABS and ESC valve block with Pumps and control unit
- Accelerator, brake and clutch pedal

Drive shafts
Fuel system and
Exhaust system

Fig. 7.1 Modules and module scopes of the chassis. (Photos on the left courtesy of © Audi AG 2018. All Rights Reserved)

dealt with in Chap. 4. The steering system and braking system are dealt with in Sects. 7.8 and 7.10. However, the characteristic feature of a chassis is the design of the suspension.

The primary task of the suspension is to establish optimum contact between the individual wheels and the road surface according to driving dynamics criteria. The optimum contact is influenced by the spatial wheel position, the wheel load distribution and road unevenness. The compensation of road unevenness requires a vertical degree of freedom from the suspension. This is provided by the design of the suspension as a system build of joints and connecting rods. When the wheel moves with an according spring travel, the system of rods and joints simultaneously influences the wheel camber, the toe angle and the track width. This effect is summarized under the term kinematics. Elasticities in the chassis, which are used in a targeted manner, especially in production vehicles, can also cause changes in the kinematic wheel position through deformation. This is referred to as compliance, which, however, plays a rather subordinate role in racing vehicles, since the central objective is to achieve the stiffest possible connection between the suspension and the structure. The support of the weight, inertia and downforce forces (see also Sect. 5.3) as well as the control of the uneven wheel movements are achieved by the interaction of the suspension geometry with body springs, shock absorbers and stabilizers.

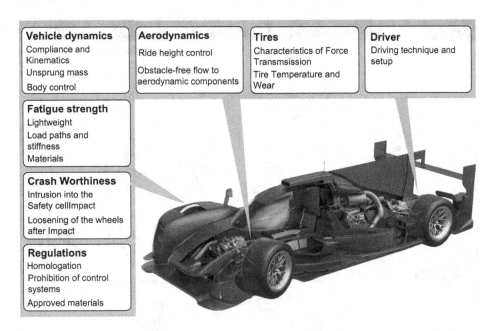

Vehicle dynamics	Aerodynamics	Tires	Driver
Compliance and Kinematics	Ride height control	Characteristics of Force Transmsission	Driving technique and setup
Unsprung mass	Obstacle-free flow to aerodynamic components	Tire Temperature and Wear	
Body control			

Fatigue strength
Lightweight
Load paths and stiffness
Materials

Crash Worthiness
Intrusion into the Safety cellImpact
Loosening of the wheels after Impact

Regulations
Homologation
Prohibition of control systems
Approved materials

Fig. 7.2 Chassis and suspension requirements of a racing vehicle. (Photo courtesy of © Dr. Ing. h. c.. F. Porsche AG 2018. All Rights Reserved)

The conceptual requirements on the chassis of a racing vehicle come from different areas, which are summarized with some examples in Fig. 7.2. Compared to production vehicles, the requirements from the field of aerodynamics are particularly important for high-performance racing vehicles, since the driving dynamic properties of a racing vehicle are very strongly dependent on the ground clearances due to the pitch sensitivity. In addition, the objective of an obstacle-free inflow of the aerodynamic components or the underbody may outweigh the classical geometric objectives. Examples are discussed in the following sections. In addition, there are other requirements from the regulations that must be observed. Among other things, the prohibition of certain materials or a limited freedom of design in the layout of pick up points for near-series racing vehicles should be mentioned in this context. With regard to the design of the suspension, a distinction is made between rigid and semi-rigid axles and independent wheel suspensions. Figure 7.3 shows the systematics of the different wheel suspensions. Figure 7.4 summarizes the overriding characteristics of the various concepts.

The defining characteristic of rigid axles is that the two wheels of an axle are firmly connected to each other and that they therefore influence each other in every driving situation. Today, rigid axles are usually only used on the rear axle. A distinction is made between driven and non-driven rigid axles. Driven rigid axles are divided into a design with a differential integrated in the axle beam and De-Dion axles, in which the differential is attached directly to the body to reduce the unsprung masses. Lateral support is provided by

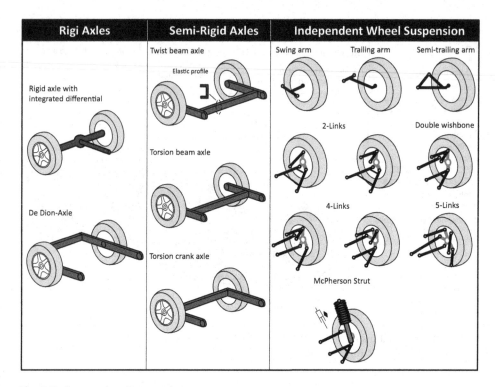

Fig. 7.3 Systematics of suspensions

a Panhard rod or a Watt linkage. Longitudinal support can be provided by two to four trailing arms or a drawbar. Despite their significant disadvantages in terms of driving dynamics, driven rigid axles are still used in various NASCAR racing series, for example, where their use is prescribed by the technical regulations.

In the case of semi-rigid axles, the fixed connection is partially eliminated by means of a specifically introduced elasticity. For this purpose, two trailing arms are welded together with an elastic torsion profile. The elasticity of the profile allows the wheels to move relative to each other – but not completely independently of each other. The position of the torsion profile defines whether the axle is a twist-beam, torsion-beam arm or torsion crank axle. Semi-rigid axles are used today exclusively on non-driven rear axles.

In the case of independent wheel suspensions, the wheels are connected to the body independently of one another via the wheels' own control arms, so that they can move relative to one another without influencing one another. The central advantages of independent wheel suspension are the low unsprung masses and the significantly better vehicle performance, which results among other things from the high potential for kinematic and compliance optimization. The optimization potential increases with the number of control arms used. The typical application for racing vehicles is the double wishbone arrangement, which consists of an upper and a lower wishbone as well as a track rod. By breaking down

Rigid Axles	Semi-Rigid-Axles	Independent Wheel Suspension
• high economic efficiency	• high economic efficiency	• no interaction of the wheels
• sideshaft without joints	• unpowered	
• high robustness and resilience	• low unsprung mass	• low unsprung mass
	• Torsion profile replaces the Stabilizer	• great possibilities to tune the kinematic and elastokinematic of the suspension (increases with the number of the links, (favorable interpretation design of roll centers and anti-geometries and wheel travel curves))
• roll center on height of the wheel center		
• no camber and track width changes during wheel travel	• favorable spring and damper ratio	
	• small track width change during wheel travel	
	• good braking anti-dive characteristics	
• High unsprung mass		• Good driving stability
• wheel interference due to bumps in the road	• adverse unfavorable lateral force support	• favorable Camber, toe angle and track width changes under wheel travel
• lack of freedom in designing the kinematic and elastokinematic characteristics	• toe-correcting bearings for avoidance of lateral force oversteer required	• Isolation of road excited vibrations
• Restrictions on designing anti-dive- and anti-squat characteristics	• stress peaks at the welded joints between trailing arms and Torsion profile	• restricted ground clearance
• Small installation space required	• limited optimization potential driving dynamics	• High complexity and costs (depending on the number of links)
		• high installation space required

Fig. 7.4 Basic properties of the suspensions [2]

the wishbones into individual track rods, this results in the 4-link or multi-link and 5-link axles. Using the double wishbone axles as an example, the essential properties of the wheel suspension and its influence on driving dynamics are discussed below. To this end, the main design forms of the double wishbone axle are first considered.

7.2   Double Wishbone and Multi-link Axles

Double wishbone axles have been used in Formula 1 since the late 1950s because they represent the best compromise between driving dynamics, aerodynamic and structural requirements. For this reason, this suspension concept can be found in different variants in almost all high-performance racing cars. The arrangement of the individual components for actuating the dampers, springs and stabilizers has been constantly evolving due to changing boundary conditions. This evolution is described in detail in [3].

Figure 7.5 shows the standard version of a double wishbone suspension, as used for example on the Aston Martin DBR9 in the late 1950s. Not shown is the track control arm or track rod, which eliminates the rotational degree of freedom of the wheel carrier. The body spring is a coil spring and is concentric around the shock absorber. The spring-damper

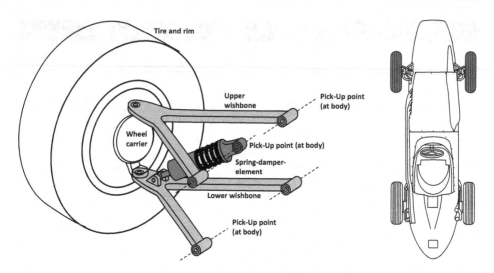

Fig. 7.5 Double wishbone with coil spring damper element on Aston Martin DBR9 (1959)

element is attached to the lower wishbone on the suspension side. During compression, the spring/damper element is compressed and during extension, it is pulled apart. This arrangement relieves the lower wishbone when absorbing lateral forces, as the spring-damper element represents a second load path. To avoid bending loads in the lower control arm, the spring-damper element must be mounted as close as possible to the wheel-side end of the lower control arm. This is also the point where the vertical movement of the control arm is greatest. This arrangement is still used today in a similar form in various production vehicles. The central disadvantage of this arrangement is that the spring-damper element is in the airflow on vehicles with open wheels, thus increasing drag. Furthermore, only a small range of variation is possible in the design of the spring and damper ratio (see also Sect. 7.3).

The two disadvantages mentioned can be overcome by the double wishbone suspension design shown in Fig. 7.6 with top rocker and internal spring-damper element. The top rocker serves here as a rocker arm, which is pivoted in its center on the body. The upward movement of the wheel during compression is translated via the top rocker into compression of the spring/damper element. This arrangement was first used on the front axle of the Lotus T21 in 1961 and was adopted for some of its successors, such as the Lotus T25 and T33. However, the upper control arm, which is designed as a top rocker, now has to support bending loads, which is why it is designed to be relatively massive. Due to the heavy top rocker the center of gravity of the chassis moves upwards. Another disadvantage of this arrangement is the missing relief of the lower wishbone.

At the beginning of the 1970s, the demands on the chassis change due to the increasing aerodynamic downforce. In particular, the support of the aerodynamic loads requires higher spring stiffnesses. In order to avoid the loss of traction caused by stiffer springs, a higher

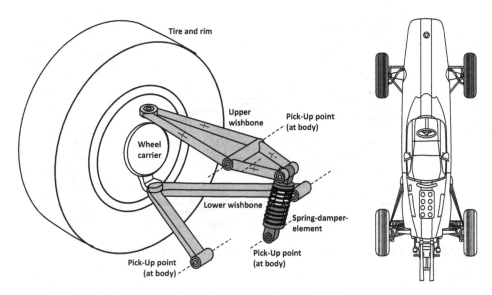

Fig. 7.6 Double wishbone with top rocker and inboard coil spring-damper element on Lotus T25 (1962)

progression for the spring rate is required. This is achieved on the Lotus T72 in 1970 by a pull-rod arrangement (Fig. 7.7). In the pull-rod arrangement, the spring and damper are actuated by a pull rod via the upper wishbone. The pull rod is mounted at the wheel end of the upper wishbone (to avoid bending stresses in the upper wishbone) and actuates a rotatable rocker on the body side. A further special feature in this case is the design of the body spring as a torsion bar, whereby the torsion bar simultaneously assumes the function of the pivot bearing for the rocker. When the wheel is deflected, the upper wishbone pulls the pull rod upwards. The pull rod also pulls the end of the rocker upwards and thus rotates the rocker around its bearing on the body side. The torsion bar opposes this rotational movement, thereby realizing the function of the body spring. The upward movement at the end of the rocker is used to actuate a telescopic shock absorber.

The use of torsion bars was mainly due to the fact that the Lotus T72 had inboard brake discs, which could not have been accommodated in the nose of the vehicle together with an inboard coil spring. The use of the pull-rod arrangement also made it possible to achieve the desired progressive spring characteristic (rising rate design). At the same time, it was possible to vary the spring and damper ratios over a wide range. The bottom-mounted rocker also had the advantage of giving the chassis a lower center of gravity. The general disadvantages of the pull-rod arrangement are the lack of relief for the lower wishbone and the heavy load on the upper wishbone due to the load peaks introduced into the upper wishbone during braking.

Figure 7.8 shows an alternative version of the pull-rod arrangement using a classic spring-damper element on the front axle of the 1974 Brabham BT44. In this arrangement,

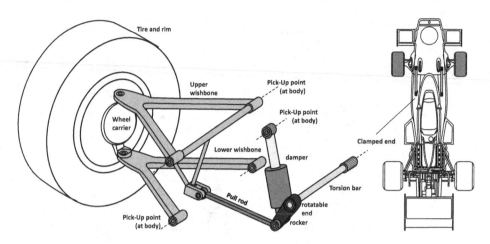

Fig. 7.7 Double wishbone with pull-rod and torsion bar on Lotus T72 (1970)

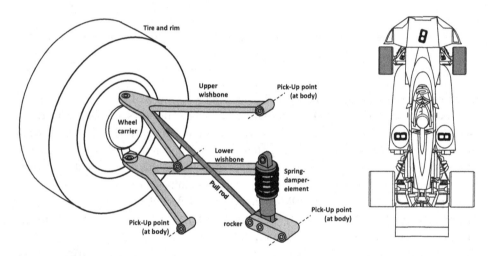

Fig. 7.8 Double wishbone with pull-rod and coil spring-damper element on Brabham BT44 (1974)

the rocker, mounted on the frame by a pivot, operates both damper and coil spring. In contrast to the Lotus T72, the Brabham BT44 had outboard mounted brake discs on the front axle, so there were no packaging problems in accommodating a coil spring damper element in the safety cell.

However, the structural disadvantages of the pull-rod arrangement became increasingly important in the following years due to the aerodynamically induced increase in lateral forces. For this reason, a push-rod arrangement was used on the 1978 Brabham BT46 (Fig. 7.9). In this, a push rod is fitted to the wheel end of the lower wishbone. On the opposite side, the push rod actuates an overhead rocker, which in turn acts on a classic

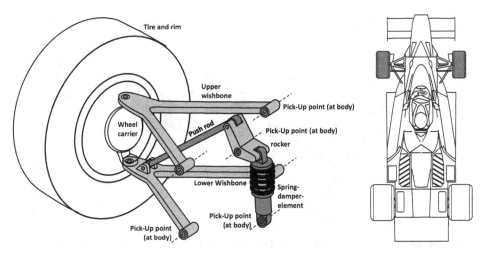

Fig. 7.9 Double wishbone with push rod and coil spring-damper element on Brabham BT46 (1978)

spring-damper element. For this purpose, the rocker is pivoted in its center on the frame. When the wheel travels upwards, the end of the rocker connected to the push rod is also moved upwards. On the opposite side, a downward movement occurs, which actuates the spring and damper. This arrangement has the known advantage of relieving the load on the lower wishbone by providing a second load path. With few exceptions, the push-rod arrangement remains the standard design for front and rear axles of F1 cars in the following decades and also establishes itself in sports car prototypes. A disadvantage compared to the pull-rod arrangement is the chassis center of gravity, which is shifted upwards due to the rocker being on top.

In the early 1990s, for example on the 1992 McLaren MP4/6, the spring-damper element was moved to a horizontal and overhead position. The axis of rotation of the rocker is rotated by about 90° for this purpose. Figure 7.10 shows this arrangement, which primarily allows a narrower and thus aerodynamically more favorable design of the monocoque. A comparison of Figs. 7.9 and 7.10 shows that the McLaren MP4/6 is significantly more compact in the area from the wing to the cockpit than the Brabham BT46. This results in improved airflow around the front wing and improved airflow to the radiators in the sidepods. Another advantage of this arrangement is the ease of access to the springs and dampers for setup adjustments. A disadvantage is that the center of gravity moves further up with the spring/damper elements. This design is still used today in many racing cars, such as the US IndyCars, on the front and rear axles. When installed on the rear axle, the spring/damper elements typically lie on the transmission housing.

At the end of the 1990s, the trend in Formula 1 began to finally replace the coil body springs with torsion bars. Figure 7.11 shows a vertically positioned torsion bar with the damper still arranged horizontally. In this case, the torsion bar also takes over the rotational guidance of the rocker. For this purpose, the torsion bar is firmly clamped at the lower end

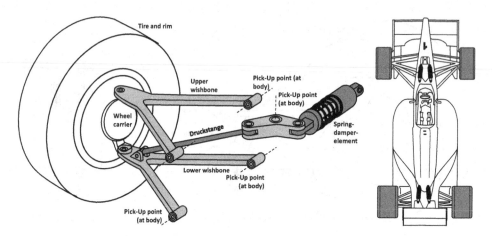

Fig. 7.10 Double wishbone with push rod and horizontally positioned spring-damper element on McLaren MP4/6 (1992)

and is rotatably mounted at the upper end (not visible in the picture). In addition to the further reduction in installation space required by the narrow torsion bars, the separation of spring and damper has the advantage that, if the spring-damper element is not arranged 100% concentrically, no friction-causing lateral forces act on the damper which would reduce its response behavior. Such an arrangement was used, for example, on the front axle of the Prost AP01 in 1998 and established by McLaren on the MP4/14 rear axle in 1999. The torsion bars and dampers are mounted on the rear axle through the gearbox, with the dampers still on top of the gearbox.

On the front axle of the Ferrari F300 from 1998, a push rod arrangement was used, in which the torsion bar assumes a horizontal position. With the torsion bar, the axis of rotation of the rocker also rotates, which in this case is in the longitudinal direction of the vehicle. The damper takes a vertical position. At the same time Fig. 7.12 shows the use of a third damper element (heave damper), the function of which is discussed in detail in Sect. 7.5. This configuration is now one of the standard front suspension designs for prototype and formula cars. Another detail is the use of a flexure joint instead of a ball joint at the body-side ends of the two wishbones, which reduces joint friction and increases joint stiffness.

The push-rod arrangement of the spring and damper actuation was the standard design of all F1 cars for both the front and rear axles until 2008, with a horizontally positioned torsion bar being used on the front axle and a vertically positioned torsion bar on the rear axle. However, in 2009, a pull-rod arrangement was again used on the rear axle of the Red Bull RB5 (Fig. 7.13), and this was adopted by all other racing teams in subsequent years. The switch to this arrangement was encouraged by a change in the technical regulations. For the 2009 season, the diffuser upsweep was only allowed to start from the rear wheel center line. Previously, the rise was allowed to start 330 mm before the rear wheel center line. In the now straightened part of the underbody it was possible to accommodate a rocker

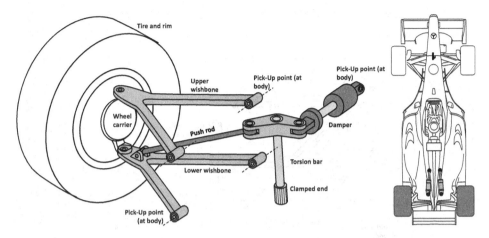

Fig. 7.11 Double wishbone with push rod and vertical torsion bar on McLaren MP4/14 (1999)

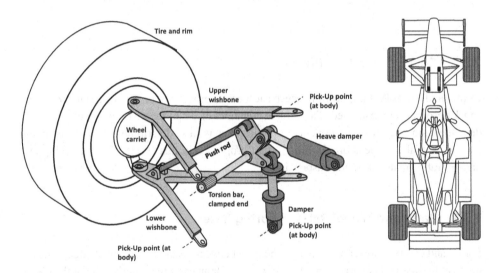

Fig. 7.12 Double wishbone with push rod, horizontal torsion bar and 3-spring damper element on Ferrari F300 (1998)

without affecting the airflow in the diffuser. This is another example of how much the suspension design of a high performance race car is affected by aerodynamics. Using a classic spring-damper element, a similar arrangement was also used on the rear axle of the 2013 Audi R18, where the spring-damper element was horizontal and parallel to the transmission.

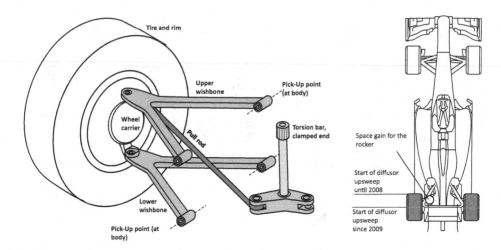

Fig. 7.13 Double wishbone with pull rod and vertical torsion bar on Red Bull RB5 (2009)

7.3 Springs and Stabilizers

In this and the following section, the basic design properties of springs, stabilizers and shock absorbers are discussed. The vehicle dynamics effect of these components is not only dependent on the pure component properties, but is also determined by the geometric properties of the suspension. Two important properties in this context are the spring and damper ratios realized by the suspension.

7.3.1 Ratio and Wheel Related Spring Rates

Spring and damper ratios are defined as the ratio between wheel travel and change in spring travel or between wheel travel and damper speed. For an exemplary derivation of the spring ratio, the kinematic ratios in Fig. 7.14 are considered.

For the relationship between wheel travel and vertical movement at the articulation point of the spring-damper element, the following applies in this case:

$$\Delta z = \Delta z_A \cdot \frac{l_R}{l_A}. \tag{7.1}$$

In addition, the following applies to the spring travel change:

$$\Delta s_F = \Delta z_A \cdot \cos(\delta). \tag{7.2}$$

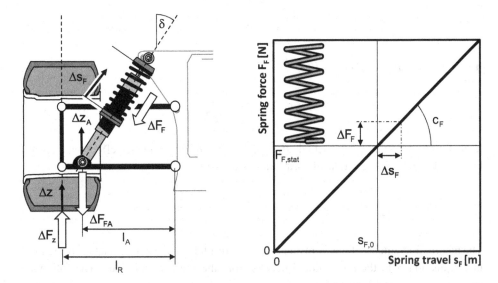

Fig. 7.14 Derivation of spring and damper ratio

From this you get the spring ratio to:

$$i_F = \frac{\Delta z}{\Delta s_F} = \frac{\Delta z_A \cdot \frac{l_R}{l_A}}{\Delta z_A \cdot \cos(\delta)} = \frac{l_R}{l_A \cdot \cos(\delta)}. \tag{7.3}$$

Since a coaxial spring-damper element is considered in the example used here, the damper ratio corresponds to the spring ratio. High damper ratios have the disadvantage that large forces must be achieved in the shock absorber to achieve high wheel-related damping, since high ratios reduce the wheel-related forces. High damping forces require high pressures in the damper, which can result in cavitation phenomena or damper noise (only relevant for production passenger cars). The spring ratio not only affects the spring travel change, but it also translates the spring force change to the lower control arm. From the moment equilibrium around the body-side bearing of the lower wishbone, the relationship between the wheel load change and the vertical force component at the point of articulation of the coil spring is obtained:

$$\Delta F_z = \Delta F_A \cdot \frac{l_A}{l_R} = \Delta F_F \cdot \cos(\delta) \cdot \frac{l_A}{l_R} = \frac{\Delta F_F}{i_F}. \tag{7.4}$$

The vertical force component F_A at the articulation point corresponds to the z-component of the spring force F_F. For the spring force, one obtains from the relationship between spring stiffness and spring travel change:

$$\Delta F_F = c_F \cdot \Delta s_F = c_F \cdot \Delta z_A \cdot \cos(\delta) = c_F \cdot \Delta z \cdot \frac{l_R}{l_A} \cdot \cos(\delta). \tag{7.5}$$

It follows from Eqs. (7.4) and (7.5):

$$\Delta F_z = c_F \cdot \Delta z \cdot \frac{1}{i_F 2}. \tag{7.6}$$

It can thus be seen that for the actual effect of a spring, its wheel-related effect must always be considered. The wheel-related spring rate is defined as:

$$c_R = \frac{\Delta F_z}{\Delta z} = \frac{c_F}{i_F^2}. \tag{7.7}$$

The natural frequencies of the front and rear axles result from the wheel-related spring rates in conjunction with the proportional masses from the static wheel loads. The following applies to the natural frequencies:

$$f_i = \frac{1}{2 \cdot \pi} \cdot \sqrt{\frac{c_{Ri}}{m_i}}. \tag{7.8}$$

Natural frequencies are another characteristic property of the chassis. Typical values for the natural frequency in the design position of various vehicles are:

- Standard passenger car: approx. 1.3–1.4 Hz at the VA, approx. 1.5 Hz at the HA
- Sports car: approx. 1.5 Hz at VA, approx. 1.7 Hz at HA
- Racing vehicles without downforce: 1.6–3 Hz
- Racing vehicles with significant downforce: 4–8 Hz

The use of body springs with higher natural frequencies or with increased wheel-related spring rates has the advantage of reducing vehicle roll, which generally improves vehicle response because the body does not move as much against the direction of the curve. Pitching movements of the body are also reduced. The required suspension travel can be reduced, which allows the vehicle to be lowered, thus reducing dynamic wheel loads. In addition, the vehicle settles later on the bump stop or the additional spring, which means that the roll stiffness distribution between the front and rear axles remains in a defined ratio for longer. In the case of near-series racing vehicles, minimizing the rolling motion also results in a reduction of the kinematically induced understeer tendency. The high natural frequencies of F1 vehicles and sports car prototypes, which can be as high as 8 Hz, result from the need to keep the ground clearance within a defined operating window by means of stiff springs, even when generating the maximum downforce, or to prevent the body from sagging and thus the vehicle from touching the road.

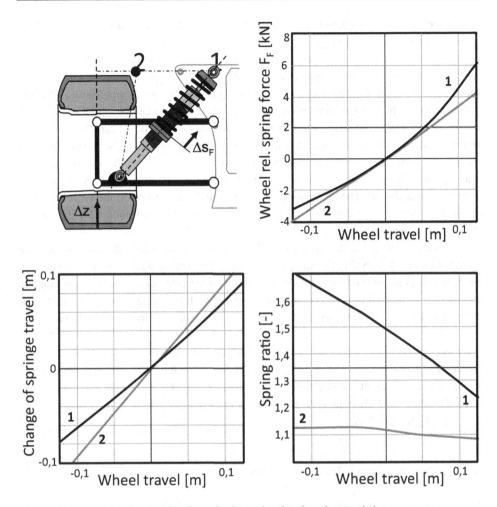

Fig. 7.15 Progressive curves related to wheel travel and spring characteristics

The wheel related spring rate is generally not constant, because the spring ratio changes over the wheel stroke according to Fig. 7.15. To maximize traction, the lowest possible vertical stiffness or a low wheel-related spring rate is required. However, such a soft suspension presents a conflict of objectives with the necessary support of aerodynamic loads. For this reason, the geometry of the spring connection is designed in such a way that a spring rate increasing over the wheel stroke or a progressive increase in the wheel-related spring force over the wheel stroke is achieved. Such a design is referred as "Rising-rate" geometry.

Another advantage of detection rising rate geometry is that a speed-dependent influence on handling can be realized for vehicles with high aerodynamic downforce. If the progression at the front axle is stronger than at the rear axle, the understeer tendency and thus the stability of the vehicle are increased with increasing speed. In this way, it is possible to partially resolve the conflicting goals of high agility in the lower speed range and high stability in the high-speed range.

By shifting the spring pivot points, the spring rate and progression can be specifically modified. In this case, the more the axis of the spring is inclined relative to the wheel movement, the greater the progression. Particularly in vehicles with suspensions in push or pull rod design, the progression and the spring or damper ratio can be varied over a wide range. In addition to the articulation point, the geometry of the rocker can also be used to influence the wheel-related curve. Figure 7.16 shows the principle influence of these two measures on the variables which are decisive for the spring curve. In addition to these variables, the change in wheel camber and track width due to the vertical movement of the wheel are also shown. The dependence of the kinematic quantities on the wheel travel is referred to as wheel lift curves. The significance of the wheel lift curves for the vehicle dynamics is considered in Sect. 7.7.

Progressive spring characteristics can also be achieved by a special design of the coil spring. Figure 7.17a, b show a conical spiral and a coil spring with variable coil spacing. Both measures allow progressive spring curves to be achieved on the component. Springs of this type were used in the Porsche 956, for example. It has already been shown in Sect. 5.3 that additional springs also enable a progressive increase in spring rate. However, the effect of such a bump stop spring only sets in when the empty distance between the damper tube and the additional spring is overcome, which is shown here for the combination with a linear rate. The progression can be specifically tuned by the geometry and the material properties. However, the contact with the additional spring leads to a strong increase in the dynamic wheel load changes on the corresponding axle. This must be taken into account when tuning the vehicle.

7.3.2 Preload, Lowering and Crossweight

Coil spring damper elements for racing vehicles are usually designed as coilover suspension. The reservoir tube and spring plate are threaded for this purpose, as shown in Fig. 7.18. In the rebound stop, i.e. when the piston rod is fully extended, the coil spring is still under compressive stress, which is referred to as preload. Turning the spring plate upwards increases the preload, turning it downwards reduces it. When the vehicle is then on its wheels, the remaining spring travel is determined by the weight of the vehicle minus the preload, with the resulting total compression of the spring being identical for all cases. Accordingly, increasing the preload results in an increase in ground clearance, while simultaneously increasing the compression travel and reducing the extension travel. Reducing the preload results in lowering the vehicle, with a corresponding reduction in compression travel and increase in rebound travel. Coilovers, then, allow high-performance race cars to selectively alter their aerodynamic characteristics by adjusting the ground clearances at the front and rear axles. Note: After changing the ride height, the toe-in and camber values of the suspension may also need to be corrected.

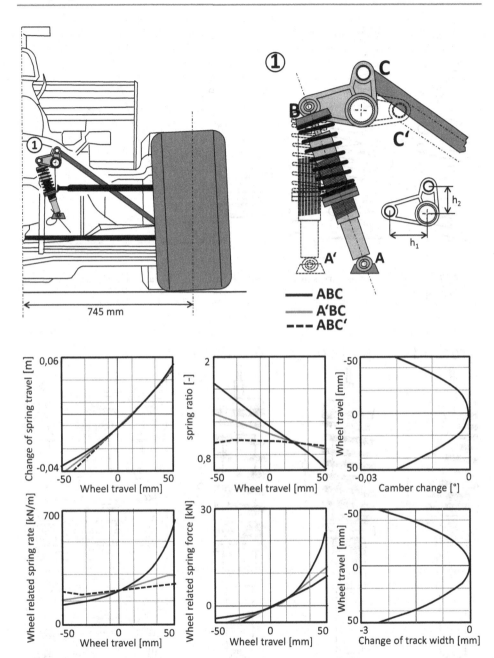

Fig. 7.16 Setting options for wheel curves on an F1 front axle

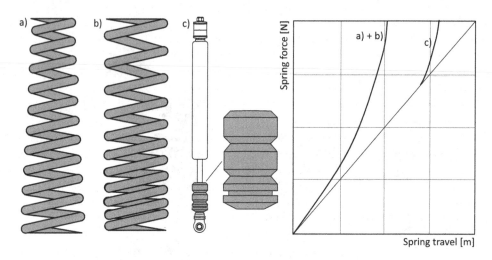

Fig. 7.17 Raising rate spring curves by (**a**) variable coil diameter, (**b**) variable coil spacing and (**c**) an additional spring

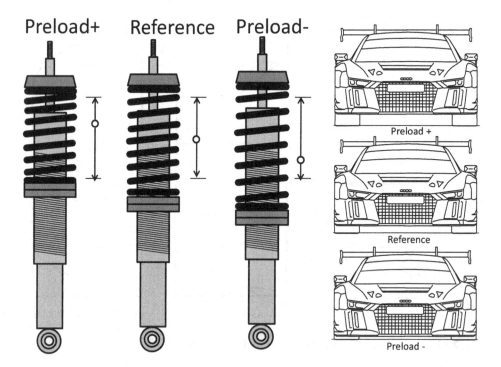

Fig. 7.18 Preload on a coilover suspension

The preload or height adjustment of the chassis is closely related to the weight distribution of the vehicle, as a change in body height alters the wheel loads across the diagonal of the vehicle. Increasing the preload or raising the body height at one wheel results in an increase in wheel load at that wheel and the diagonally opposite wheel. At the same time, the wheel loads on the opposite wheel of the same axle decrease by the same amount, since the mass distribution between the front and rear axles and between the left and right sides cannot be changed. This situation is illustrated in Fig. 7.19. Graphically, this is the same effect that occurs when adjusting the height of table legs. The further the table leg is extended, the higher the load on this table leg. This effect must also be taken into account in connection with the steering geometry.

The weight distribution of the vehicle is thus characterized by three factors:

$$f_{VH} = \frac{m_1 + m_2}{m_1 + m_2 + m_3 + m_4},$$
(7.9)

$$f_{LR} = \frac{m_1 + m_3}{m_1 + m_2 + m_3 + m_4},$$
(7.10)

$$f_{CW} = \frac{m_2 + m_3}{m_1 + m_2 + m_3 + m_4}$$
(7.11)

The diagonal weight distribution f_{CW} is referred to as crossweight and can be specifically influenced by adjusting the height positions. In the example shown, it is 55%, which causes the vehicle to behave noticeably differently in left-hand and right-hand bends. On the usual circuits, a diagonal weight distribution of 50% is aimed at for this reason. In this case, this is achieved by increasing the wheel load on the right front and left rear wheel by 32 kg. For racing vehicles, one tries to achieve a deviation from the ideal diagonal distribution of maximum 0.1% [4]. In contrast, the factors for weight distribution between the front and rear axles f_{VH} and between the left and right sides of the vehicle f_{LR} always remain constant. They are fixed by the vehicle concept. On the US ovals, which have only one curve direction, an asymmetrical weight distribution can be advantageous, as in this way a positive interaction with the dynamic wheel load changes is achieved.

Figure 7.20 shows two crossweight settings for driving on the Indianapolis Speedway oval. By shifting the weight to the front wheel on the outside of the curve and the rear wheel on the inside of the curve, the traction or understeer tendency and the stability of the vehicle are increased in left-hand curves, since the dynamic wheel load difference on the rear axle is lower in this configuration. This measure increases the adhesion potential at the rear axle when cornering.

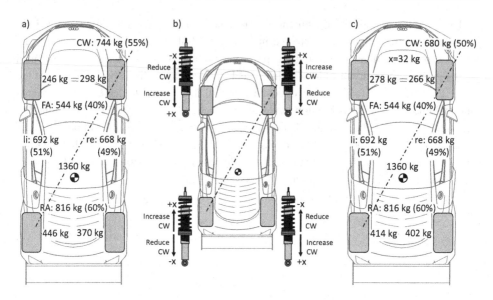

Fig. 7.19 Crossweight. (**a**) Parameters of the static wheel load distribution, (**b**) Wheel load setting on the coilover suspension, (**c**) Optimized wheel load distribution

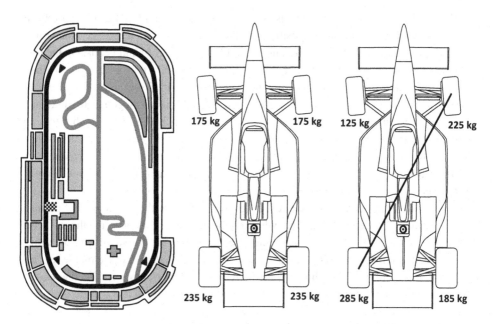

Fig. 7.20 Increase of understeer tendency by crossweighting in the Indianapolis Speedway driven through on the left side

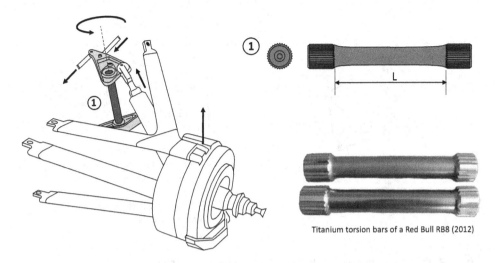

Titanium torsion bars of a Red Bull RB8 (2012)

Fig. 7.21 Actuation of the torsion bar and stiffness-determining properties

7.3.3 Torsion Bars

In various applications for high-performance racing vehicles, the classic coil spring has been replaced by a torsion spring or torsion bar. In push-rod and pull-rod chassis, the torsion bar is firmly clamped at its lower end via serrations. At the upper end, the rocker is placed. The upper end is also supported in such a way that the torsion bar can only twist, but not bend. In this way, the torsion bar acts as a body spring and pivot bearing for the rocker. This arrangement is shown in Fig. 7.21 using the example of a push-rod design, on whose rocker the connecting rods for actuating the body dampers and the stabilizer can also be seen. When the wheel moves upwards, a rotational movement of the rocker around the torsion bar is initiated via the push rod. This rotational movement is opposed by the torsion bar with a resistance proportional to the angle of twist. The stiffness of the body spring is accordingly also defined by the torsional stiffness of the torsion bar, for which the following applies approximately:

$$c_T = \frac{M_T}{\varepsilon} = \frac{G \cdot I_p}{L}. \tag{7.12}$$

The torsional stiffness is thus determined in the form of the shear modulus G by its material properties, and in the form of the polar moment of inertia I_p by its geometry and by its restrained length l.

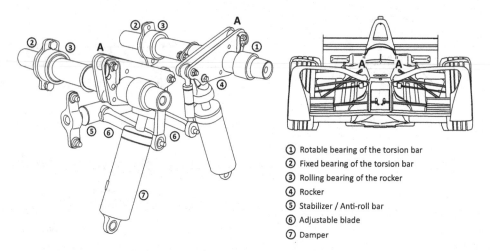

① Rotable bearing of the torsion bar
② Fixed bearing of the torsion bar
③ Rolling bearing of the rocker
④ Rocker
⑤ Stabilizer / Anti-roll bar
⑥ Adjustable blade
⑦ Damper

Fig. 7.22 Push rod front axle with horizontal torsion bar of a Formula E vehicle

Figure 7.22 shows the front axle of a Formula E vehicle. The horizontal torsion bars are not directly visible, as they are each guided through a hollow shaft. The rocker is mounted on this hollow shaft and is operated by a push rod. The hollow shaft is rotatably mounted at both ends via roller bearings. At the front end, the hollow shaft receives the torsion bar via a serration. At this end, the torsion bar and hollow shaft have the same torsion angle. At the opposite end, the second torsion bar end is firmly clamped. For this purpose, a stationary bearing is screwed against the structure, which receives this torsion bar end with a serration. The torsion bar is thus twisted by a rotation of the hollow shaft. The bearing of the hollow shaft keeps bending loads away from the torsion bar. In some applications, the bearing with the fixed end of the torsion bar is provided with an adjustment mechanism so that this bearing can be externally rotated to adjust the ground clearance. Such a "Ride Height Adjuster" can be seen in Fig. 7.39.

7.3.4 Stabilizers

The basic function of the stabilizer has already been explained in Sect. 5.3. Figure 7.23a shows again the principle illustrated there. The stabilizer is designed as a torsion spring and has a bend with holes at both ends. The connecting rods are accommodated at these bores. The stabilizer bearings, in which the stabilizer is rotatably mounted, are attached to the body. The connecting rods are attached to the wheel carrier, to a link or to the spring/ damper element (see also Fig. 7.1). This is the typical design of anti-roll bar for production

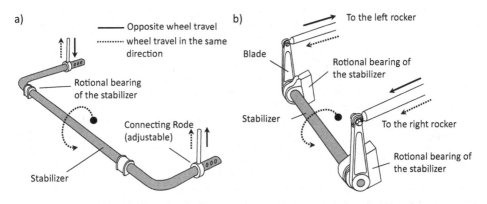

Fig. 7.23 Construction methods of U-shaped stabilizers (**a**) in production passenger cars or near-production racing vehicles and (**b**) in racing vehicles

passenger cars and near-production race cars. When both springs moves in the same direction, the connecting rods perform the same vertical movement and the anti-roll bar rotates without resistance in the anti-roll bar bearings. A spring movement of the wheels in the opposite direction, as occurs particular when cornering, also leads to a movement of the connecting rods in the opposite direction, and the two stabilizer ends are twisted by the connecting rods in opposite directions. This twisting is resisted by the stabilizer. The higher this resistance, the greater the roll stiffness of the vehicle. The roll stiffness of the stabilizer is determined on the one hand by the torsional stiffness of the torsion bar, and thus by the properties given in Eq. (7.12), and on the other hand by the lever arm which the connecting rods have around the torsion bar. The longer this lever arm, the lower the roll stiffness of the stabilizer. Another way of adjusting the roll stiffness is to use adjustable connecting rods (not shown), which can be varied in length. By changing the angular position, this leads to a change in the effective lever arm and thus the roll stiffness.

Figure 7.23b shows the typical design of a U-shaped stabilizer as used in connection with push-rod and pull-rod suspensions. In these cases, too, the stabilizer is designed as a torsion spring which is actuated by two coupling rods. The coupling rods are connected to the bell crank (see Fig. 7.21).

Another typical design for high-performance racing vehicles is the T-shaped stabilizer, as shown in Fig. 7.24 using the rear axle of a Formula E vehicle as an example. In the T-shaped arrangement, the anti-roll bar performs a tilting rotation in the event of a spring movement in the same direction, whereby the anti-roll bar is not twisted. A spring movement in the opposite direction leads to an opposite rotational movement of the rocker

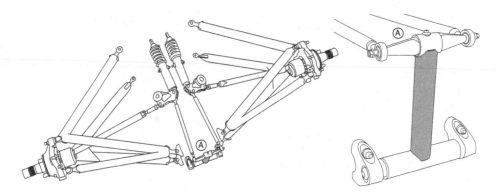

Fig. 7.24 Adjustable T-stabilizer on the push-rod rear axle of a Formula E vehicle

and thus to an opposite movement of the coupling rods. The stabilizer is twisted. The lever arms on the T-shaped stabilizer are subjected to bending stress during a movement in the opposite direction. The higher the bending stiffness of the lever arms, the higher the roll stiffness of the stabilizer. In this case, lever arms are used in which the bending stiffness can be adjusted by twisting the lever arm. The lever arm is rotatably mounted in its connection to the stabilizer and has an I-shaped profile. Rotating the lever arm changes the effective surface moments of inertia, which counteract the bending stress. In the "stiff" setting the high side counteracts the bending stress, in the "soft" setting the flat side. In Fig. 7.24, the stabilizer is thus in the "soft" setting. The deformation of the lever arm reduces the twist in the stabilizer and thus reduces the resulting roll stiffness of the corresponding axle. The U-shaped stabilizer in Fig. 7.22 also has such an adjustment option. There the stabilizer is in the "stiff" setting.

An alternative to the design of the stabilizer as a torsion bar is shown in Fig. 7.25. The function of the stabilizer is realized there by two bending arms coupled to each other. Both bending arms are attached to their respective bell rockers so that they cannot rotate. At their endings the bending arms are rotatably mounted in the coupling element. When the wheel moves in the same direction, the ends of the two bending arms move in the same direction. This movement is not opposed by any resistance. When the wheel moves in the opposite direction, one bending arm moves upwards and one downwards, which leads to a bending stress in the bending arms. The higher the bending stiffness, the greater the resistance to vehicle roll. Such a stabilizer construction was used among others in the Minardi PS01 (2001) and in the McLaren-Mercedes MP4–26. The advantage of this design is the more flexible package.

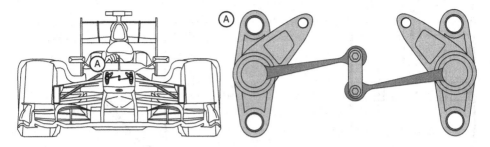

Fig. 7.25 Stabilizer with bending arms in a McLaren-Mercedes MP4–26 (2011)

7.4   Shock Absorbers/Dampers

7.4.1   The Function and Types of Telescopic Dampers

The suspension of a vehicle has the task of absorbing road unevenness and supporting the weight, inertia and downforce forces of the body. The energy used to deflect the spring is stored in the spring. An undamped system would release this energy in the form of a relatively long-lasting oscillation, which would lead to vibrations of the body and the wheel on a vehicle with corresponding wheel load fluctuations. The task of the shock absorber is to reduce the energy supplied to the spring and to dissipate the energy absorbed in the spring. In this way, the shock absorber as shown in Fig. 7.26 ensures that the vibration excited by a deflection of the spring decays more quickly. The spring is deflected by inertial forces and aerodynamic forces in addition to road unevenness. The damper provides controlled movement of the body and wheel and ensures the best possible contact between the tire and the road. In high performance race cars, the control of the body by the damper is closely related to the aerodynamic characteristics due to ground effect and pitch sensitivity. The excitations caused by ground unevenness, inertial forces and aerodynamic forces take place in different frequency ranges, which must be taken into account when designing the damper.

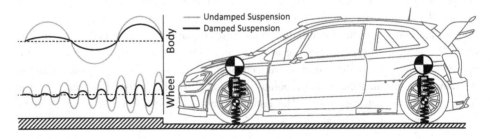

Fig. 7.26 Vibration of wheel and body masses

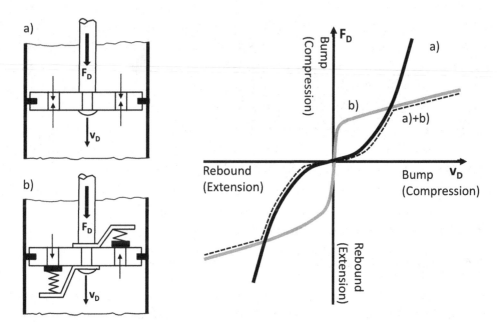

Fig. 7.27 Functional principle of hydraulic damping (**a**) with bleed holes and (**b**) with spring-loaded valves

Modern shock absorbers for road and racing vehicles all use the principle of hydraulic damping. The effect of hydraulic damping is based on the fact that when flowing through hydraulic lines, the flow energy is irreversibly converted into heat by friction and vortex processes. Figure 7.27 shows the operating principle of a hydraulic telescopic damper. A piston guided on a rod moves in a working chamber filled with hydraulic oil. During its movement, the piston must displace the hydraulic oil. In the example shown, the displacement takes place through holes in the piston. These holes represent the hydraulic lines.

Bores are distinguished between uncovered bores, also referred to as bleed holes or pre-opening cross-section, and spring-loaded valves. Both types of bores produce different damping characteristics [5]. The bleed holes produces a damping force that increases sharply with piston velocity. At very slow piston speed, the damping force is zero. As the speed increases, the viscosity of the oil makes it more difficult to displace the oil through the bores, and the force required increases quadratically with piston speed. The result is a sharp progressive increase in damping. When the damper compresses ("compression" or "bump") or the piston rod retracts, it is referred to as the compression stage of the damper, and when the piston rebound ("rebound") or the piston stages extend, it is referred to as the rebound stage. The compression and rebound stages generally have different damping force characteristics.

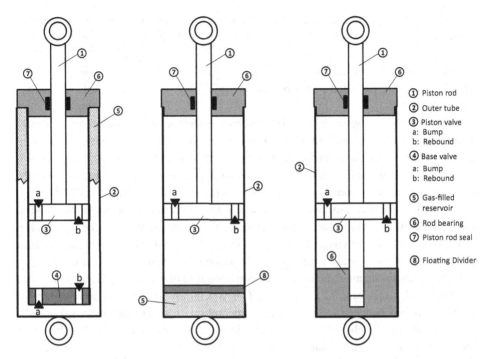

Fig. 7.28 Types of telescopic dampers: (**a**) Twin-tube damper, (**b**) Monotube damper, (**c**) Through-rod damper

Spring-loaded valves function like a pressure relief valve. By preloading the spring or a preload, a defined opening pressure can be set, from which the valve lifts off the bore. The further force-velocity curve, which is shown linearly at this point, depends on how the opening geometry changes over the valve stroke. The progression can be tuned from a degressive to a linear to a progressive characteristic. The opening of the spring-loaded valves marks the transition from low piston speeds ("low-speed" range) to high piston speeds ("high-speed" range).

The combination of bleed holes and spring valves results in the force-velocity characteristic of shock absorbers. To simplify, it can be said that bleed holes mainly influence damping in the "low-speed" range and thus control of body movement (or sprung masses) and dynamic wheel load changes, while spring valves are mainly responsible for damping in the "high-speed" range and thus wheel control (unsprung masses). The geometry of the bleed holes and the characteristics of the valve springs can be designed differently for compression and rebound. Since the "low-speed" and "high-speed" characteristics in rebound and compression can thus be designed independently of each other, this is also referred to as "four-way" damping. By varying the pre-opening cross-sections and different valve springs, the driving dynamic properties of a damper can be tuned over a wide range.

Figure 7.28 shows the types of telescopic dampers used in automotive engineering. A distinction is made between twin-tube, monotube and through-rod dampers. In the case of twin-tube and single-tube dampers, the retraction of the piston rod results in more oil having to be displaced below the piston than can be absorbed above the piston. This difference in volume, which just corresponds to the volume of the piston rod in the working chamber, must be compensated for by design. The difference in volume between the fully retracted and fully extended piston rod can be more than 10% of the working chamber volume [6].

In the *twin-tube damper*, the required volume compensation is achieved by a compensation chamber arranged concentrically around the working chamber, which gives the twin-tube damper its name. In the vertical position of the damper, the compensation chamber is filled approximately half with oil and half with a gas. The working chamber and the equalizing chamber are connected to each other via the base valve. The main disadvantage of the twin-tube damper is its limited inclined position, since at high inclination the oil level on one side drops sharply and thus gas may be sucked in through the base valve. Furthermore, it must be ensured that the working chamber is always completely filled with oil, even in the event of temperature-related fluctuations in the oil volume.

In the *monotube damper*, the volume differences due to ring volume and temperature are compensated by a pressurized gas volume. This is achieved by means of an impermeable separating piston between the gas-filled compensation chamber (gas reservoir) and the oil-filled working chamber, which are located in a common container tube. The damper is under permanent internal pressure, which causes the piston to extend in the unloaded state. In the simplest case, the characteristics of the rebound and compression stage are completely realized via the piston valves.

The main advantages of the monotube damper with pressure loading of the oil column are:

- Good cooling due to an external boundary of the working chamber
- Larger piston area for the same diameter (allowing lower internal pressures)
- Lower tendency to oil foaming or cavitation
- Freely selectable mounting position

The gas reservoir can also be arranged outside the housing, as shown in Fig. 7.29. This design is used for numerous applications in motor sports in order to create a quick adjustment possibility by using a valve in the supply line to the external gas reservoir and to reduce the necessary installation length. The operation of adjustable valves is explained in Sect. 7.4.3.

The distinguishing feature of the *through-rod damper* is a piston rod running through the entire working chamber. In the field of body damping, they are only used on vehicles with relatively small spring travels because otherwise the piston rod would have to be routed through both ends. Since the piston rod volume does not need to be balanced, through-rod dampers require only a small volume of gas to compensate for fluctuations in oil temperature. Combined with small spring travel, this gives package advantages over monotube and twin-tube dampers on race cars. The piston surfaces on the rebound and compression sides are identical, so that the internal pressure does not generate any

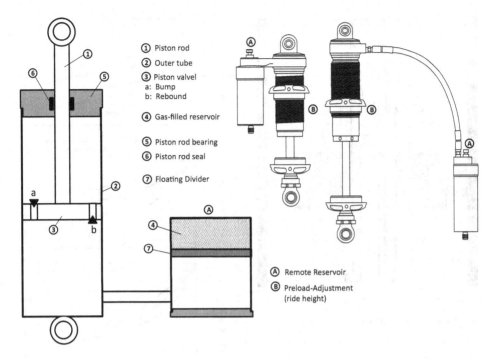

Fig. 7.29 Monotube damper with external reservoir

extension force. Temperature-related fluctuations in the filling pressure do not affect the damping behavior. The distance to support lateral forces and bending stresses remains constant. When using through-rod dampers, the friction caused by a concentrically arranged spring can thus be significantly reduced.

7.4.2 Valve Systems

Modern telescopic dampers for series production and motorsport applications use a combination of different valve systems, which are illustrated in Fig. 7.30 using the example of various piston valves. Accordingly, one can distinguish between three basic forms of bores or valves:

- Bleed holes
- Check valves
- Spring valves (pressure relief valve)

A needle valve in combination with a bore through the piston rod represents the bleed hole here. The opening cross-section can be varied by different positions or geometries of the needle valve as well as directly by the bore diameter. The bleed hole essentially determines the behavior at low piston speeds. The larger the bore cross-section, the lower the damping

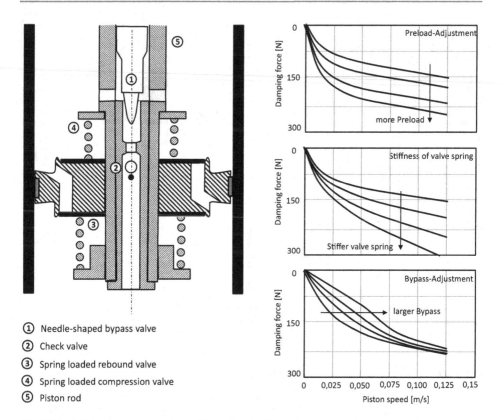

Fig. 7.30 Schematic representation of the valve setting options

① Needle-shaped bypass valve
② Check valve
③ Spring loaded rebound valve
④ Spring loaded compression valve
⑤ Piston rod

force. The damping behavior at high speeds is determined by the spring valves and is only influenced to a small extent by the bleed hole. A check valve ensures that the bleed hole integrated into the piston rod is only effective in the rebound stage. In the compression stage, this opening is closed. A separate bypass channel is therefore required for the compression stage.

In the example shown, the spring valves are equipped with helical springs. The characteristics of the spring valve are determined by the bore, spring preload force and spring stiffness. The rebound and compression stages are usually tuned differently. At low piston speeds, the spring valves remain closed. Only when at higher piston speeds the pressure acting on the valve surfaces overcomes the preload force does the valve open. This behavior is marked in the force-velocity diagram by the degressive buckling. This characteristic point of the damping curve represents the transition to the range of high piston speeds. Increasing the preload shifts this kink to higher damping forces. The slope of the force increase after overcoming the preload remains almost constant. In the range of low speeds, the properties of the damper are ideally only slightly affected. The stiffness of the valve spring determines the increase in the opening cross-section as a function of the

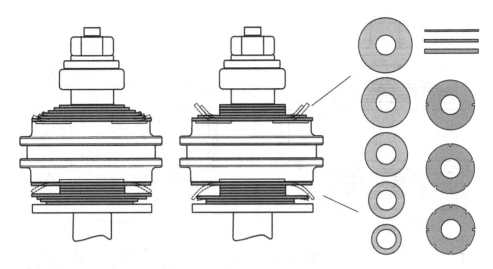

Fig. 7.31 Variation of the valve disc arrangement

pressure increase. The stiffer the valve spring, the smaller the increase in opening cross-section and the higher the damping force. In the force-velocity diagram, the slope of the damper characteristic curve is thus essentially determined by the valve spring at high piston speeds. The stiffer the valve spring, the higher the speed-dependent force increase and the steeper the slope of the characteristic curve.

Often the damper valves are not designed as helical springs, but as annular valve discs with defined elastic properties, as exemplified in Fig. 7.31. The valve disc here is an annular bending spring. The desired opening behavior is set by arranging different valve discs one above the other. The properties of the valve discs can additionally be varied by changing their thickness, diameter or shape.

The behavior of the damper can also be influenced by different designs of the bore geometry, as Fig. 7.32 shows using the example of various piston valves from the Penske company. In this case, the arrangement of the valve discs is identical for all three configurations. The behavior can be adjusted from degressive to linear to progressive.

The tuning of the driving behavior can therefore – given the design of the damper – be carried out by adjusting the following characteristics:

- Geometry of the bleed holes
- Geometry of the opening cross sections for the spring-loaded valves
- Preload on the valve springs
- Stiffness of the valve springs

These properties can often only be changed by time-consuming dismantling and disassembly of the damper with subsequent re-fitting. Adjusting the damper in the time-limited training sessions of a racing event is not practicable in this way. Mechanisms have

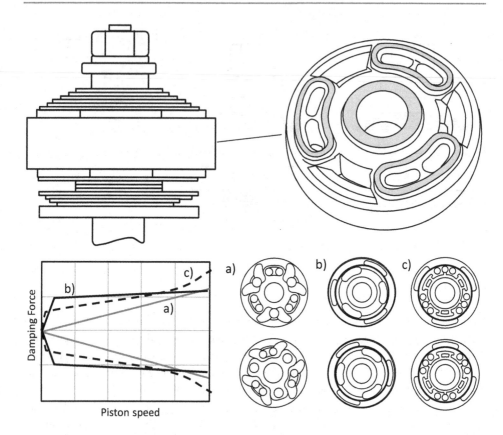

Fig. 7.32 Piston valve with valve discs

therefore been developed for changing the pre-opening cross-sections and the spring preload, which also allow the damper to be fine-tuned at short notice.

7.4.3   Shock Absorbers for Motor Sport Applications

For dampers used in motor sports, the general development goals of optimized size and low component weight as well as low friction and good response behavior can be formulated first. An optimized size and a flexible installation position increase the freedom in the aerodynamic design of the outer contour. To reduce the overall weight, which is only about 200–600 g for dampers for high-performance racing cars, many components are made of aluminum and titanium. The realization of a good response is a design challenge, especially for high performance race cars, because they have only small strokes. The total stroke of the telescopic dampers on a 2007 F1 car is only about 30–50 mm for aerodynamic reasons. At the rear axle, the vehicle deflects up to 15 mm at top speed due to aerodynamic downforce. The dynamic strokes during driving are even only ±3 mm at the front axle

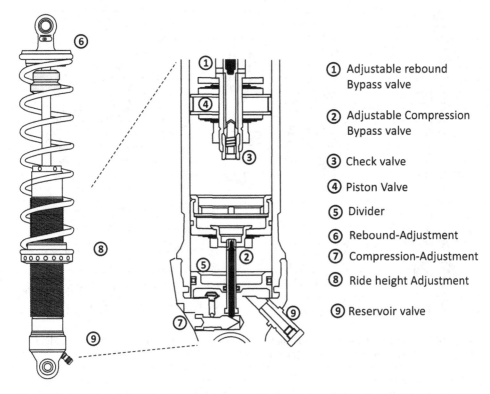

Fig. 7.33 Double adjustable monotube damper with base valve. (According to documents of the company Penske Racing Shocks)

and ± 6 mm at the rear axle [7]. In addition to these development goals and the demand for high reliability by avoiding cavitation phenomena, the adjustability of a damper in particular plays a decisive role. In this context, adjustability means being able to change the damper's characteristics quickly from the outside without having to disassemble and reload the damper in a time-consuming procedure. Such adjustability allows the tuning of the vehicle to be adapted to the current conditions at short notice.

Figure 7.33 shows the PS-7500-DA ("Double Adjustable") monotube damper from Penske Racing Shocks. Double adjustable in this case means that the low-speed characteristics of the damper can be adjusted independently for compression and rebound. The high-speed characteristics of the rebound and compression stages are determined by the configuration of the spring-loaded piston valves. In the rebound stage, adjustment is made via a needle valve incorporated in the piston rod, which can be moved up and down via a thread and an adjustment screw, allowing the opening cross-section to be varied. In the PS-7500-DA there is an additional base valve which is responsible for the adjustment in the compression stage. The volume displaced by the retracting piston rod flows through the bottom valve. The opening cross-section thus determines how quickly this volume can be compensated by the separating piston and gas reservoir. The smaller the opening cross-

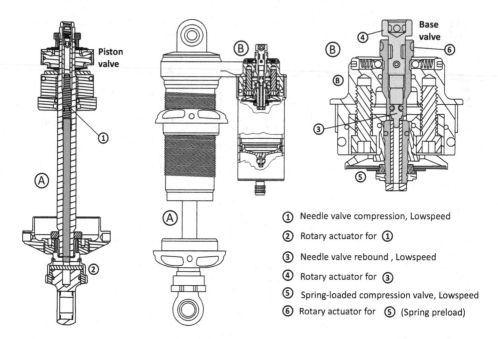

1. Needle valve compression, Lowspeed
2. Rotary actuator for ①
3. Needle valve rebound , Lowspeed
4. Rotary actuator for ③
5. Spring-loaded compression valve, Lowspeed
6. Rotary actuator for ⑤ (Spring preload)

Fig. 7.34 Triple adjustable monotube shock absorber with external reservoir. (According to documents of the company Penske Racing Shocks)

section, the higher the damping. The bleed valve is also realized by an adjustable needle valve, which allows 40 different adjustment positions. The independence of rebound and compression damping is ensured by a check valve on the piston side. In the high-speed range of the compression stage, the piston rod volume additionally flows through a non-adjustable disc spring valve. The basic tuning of the damper must be done via the classic piston valves, the adjustment of the needle valves should only be used for fine tuning of the vehicle.

Figure 7.34 shows the PS-8760 monotube damper with external reservoir from Penske Racing Shocks. This damper allows a triple adjustment. The low-speed characteristics of the rebound and compression damping are also realized with this damper by adjustable needle valves. The needle valve of the rebound stage is located in the hollow-bored piston rod just like on the PS-7500-DA. However, the bottom valve on this damper is located in the external gas reservoir, which allows for a more elaborate valve design without increasing the overall length of the damper. The bottom valve is designed to increase the preload on the valve springs, which allows for a change in the high-speed characteristics in the compression stage of the damper, making it the third adjustment option.

ZF Sachs Race Engineering GmbH has developed the Formula Matrix damper for high-performance racing vehicles. The functional principle and valve system of this damper are shown in Figs. 7.35 and 7.36. The Formula Matrix damper is offered in different variants:

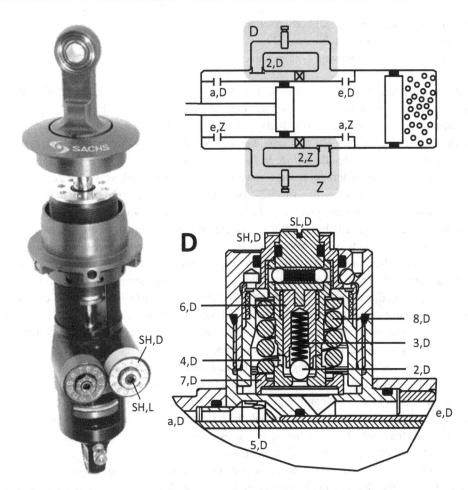

Fig. 7.35 Formula matrix dampers from ZF Sachs. (Courtesy of © ZF Friedrichshafen AG 2018. All Rights Reserved)

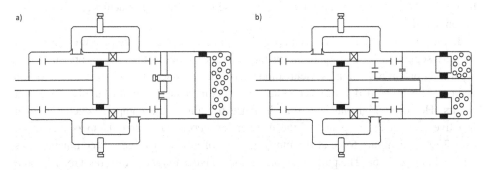

Fig. 7.36 Formula Matrix damper (**a**) as a monotube damper with bbase valve and (**b**) as a through-rod damper with base valve. (Courtesy of © ZF Friedrichshafen AG 2018. All Rights Reserved)

- Monotube damper without bottom valve
- Monotube damper with bottom valve
- Through-Rod damper with bottom valve

All variants feature a twin-tube design with two external valves, which allow a four-fold power adjustment (from product brochure: Formula Matrix damper). The twin-tube design creates a channel system in which the oil flow in compression and rebound can be influenced independently. In this case the damper piston can be designed as a pure displacer or with a Belleville spring. In the compression stage, the oil is forced through the bore eD via the compression stage valve D and the open check valve 2D through the bore aD. The closed non-return valve 2Z prevents oil flow via the rebound stage valve Z. The displaced oil initially presses against the ball 2D, and as soon as the spring force of the spring 3D is exceeded, the ball lifts off and the oil flows through the bores 4D as well as 5D to the opposite side of the working chamber. When the piston reverses direction, the ball acts as a check valve. At slow piston speeds, the oil flow takes place completely through bore 4D. The opening cross-section of this bore can be regulated by turning the slide 6D. The adjustment is made at the associated screw SLD. There are 16 different settings possible which influence the low-speed behavior of the damper. At higher piston speeds and increasing pressures, in addition to the ball, the valve body 7D, which is preloaded by the helical spring 8D, lifts off and also releases the oil flow through the bore 5D. The preload of the spring 8D can be varied in 12 steps by means of the adjusting screw SDH, which allows the high-speed characteristics of the damper to be tuned.

In the rebound stage, the conditions are reversed. The check valve 2D is now closed and the oil is displaced via the bore eZ through the rebound valve Z and the open check valve 2Z through the bore aZ. The rebound and compression stages can be adjusted independently of each other, so that the aforementioned four-fold power adjustment is possible. This basic principle is identical for all variants of the Formula Matrix damper. The main advantage of the double-tube arrangement is that the displaced oil volume practically always flows completely through the valves. With the previously shown dampers, only the piston rod volume flows through the bottom valve in the compression stage. Due to the higher flow rate, a very precise and fast response of the damper is achieved even at low piston speeds.

In the conventional monotube damper, the working pressure in the pressure stage must be fully supported by the separating piston and gas volume. At very high pressures, the separating piston may initially yield somewhat, which worsens the initial response of the damper. This is avoided if the gas pressure is greater than the maximum damping force that occurs. However, high gas pressures increase friction in the damper, which can then negatively affect the response of the damper. For Formula Matrix dampers that reach over 2000 N damping force at 0.5 mm/s, the use of a bottom valve is recommended, as outlined in Fig. 7.36a. The purpose of the bottom valve is to relieve the pressure generated in the damper such that the separating piston is displaced only by the displaced oil volume of the piston rod. The gas pressure can thus be kept smaller and the damper responds well even at high forces.

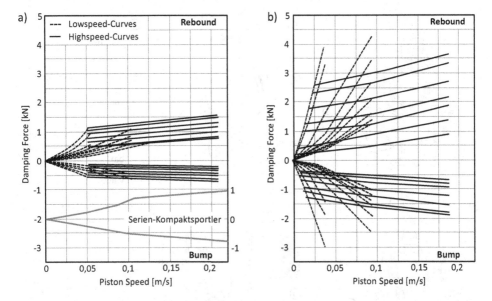

Fig. 7.37 Examples of characteristic variation of a Formula Matrix damper (**a**) with large bleed valve cross-section and (**b**) with small bleed valve cross-section. (Courtesy of © ZF Friedrichshafen AG 2018. All Rights Reserved)

In Fig. 7.36b the Formula Matrix damper is designed as a through-rod damper with bottom valve. Since there is no difference in displaced volume between rebound and compression in this variant, the gas reservoir is only used to compensate for temperature fluctuations. Therefore only a low gas pressure is needed, which reduces friction and results in a lower temperature sensitivity. During compression, the separating piston is not displaced and the damper can be operated with higher damper capacities.

Formula Matrix dampers are used in Formula 3, GT vehicles, various touring car series, sports car prototypes, IndyCar and Formula 1, among others. A wide portfolio of pistons, valve slider geometries, valve discs and springs is offered for equipping the Formula Matrix damper, so that the damping characteristics can be optimally matched to the respective application. Figure 7.37 shows two setting examples for the Formula Matrix damper in the form of the characteristic curve sheets. Note: For the determination of the actual damping forces at the wheel the damper ratio has to be considered. The strongly different damping level of the variants a and b results from a different pre-opening cross-section in the piston. The smaller the cross-sectional area of the piston bore, the higher the basic damping level. The adjustment of the valve spool then produces the characteristic curves shown for the low-speed range. For the adjustment of the high-speed range, the preload of the valve spring is adjusted, resulting in the characteristic curve of the high-speed range. The slope in the high-speed range can only be changed by using a different valve spring. Essential characteristics of the damper, such as basic damping and gradient, must already be determined in the basic tuning, so that the fine tuning of the damper can

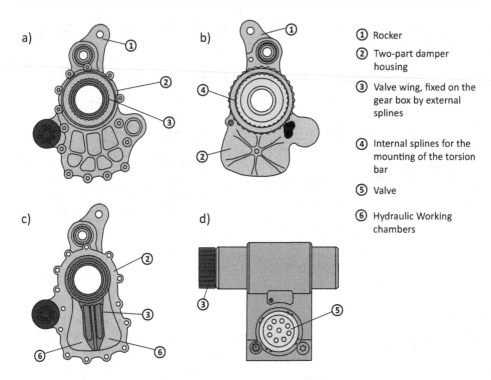

①	Rocker
②	Two-part damper housing
③	Valve wing, fixed on the gear box by external splines
④	Internal splines for the mounting of the torsion bar
⑤	Valve
⑥	Hydraulic Working chambers

Fig. 7.38 Rotary damper in view (**a**) from the front, (**b**) from the rear and (**c**) from the front without housing cover, (**d**) valve blade

take place via the given adjustment possibilities. In Fig. 7.37a the damping characteristic curve of a standard compact sports car is entered in grey, whereby the original characteristic curve sheet has a piston speed range of up to 1 m/s.

7.4.4   Rotary Dampers

For the 2003 season, Ferrari used a rotary damper developed by ZF Sachs on the rear axle for the first time. Two rotational dampers each replaced the classic telescopic dampers, with the rotational damper also taking over the function of the rocker.

Figure 7.38 illustrates the function of the rotary damper. It essentially consists of the two-part damper housing and the valve vane. The valve vane is flange-mounted on the gearbox via external serration so that it cannot move. The piston valves are located in its vane. The two-piece housing is rotatably mounted on the valve vane. The housing contains the working chambers and the receptacles for the suspension elements to be controlled. Also incorporated in the housing is an internal spline which receives the torsion bar. By rotating the damper housing around the valve vane, the volume of the working chambers changes and ensures a compensating flow through the piston valve. The basic mode of operation is therefore the same as that of a telescopic damper. Since the valve vane is always fully seated in the housing, the oil volume from both working chambers is constant and no balancing volume is required.

Due to the high degree of functional integration, the chassis design with a rotaryl damper builds lighter than a classic rear axle suspension with telescopic dampers. The use of a rotational damper reduces the number of bearings required to support the damper, which increases the overall rigidity of the system and thus improves the response of the chassis to high-frequency excitations. However, the central advantage of a rotational damper is the small installation space it requires, which allows for an aerodynamically optimized package in front of the rear wing. The disadvantages of the rotary damper are the small working travel and the relatively high friction, which worsens the overall response. The high friction results from the compact design because higher pressures are required to generate corresponding damping forces due to the small piston area. For this reason, the seal must be pressed against the housing with a relatively high preload, which leads to a corresponding increase in friction. The rotary damper does not offer an externally accessible adjustment option. The cost of a rotational damper is about 15,000 euros per piece [8], which is a not insignificant cost factor, especially for smaller teams. Besides Ferrari, only a few teams, including Toyota and Brawn GP, have used a rotational damper. However, Formula 1 has since returned to telescopic dampers. Outside of Formula 1, rotational dampers were used on the rear axle of the Peugeot 908 HDi FAP (LMP1), for example.

7.5   3Heave-Spring/Damper and FRIC Systems

Figure 7.39 shows schematically the front axle of modern Formula 1 cars (see also [9]). The double wishbone axle is designed as a push rod system with a horizontally positioned torsion bar (4). The torsion bar (4), the body dampers (6) and, via two connecting rods (8), the stabilizer (7) are actuated by the rocker (2). In addition, the front axle has a 3-spring damper element (3) also called heave spring damper, which consists of a heave spring, a heave damper and, if necessary, an inertia damper. These so-called heave-spring damper systems became established in Formula 1 after active suspension systems were banned for the 1994 season and are now also used in other high-performance racing cars. They offer further possibilities for controlling the movement of the body or its aerodynamic behavior.

Figure 7.40 explains the basic operation of the heave-spring damper element. *Case a* shows the static equilibrium case. The distance between the two heave spring supports is d_a. During roll, shown as *case b* in Fig. 7.40b, the wheels travel in opposite directions. In this example, the right wheel mpves upwards and the left wheel moves downwards, which causes the pivot points on the rocker to move in the same direction. The distance between the two support points does not change ($d_b = d_a$). This means that there are no reaction forces in the heave spring, heave damper and inertia damper. In this arrangement, the heave-spring damper system has no influence on roll stiffness and roll damping.

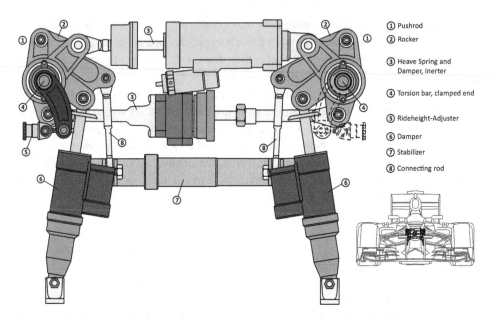

Fig. 7.39 Components (from cockpit view) of the front axle suspension of a modern F1 car

Case c shows the same directional deflection of the wheels, which occurs during pitching or due to the increase in aerodynamic forces above the driving speed. Due to the counter-rotational movement, the pivot points of the rocker move towards each other. The distance between the contact points decreases ($d_c < d_a$), resulting in a corresponding increase in spring and damper force. The heave-spring damper system thus increases vertical stiffness and vertical damping without affecting roll stiffness and roll damping. It thus has the reverse effect of a stabilizer. When using the heave spring, a progressive additional spring can be used in the same way as the body spring. The insertion point of the heave spring can be delayed by an idle travel. For example, lower spring stiffnesses can be used to optimize traction at low vehicle speeds. The vertical stiffness required as the downforce level increases is achieved by inserting the heave spring and the progressive auxiliary spring.

The rotational movements of the rocker caused by different wheel movements can also be used for the application of a roll damper. In this case, as shown in Fig. 7.41, a further pivot point is used below the torsion bar. During roll, the distance between the two articulation points of the roll damper increases and a damping force is created which counteracts the roll movement. In contrast, the distance between the two pivot points remains constant when the spring compresses or extends in the same direction. In this way, the roll damping is increased without affecting the vertical damping. A roll damper of this type was used, for example, on the Lotus E21 in the 2013 season, which could be seen from an asymmetrical bulge for the rocker on the top of the cockpit.

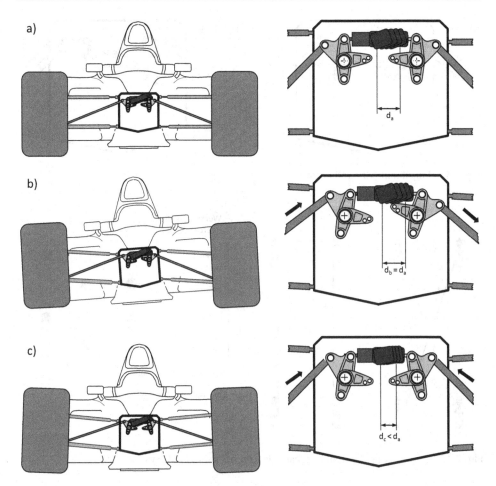

Fig. 7.40 Kinematics of the heave-spring damper element: (**a**) in static equilibrium, (**b**) when rolling and (**c**) when pitching or compressing in the same direction

Another way of influencing the body's heave, roll and pitch movements largely independently of one another is offered by the so-called front-rear interconnected systems (FRIC for short). In FRIC systems, the front and rear body dampers are interconnected via additional lines and valves. A distinction is made between a left-right-left-right and a left-left-right-right connection. Both forms are summarized in Fig. 7.42.

The operating principle for influencing the various forms of movement of the body is shown in Fig. 7.43 for a left-right-left-right connection.

A FRIC system can be operated in two modes. In the first mode, the pressure chambers of the opposing front and rear axle dampers are connected in *parallel. As* shown in Fig. 7.43, this results in the oil flowing out of a pressurized chamber into a depressurized chamber during roll. This compensating movement takes place without the build-up of additional damping forces. Also during pitching, the body dampers on the front and rear axle behave in opposite directions, so that with a parallel connection, the pitch damping

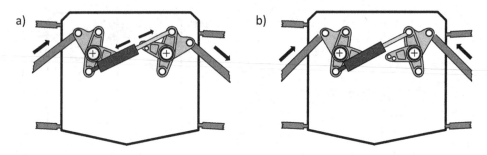

Fig. 7.41 Functioning of a roll damper

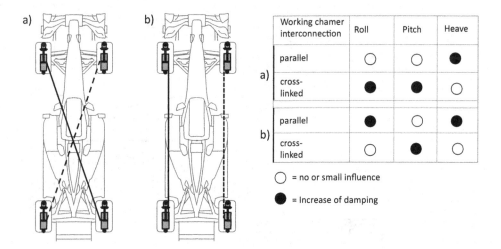

	Working chamer interconnection	Roll	Pitch	Heave
a)	parallel	○	○	●
	cross-linked	●	●	○
b)	parallel	●	○	●
	cross-linked	○	●	○

○ = no or small influence

● = Increase of damping

Fig. 7.42 Cross-linking the front and rear axle dampers using FRIC systems

also remains largely unaffected. In contrast, during compression in the same direction, which in this case corresponds to a pure lifting movement of the body, both connected chambers are under pressure. Therefore, no volume compensation can take place via the connecting line, and the vertical damping increases.

In the case of a *crossover connection, the* conditions are reversed, as the opposing dampers are each connected to each other at different chambers. In this case, volume compensation is prevented during roll, which is accompanied by a corresponding increase in roll damping. Even when the vehicle is pitching, two pressurized chambers are connected in this way, so that a crossover connection also increases pitch damping. In contrast, compression in the same direction is not additionally damped.

These interactions are summarized in tabular form in Fig. 7.42. Analogously, these cases can be derived for the left-left-right-right system (cf. Fig. 5.24). FRIC systems were banned in Formula 1 for the 2015 season. The Audi R18 (LMP1-H) is said to have used such a system in the 2015 24 Hours of Le Mans. However, the coupling of the dampers

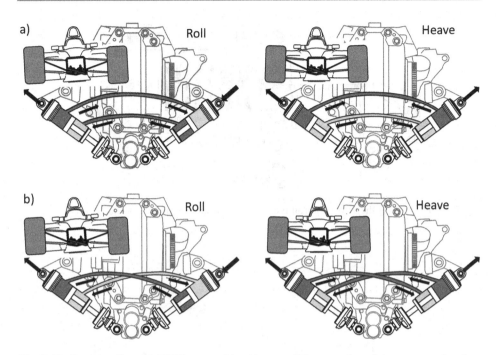

Fig. 7.43 Functionality of a FRIC system (**a**) with a parallel connection of the pressure chambers and (**b**) with a cross connection of the pressure chambers

does not necessarily have to take place between the front and rear axles. Coupling of the opposite wheels of an axle is also possible. However, the movements of the body are not only determined by the suspension, damping and stabilization of the vehicle, but are already influenced in a significant way by the geometry of the wheel suspension. This is explained in Sect. 7.7.

Figure 7.44 shows another heave spring and roll damping system, which was used in a comparable form in the Porsche 919 [10]. The system is based on a push rod arrangement in which the stabilizer is actuated via a Watt linkage and a central rocker. When the outer rockers move in the opposite direction – i.e. when both wheels travels in the same direction – the central rocker rests. The two connecting rods cause a rotary movement of the Watt linkage, whereby the heave spring and heave damper are actuated. The movement of the outer linkage in the same direction, as occurs when the body is rocking, causes the central linkage to rotate due to a displacement of the pivot bearing on the Watt linkage. This rotation causes the stabilizer and the roll damper to be actuated. Porsche's solution dispenses with the usual symmetrically arranged body dampers. These two dampers always act on both heave and roll movements. With the Porsche solution, on the other hand, both forms of movement can be damped completely independently of each other, which creates further freedom in chassis tuning. Since the roll inertia of a vehicle is usually significantly lower than its pitch inertia, the roll motion is often overdamped. The suspension design

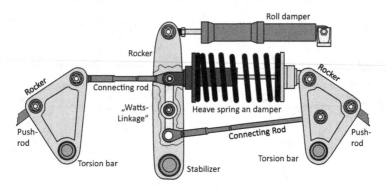

Fig. 7.44 Decoupling the heave and roll motion on the front axle of the Porsche 919

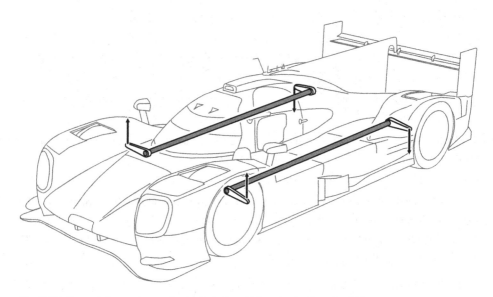

Fig. 7.45 Potential use of a mechanical pitch stabilizer in the Porsche 919

shown therefore uses only two dampers, which provides an additional weight advantage. This approach is also possible for the vertical and roll stiffness if the mounting of the rockers by torsion springs can be dispensed with.

To further decouple the heaving, rolling and pitching motion, a mechanical pitch stabilizer was considered to be used in the Porsche 919, the operating principle of which is outlined in Fig. 7.45. Here, the front and rear axles of one side of the vehicle are each coupled to each other via a torsion bar. The mode of operation is identical to that of a classic stabilizer. The movements of the individual wheels are transmitted to the lever arms of the pitch stabilizer via a connecting rod (not shown in the picture). If, as in the case of braking, the front wheels move upwards and the rear wheels moves downwards, then the movement of the lever arms in opposite directions leads to a twisting of the torsion bar, which in turn provides resistance to the pitching movement. Strictly speaking, therefore, this is a mechanical FRIC system. To optimize weight, the torsion bar in the Porsche 959 were designed as carbon-fiber-reinforced hollow plastic profiles, which would have resulted in a weight advantage over a hydraulic solution.

7.6   Inertia Dampers

Inertia dampers are a relatively new element of the chassis, which is used exclusively in racing cars. The inertia damper was developed in the early 2000s as part of a cooperation between the McLaren F1 team and the University of Cambridge. It was first used in 2005 in the McLaren-Mercedes MP4/20 at the Spanish Grand Prix in Barcelona.

Figure 7.46 shows the structure of an inertia damper. This essentially consists of three elements: Housing, threaded rod and ball nut with rotating body. The housing and the threaded rod are each connected, like the heave spring and heave damper, to one of the two rockers. The threaded rod retracts into the housing when the spring moves in the same direction. The ball nut converts the translational movement of the threaded rod into a rotational movement of the rotating body, which is mounted with low friction in the housing for this purpose. The rotational inertia of the rotating body opposes the movement of the threaded rod with a resisting force proportional to the acceleration. The moment of inertia can be used in particular to dampen high-frequency vibrations, such as those that occur when driving over curbs.

7.7   Chassis Geometry

The suspension transmits the longitudinal, lateral and wheel contact forces acting on the tires to the body. Due to the vertical degree of freedom and the required compliance of the body springs, these forces as well as the inertia forces arising on the body – caused by

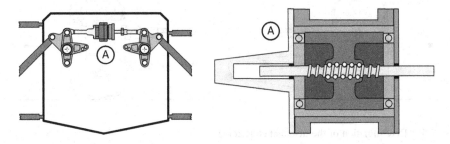

Fig. 7.46 Functional principle of an inertia damper (inerter)

deflection and rebound movements – cause a relative movement between wheels and body. The changes in the spatial wheel position caused by spring compression and extension movements, described by toe-in and camber, among other things, are referred to as chassis kinematics. This dependence of the wheel position on the spring travel is described by the so-called wheel lift curves. The forces acting on the body springs and thus also the movement of the body are decisively influenced by the geometry of the suspension. These relationships are discussed in this section.

7.7.1 Instantaneous Center and Roll Axis

The first characteristic properties of the chassis to be considered are the roll centers at the front and rear axles and the resulting rolling axis of the vehicle. The roll center describes the point around which the body rotates in its current position. Figure 7.47 shows how the position of the roll center on a double wishbone axle is determined by the geometry of the control arms. The body, control arms, wheel carrier and wheels, and roadway are considered here as an system of rods and joints. In the first step, the instantaneous centers of the two wheel carriers are determined in relation to the body. The term instantaneous center originates from mechanics. The instantaneous center determines the point around which the motion of a body can be represented as a pure rotary motion. In the double wishbone axle shown here, the instantaneous center of a wheel carrier results from the intersection points of the extended connecting lines of the wishbone bearings. In the second step, the wheel contact points are considered as bearings between the road surface and the wheel or wheel

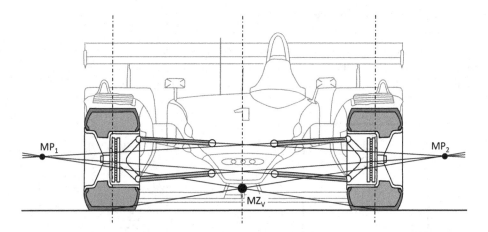

Fig. 7.47 Determination of the instantaneous center

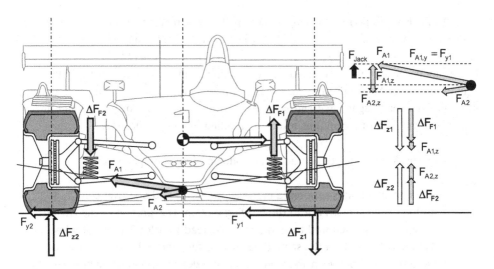

Fig. 7.48 Supporting the wishbone forces

carrier. The connecting line between the instantaneous center belonging to the wheel and its contact point represents a virtual control arm. The intersection points of the extended connecting lines form the roll center of the axle. The body rotates around this point. As the position of the wheel and the wishbones change during compression and extension, the position of theroll center also changes due to compression and extension movements.

Figure 7.48 shows how the geometry of the control arms affects the support of the body forces. For simplicity, it is postulated that the body springs are actuated directly by the lower wishbones and that the vertically arranged spring only absorbs vertical forces and no lateral forces. Thus, the lateral force acting on the respective wheel is transmitted exclusively by the wishbones. The resulting forces F_{A1} and F_{A2}, which are absorbed by the two wishbones of the associated wheel, act on the connecting lines between the wheel contact points and the instantaneous centers. In addition to the lateral forces F_{y1} and F_{y2}, the wishbones absorb the vertical force components $F_{A1, z}$ and $F_{A2, z}$. These build up at the same time as the lateral forces, which is why they are also referred to as *direct wheel load transfer*.

The lateral forces and vertical force components give rise to the resulting wishbone forces F_{A1} and F_{A2}. The resulting wishbone forces act on the body side at the roll center and therefore have no leverage effect. They do not contribute to the body roll. Lateral forces of different magnitudes are applied to the wheels due to the dynamic wheel load changes, which results in different magnitudes of vertical force components acting from the control arms at the roll center. The difference between these forces represents "jacking force" F_{Jack}.

It causes the body to perform a heaving motion. The difference between the vertical force components in the control arms and the dynamic wheel load changes ΔF_{z1} and ΔF_{z2} is absorbed into the body springs, which is referred to as *indirect wheel load transfer* because it does not occur until the body moves. The resulting spring force changes (from the static equilibrium state) ΔF_{F1} and ΔF_{F2} cause a moment about the roll center, supporting the roll moment from centrifugal force and lever arm between the center of gravity and the roll center. The yielding of the springs results in a rolling motion of the body.

▶ The wheel load changes are therefore made up of two components:

- A proportion that is supported in the control arms and does not contribute to the rolling motion but to the heaving motion. This portion of the wheel load changes is referred to as direct wheel load transfer. The direct wheel load transfer starts at the same time as the lateral forces build up. It cannot be distributed between the front and rear axles by springs, dampers and stabilizers;
- A proportion which is absorbed in the body springs, stabilizers and dampers. Due to the resilience of the springs, stabilizers and vibration dampers, this proportion causes body roll. It is referred to as indirect wheel load transfer. Indirect wheel load transfer takes place only after the body has moved. By adjusting the stiffnesses and damping constants, this proportion of the wheel load changes can be distributed between the front and rear axles and used to tune the handling in the short term.

With regard to the position of the roll center, a distinction can be made between two extreme cases, which are shown in Fig. 7.49. In the first case, the roll center is at the level of the road surface, which means that the wishbones do not absorb any vertical forces. The wheel load changes are fully absorbed by the springs, stabilizers and dampers, which therefore have to support the entire roll moment. The high spring forces lead to a correspondingly high body roll. However, there is no supporting force or jacking force. The wishbone forces do not cause any heaving movement of the body.

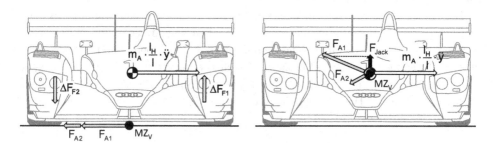

Fig. 7.49 Influence of instantaneous center on roll and jacking

In the second extreme case, the roll center (or roll axis) is at the height of the body center of gravity, which means that the inertia force has no lever arm around the roll center. In this case, the forces are supported solely by the control arms. No rolling motion occurs, but the axle exhibits an extreme jacking effect, the effect of which must be considered in conjunction with the kinematic behavior of the suspension.

The significance of the roll center for the kinematic behavior of the suspension is considered using three different variants of the double wishbone axle, which are shown in Fig. 7.50. *Variant a* has wishbones of equal length and parallel arrangement. The roll center of this variant is on the road surface in the static equilibrium case. The parallel and equal length wishbones cause the camber to remain constant during compression and extension. This means that the contact patch is not affected during high longitudinal accelerations and decelerations. The track width is reduced during compression and rebound. These lateral movements of the tire cause scrubbing over the road surface and increase tire wear. The dependence of wheel camber and track width (as well as other chassis parameters) on suspension travel is captured in wheel travel curves, such as those shown as examples in Fig. 7.51. Wheel travel curves show the relative change in suspension parameters to a fixed body. During roll, parallel control arms of equal length cause the respective camber angles at the wheels to be approximately equal in magnitude to the roll angle. (The wheel center lines remain parallel to the vehicle center line.) The result is a negative camber development at the inside wheel and a positive camber development at the outside wheel.

The actual camber angle at the wheel results from the sum (outside curve) or the difference (inside curve) between the roll angle and the relative camber angle to the body measured in the wheeltravel curve. The rolling motion of the vehicle therefore tends to result in a loss of camber. The positive camber development at the outer wheel of the curve reduces the transferable longitudinal and lateral forces (cf. Fig. 4.15). At the limit, however, the outer wheels are mainly responsible for the lateral control of the vehicle, as they have significantly higher wheel loads at high lateral accelerations than the inner wheels. At the outer wheel, the positive camber therefore has a significant negative influence on the adhesion potential of the axle. The negative camber at the inside wheel also reduces longitudinal and lateral forces, but due to the low wheel loads this effect plays a rather minor role. One of the most important tasks of the wheel suspension is therefore to prevent positive camber at the outside wheel of the curve.

Elasticities in chassis components and connections to the frame of the vehicle can be another cause of positive camber at the outside wheel of the curve. Figure 7.52 shows from the free section of a cornering outer rear wheel that the lower wishbone is loaded in compression and the upper wishbone in tension. Due to deformations of the control arms, the wheel carrier and the frame, the wheel thus turns into positive camber under the effect of a lateral force. This effect results in the demand for a high stiffness of the chassis components as well as the highest possible local body stiffness at the connection points to the chassis.

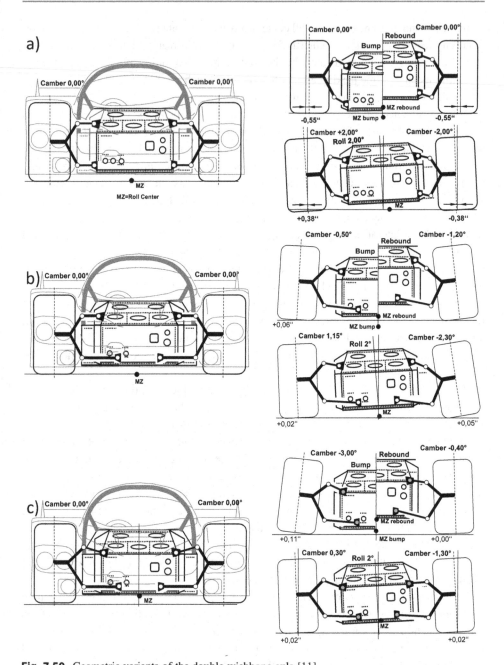

Fig. 7.50 Geometric variants of the double wishbone axle [11]

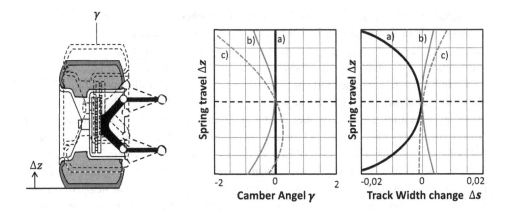

Fig. 7.51   Wheel travel curves for wheel camber and track width change

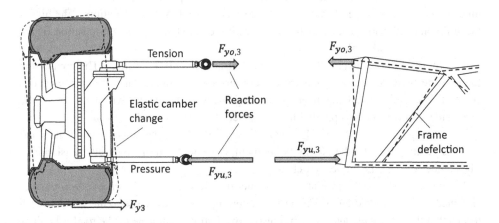

Fig. 7.52   Wishbone forces of a curve outer rear wheel

Positive camber at the outer wheel can be avoided or reduced if the wheels go into negative camber relative to the body during compression. Such a behavior is shown for example by the variants b and c from Fig. 7.50.

Variant b has parallel and unequal long wishbones. The roll center is also at the height of the road in this case. During compression, the short upper wishbone pulls the wheel into negative camber. This camber movement of the wheel simultaneously reduces the change in track width. When cornering, the negative camber reduces the camber loss resulting from the rolling motion. Compared to variant a, the positive camber at the wheel on the outside of the curve is now only 1.15 instead of 2°. On the other hand, the grip-reducing negative camber even increases on the inside wheel, which, however, as already described, is of rather minor importance due to the low dynamic wheel load. When rolling, the track changes with variant b are smaller. When driving straight ahead, however, the negative camber development reduces the transmissible longitudinal forces during braking and acceleration. The optimum compromise must be found for this conflict of objectives.

Variant c has non-parallel and unequal long handlebars. The roll center is now clearly above the road surface level. The adjustment of the upper short wishbone increases its effect in generating negative camber during compression. Due to the increased camber development, the change in track width increases accordingly compared to variant b. Likewise, the detrimental effect of camber during braking and acceleration is increased compared to variant b. However, it is advantageous that practically no positive camber occurs any more when rolling at the wheel on the outside of the bend. The negative camber at the inside wheel reduces the transferable longitudinal forces, but the lateral force potential increases. The effects on the inside wheel are of secondary importance.

By shifting the control arm points, an infinite number of variants of double wishbone geometries can be created, especially if one considers that only a simplified two-dimensional consideration was carried out at this point. An important – but not unrestrictedly valid – rule of thumb can be derived from these considerations: The higher the roll center, the lower the camber loss. In this context, however, the position of the instantaneous pole also plays a decisive role.

Figure 7.53a shows again the case of parallel and equally long links. The connecting lines of the links run parallel and therefore intersect at infinity. Since the instantaneous center lies at infinity, all the lines connecting the wheel and the wheel carrier to the instantaneous pole are also parallel to each other. This results in the instantaneous center lying on the plane of the road. All speed vectors are right-angles to connecting lines and thus are parallel to each other. So at the moment, the wheel will move purely translational. The camber of the wheel does not change.

Figure 7.53b shows the instantaneous center determination for the case of non-parallel and unequal-length handlebars. The instantaneous center is now close to the wheel, which means that the instantaneous center has also moved above the plane of the road. The lines connecting the instantaneous center to the wheel and wheel support open a fan. The horizontal speed components of the speed vectors standing perpendicular to the connecting lines show different amounts at the lower and upper wheel centers and in this case even opposite signs. Thus, the wheel performs a rotational motion relative to the body, resulting in a negative change in wheel camber during compression. In Fig. 7.51 it can be seen that this geometry already leads to a strong negative change in camber from the zero position. When the wheel is deflected, it initially goes into positive camber. Only with large deflection movements does the negative camber shown in Fig. 7.50c occur.

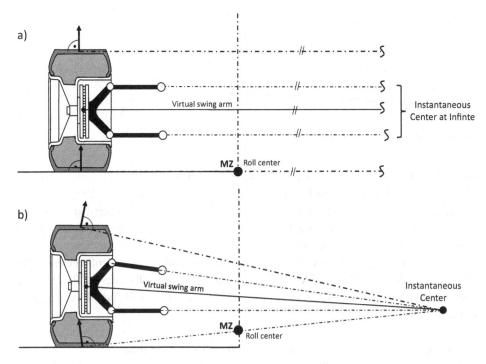

Fig. 7.53 Influence of the instantaneous center position on the camber change

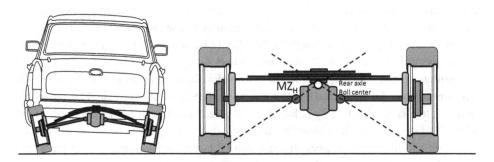

Fig. 7.54 "Tuct-under" effect on the pendulum arm axle of a Herald Triumph

If the roll center is very high, as was the case with the pendulum arm axle of the Herald Triumph (Fig. 7.54), two effects occur: Firstly, very high jacking forces occur when cornering, which push the body upwards relative to the chassis. At the same time, the

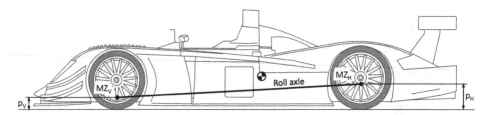

Fig. 7.55 Definition of the roll axis

high roll center causes a very high positive wheel camber during this deflection movement. The contact patch is pulled under the vehicle, which is also known as the "tuct-under" effect. The high camber results in a large reduction in the adhesion potential at the outside cornering wheel. Because this effect occurs on the rear axle, the vehicle tends to become unstable during braking and cornering.

For the driving dynamic properties of a vehicle, the position of the roll centers at the front and rear axle must be considered. The line connecting the roll centers of the front and rear axles forms the rolling axis shown in Fig. 7.55. Typical values for the roll center height of racing vehicles are:

- Front axle: −2.6 cm (below the road) to +5.6 cm (above the road)
- Rear axle: 7.6–15.3 cm (above the road surface)

These values generally result in a rolling axis that rises from front to rear. Assuming that the above rule of thumb is valid, this results in a setup that tends to understeer, because the camber loss on the rear axle is less than on the front axle. The adhesion potential of the rear axle is therefore reduced less than that of the front axle. It should be noted, however, that raising the roll centers increases the direct wheel load transfer, which may need to be compensated for by adjusting the roll rate distribution. In some modern sports cars, the roll axis is nearly parallel to the road due to a lowered roll center at the rear axle. The aim of this design is to increase traction by reducing direct wheel load transfer. The influences of the wheel load transfer and the kinematics overlap, which must be taken into account when designing and tuning the chassis. It must also be taken into account that the kinematic effects become weaker when the roll stiffness of the vehicle is increased. Under certain circumstances, this can lead to expected effects not occurring due to a changed roll rate distribution. The position of the roll axis additionally determines the lever arm h' and thus the amount of roll moment or the proportion of support provided by the control arms, as discussed earlier. Therefore, to avoid centrifugal force-induced roll, the roll axis (rather than a single roll center) must intersect the center of gravity. Which, however, brings with it the well-known disadvantages of the jacking effect.

In today's F1 cars, however, the geometry, especially that of the front axle, often has to submit to aerodynamic requirements, so that the so-called keel design has changed considerably over the last two decades. The associated development stages are summarized in Fig. 7.56. Until the early 1990s, a low-slung and relatively bulky nose, which divides the

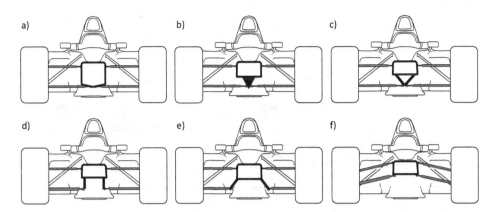

Fig. 7.56 Keel designs on F1 cars: (**a**) "early keel", (**b**) "single keel", (**c**) "V keel", (**d**) "twin keel", (**e**) "McLaren twin keel", (**f**) "zero keel"

front wing into two halves, was typical for Formula 1 cars. The upper and lower wishbones are therefore tied directly to the monocoque structure, resulting in a nearly parallel control arm arrangement. This classic design is known as the "early keel". Starting with the Benetton B191, the concept of a high nose and a continuous front wing began to gain acceptance from 1991 onwards. As a result, the monocoque structure is much higher and a more efficient use of the front wing is possible. In the middle below the monocoque is a rail to which the lower wishbones are connected. With this "single keel", the suspension geometry can be retained almost unchanged.

Optimizing the underbody and front wing inflow by minimizing or removing flow obstacles became even more significant with the 2001 season, as the efficiency of the front wing was significantly impaired by a repeated increase in the prescribed ground clearance. The volume of the nose has since been further reduced. The previous solid "single keel" can be designed as a permeable "V keel" for further flow optimization. The "V keel" can be dissolved, in particular to improve the flow to the underbody, so that the "twin keel" or, with a slightly modified geometry, the "McLaren twin keel" is created. The disadvantage of this design is that the struts to which the lower control arms are attached are subjected to bending stresses and must therefore be of correspondingly stiff dimensions. With the introduction of the "zero keel", as it is still used today in Formula 1, all chassis-related flow obstacles are consistently removed. By raising the control arms, the vortices generated at the end plate of the front wing, as well as the Y-200 vortex, can be directed under the chassis. Otherwise, the vortex hitting the lower control arm would result in a sensitive disturbance of the flow in this area [12]. The aerodynamic advantages of this concept outweigh the kinematic disadvantages of the high roll center.

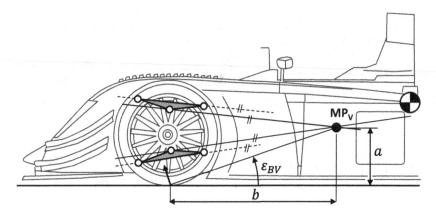

Fig. 7.57 Determining instantaneous center on the front axle

7.7.2 Pitch Center and Anti-geometries

During braking and acceleration, the chassis geometry also determines how the support of the inertial forces is distributed between the compliant springs and dampers and the control arms. The inertial force acting at the center of gravity causes a pitching motion that usually results in front axle compression during braking and rear axle compression during acceleration. In high performance race cars, these pitching motions cause a change in aerodynamic balance. For this reason, attempts are made to reduce the pitching movements by using anti-geometries. A suspension geometry that counteracts deflection under braking is called an anti-"dive" geometry. The deflection of the rear axle during acceleration is minimized by an anti-"squat" geometry.

The position of the instantaneous center in the x-z plane of the vehicle is decisive for the functioning of an anti-geometry. Figure 7.57 uses the example of a double wishbone front axle to show how the instantaneous center results from the control arm geometry. The instantaneous center is the point of intersection of the two straight lines running parallel to the axis of rotation of the wishbones and through the wheel carrier-side connections. The line connecting the instantaneous center and the wheel contact point includes the brake support angle ϵ_{BV} with the road surface. The arrangement of the control arms shown here already corresponds to an anti-"dive" geometry, which will be justified below. The brake support angle determines the direction of motion of the wheel contact point during compression. The velocity vector during compression is perpendicular to the line connecting the instantaneous center and the wheel contact point. The brake support angle of an anti-"dive" geometry causes the wheel to move forward during compression. This means that the wheel does not avoid a bump in the road by moving rearwards, but braces itself against it. Such behavior reduces the absorption capacity of the axle and is therefore a limiting factor for the design of an anti-dive geometry.

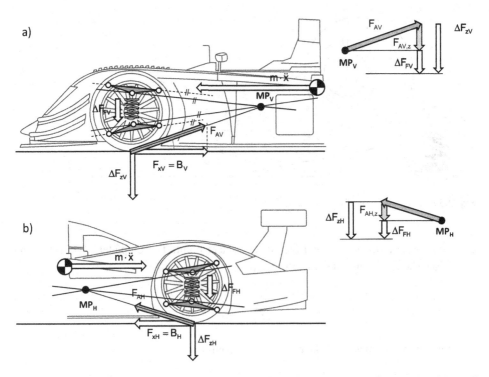

Fig. 7.58 Force support (**a**) by an anti-"dive" geometry at the front axle, (**b**) by an anti-"squat" geometry at the rear axle

Analogous to the procedure for the roll center, Fig. 7.58 considers the support of the forces by the two wishbones. Part a shows the anti-"dive" geometry on a front axle, part b the anti-"squat" geometry on a rear axle. In both cases, it is postulated that the body spring absorbs vertical forces only. The resulting wishbone force runs along the line connecting the instantaneous center and the wheel contact point. In the case of braking, the longitudinal component of the resulting wishbone force F_{xV} just corresponds to the braking force B_V on the front axle. At the same time, the wishbones absorb the vertical force component $F_{AV, z}$. This vertical component compensates proportionally for the dynamic wheel load change ΔF_{zV} arising on the front axle, and only the remaining proportion must be supported in the body spring. The proportional absorption of the dynamic wheel load changes in the control arms reduces the pitching motion of the vehicle.

The parallel and horizontal arrangement of the control arms represents an extreme case. In this configuration, the links cannot absorb any vertical force component and the inertial forces are fully supported by the spring. In this case, the anti-"dive" effect is 0%. For a complete support of the inertia force in the control arms or a 100% anti-dive effect, the resulting force vector of braking force and dynamic wheel load change must lie on the line

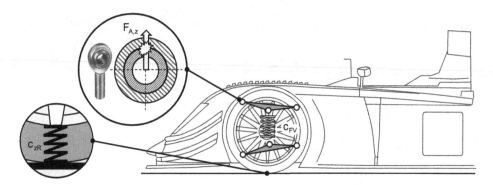

Fig. 7.59 Tire vertical stiffness and joint friction

connecting the moment pole and the wheel contact point. To completely avoid a spring deflection, the following must therefore apply:

$$\frac{B_V}{\Delta F_{zV}} = \frac{a}{b}.$$ (7.13)

The percentage compensation level is derived from this:

$$\text{Anti} - \text{"dive"} - \text{compensation level} := \frac{B_V}{\Delta F_{zV}} \cdot \frac{b}{a} \cdot 100\%$$ (7.14)

Figure 7.58b shows the force support of an anti-"squat" geometry, which works in exactly the same way as an anti-"dive" geometry. The only difference is that the resulting force vector now results from the dynamic wheel load changes and the driving force. The drive force is applied to the center of the wheel because the drive torque is supported by the driveshaft on the body, not the suspension. An anti-"squat" geometry requires a positive brake support angle. This is also advantageous for the rear axle's absorption capacity, because the wheels deflect backwards when the suspension compresses.

Two comments are important regarding the use of antigeometries: Firstly, the tire also has a vertical compliance. As shown in Fig. 7.59, the tire and the body spring represent a series connection. In high performance race cars, the tire vertical stiffness and the body spring stiffness are approximately equal. This graphically means that half of all body movements result from the compliance of the tire alone. Pitch and roll movements can thus never be completely suppressed, regardless of the position of the moment pole. Secondly, the vertical force components absorbed by the control arms lead to an increase in friction in the joints. Especially in vehicles with high braking and traction forces, this can lead to a significant increase in frictional forces. Frictional forces in the moving chassis components mean a delayed response of the suspension and damping and thus an increase

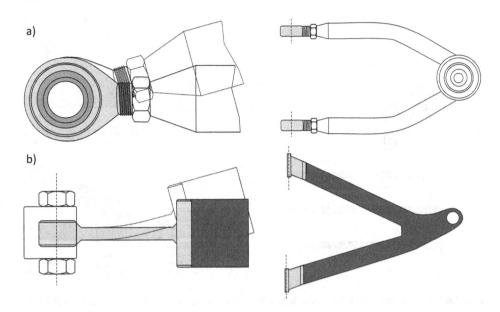

Fig. 7.60 Wishbone (**a**) with conventional bearings, (**b**) with bending elements

in wheel load fluctuations. Due to this effect and the kinematic interactions, the degrees of compensation of antigeometries are rarely more than 50%.

The wheel load fluctuations caused by joint friction can be significantly reduced if the ball or unibal or swivel joints usually used are replaced by flexure joints (see also Fig. 7.12). This is possible if only relatively small spring deflections have to be provided, which is especially true for vehicles with significant aerodynamic downforce. Figure 7.60 shows a wishbone that has conventional bearings and the wishbone of an F1 car that has flexure joints at its body-side ends. Such wishbones are manufactured, among others, by the company Kaiser Werkzeugbau GmbH, which indicates two possible implementations for the integration of these flexible elements. Firstly, rigid metal connections are glued to the ends (usually made of titanium), which ensure the spring travel through their defined bending. This possibility is sketched in Fig. 7.60b. In the alternative solution, these flexible areas are fully integrated and made of carbon, and only the bolting is represented by reinforcing plates.

7.7.3 Bump Steer

In racing, "bump steer" is understood to be the change in wheel steering angle or toe angle caused by wheel travel. For racing vehicles, the aim is often to achieve a small change in toe-in over the wheel travel. The influence of the suspension geometry on the "bump steer"

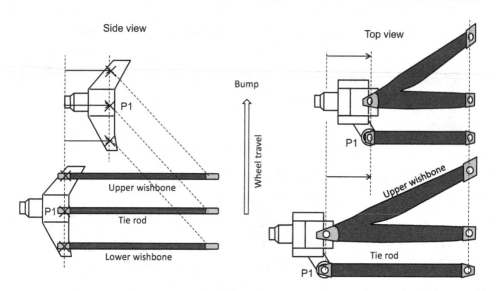

Fig. 7.61 "Bump-steer" behavior of a double wishbone suspension with parallel and equally long control arms

behavior is shown by the examples outlined in Figs. 7.61, 7.62, 7.63 and 7.64. In each case, a double wishbone axle with associated track rod is considered.

In the first case, it is a double wishbone suspension with equal length and parallel control arms. The track rod has the same length as the wishbones and is also arranged in parallel. In the side view, it can be seen that the joints on the wheel carrier side move on a circular path, the radius of which corresponds to the length of the control arm. Due to the parallel and equal-length arrangement, all wheel-side joint points move upward and inward by the same amount. The wheel experiences no change in camber. In the top view, it can be seen that the wheel carrier does not rotate due to the identical movement of the wheel-side wishbone bearings and the tie rod joint. The wheel steering angle or toe-in angle remains unchanged during compression.

In the second example, the tie rod is shortened and the pivot point on the wheel carrier is modified accordingly. The result of this modification is that the track rod joint on the wheel carrier side now moves on a smaller circular radius. For the same spring deflection movement, this bearing point therefore covers a longer distance inwards. The paths of the pivot points of the lower and upper wishbones remain unchanged. In the top view it can be seen that the wheel carrier is turned due to the movement of the outer track rod joint. The toe-in of the wheel changes due to the compression movement. In production passenger cars, this kinematic behavior is deliberately used to optimize the vehicle's stability by increasing the understeer tendency.

Changes in wheel position resulting from the geometry of the suspension are summarized under the term "kinematics". "Compliance" refers to changes in wheel

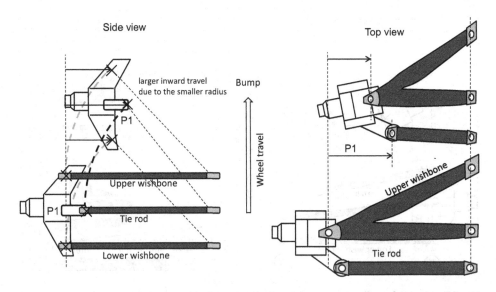

Fig. 7.62 "Bump-steer" behavior with shortened track rod

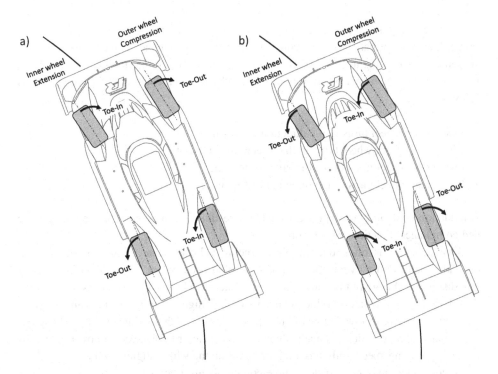

Fig. 7.63 Kinematic and elastokinematic wheel movements (**a**) to increase the understeer tendency or stability and (**b**) to reduce the understeer tendency or stability

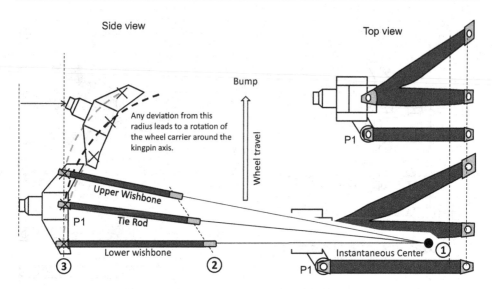

Fig. 7.64 General condition to avoid bump steer

position resulting from elastic deformations in the suspension components and on the frame. Elastic deformations occur primarily when longitudinal and lateral forces act on the suspension. According to Fig. 7.63, the understeer tendency or the stability of the vehicle increases when the wheel position is influenced by kinematic and compliance effects as follows:

- Inside front wheel turns in the direction of toe-in
- Curve outer front wheel turns in the direction of toe-out
- Inside rear wheel turns in the direction of toe-out
- Inside rear wheel turns in the direction of toe-in

A reduction in stability or understeer tendency results from a reversal of these kinematic and compliance directions of rotation.

Such behavior is often undesirable in racing vehicles. One reason for this is that one-sided deflection movements, such as those that occur when driving over curbs, lead to a disruptive change in the vehicle's course. Racing vehicles are more sensitive to such disturbances than production vehicles because of their high cornering stiffness and low yaw moments of inertia. A low degree of "bump steer" also has the advantage that a change in ground clearance, e.g. due to an adjustment of the setup or the increasing downforce with increasing driving speed, only has a minor effect on the wheel aligning values.

To completely avoid bump steer, three conditions must be met:

- The virtual extension of the tie rod intersects the instantaneous center of the wheel carrier.
- The inner tie rod joint lies on the connecting line of the inner wishbone bearings.
- The outer tie rod joint lies on the connecting line of the outer wishbone bearings.

This generalized case is shown in Fig. 7.64. It should be noted, however, that this statement is only sufficient for the two-dimensional consideration of the geometry carried out here.

7.8 Steering System

7.8.1 Steering Geometry

A central component of the chassis is the steering system. The steering system comprises the steering wheel, steering column and steering gear with the two track rods. To implement the steering function, the wheel carrier of the front axle of a double wishbone axle is rotatably mounted in the two wishbones. The connecting line of the wheel-side control arm bearings forms the kingpin axis. The wheel carrier rotates around the kingpin axis. The outer tie rod joint has a defined lever arm to the kingpin axis, so that a translatory movement of the tie rods leads to a rotation of the wheel carriers and thus to a steering movement of the wheels. This interaction is illustrated in Fig. 7.65 using the example of the steering system of an F1 vehicle. The steering gear generates the movement of the track rods by converting the rotational movement of the steering wheel and steering column into a translational movement of the rack and transmitting this to the inner track rod joint.

When steering, the wheel performs a three-dimensional movement, which means that in addition to the toe-in of the wheels, other kinematic variables such as camber change. This behavior depends on the steering geometry that is executed. The main geometric parameters of the steering system are also summarized in Fig. 7.65. The spatial position of the steering axis in relation to the vehicle and the wheel is determined by the kingpin inclination σ, the spindle length r_σ, the caster angle τ and the caster offset n_τ. These four parameters determine, among other things, the change in camber angle as a function of the steerangled wheels. To illustrate this, imagine a steering axle vertical to the road. In this case, the wheel rotates around the kingpin axis without changing the camber.

> The movement of the wheel can be represented by a bending straw. The long tube of the straw forms the axis of rotation, the short tube represents the distance from the kingpin axis to the center of the wheel.

The effect of the kingpin inclination is shown in Fig. 7.66. Increasing the inclination results in a positive camber angle on both wheels. At the same time, in conjunction with a positive spindle length, the center of both wheels moves downward, that is, the body lifts upward. The lifting increases with the length of the spindle. The negative effects of positive

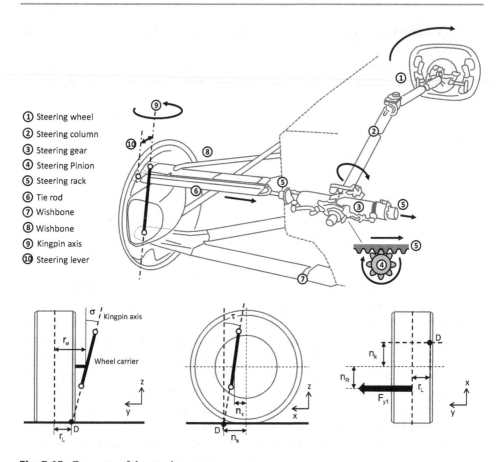

Fig. 7.65 Geometry of the steering system

camber on the outside wheel of the curve were explained in the previous sections. This effect is predominantly seen in tight curves as the camber change is relatively small. Raising the body changes the aerodynamic characteristics of the vehicle during steering. Due to the aforementioned effects, the kingpin inclination is kept as small as possible on racing vehicles.

As Fig. 7.67 shows, increasing the caster angle leads to a change in camber and height at the wheels in opposite directions. A negative camber occurs on the outer wheel of the curve and the body lowers on this side. The wheels on the inside of the bend go into positive camber, and the body is raised there. Lifting the body on the inside wheel, as described in Sect. 7.3.2, changes the diagonal weight distribution. The wheel load on the inside front wheel and on the outside rear wheel increase, which tends to lead to a stronger tendency to oversteer [13], since the dynamic wheel load changes on the front axle are reduced and increased on the rear axle. The negative camber created at the outside wheel during steering can be used to compensate for the positive camber caused by body roll and the kingpin inclination. This compensation is higher the larger the caster angle is selected.

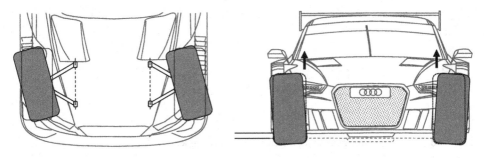

Fig. 7.66 Influence of the spread angle on camber and body position

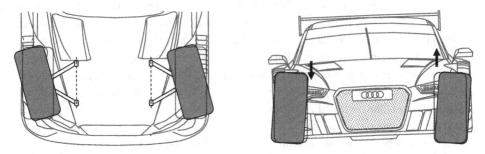

Fig. 7.67 Influence of the caster angle on camber and body position

The caster angle increases the mechnaical trail and thus the steering torques to be applied by the driver, since the caster attempts to return a wheel rolling under slip angle to straight-ahead position. It thus causes an automatic steering return and good straight-line stability. At point D the kingpin axis intersects the road surface. The distances from point D to the wheel center determine the mechanical trail n_K and the scrub radius r_L. The mechanical trail together with the pneumatic trail determines the moment effect of the lateral forces around the kingpin axis. The scrub radius determines the moment effect of the longitudinalforces around the kingpin axis. The moment effect of the scrub radius is also responsible for the occurrence of disturbances due to asymmetrical drive forces, such as those caused by bumps due to uneven surfaces. The aligning torque of the steering system results approximately to:

$$M_R = M_{L,Zst} + M_{Servo} = \left(F_{y1} + F_{y2}\right) \cdot \left(n_k + n_R\right) + \left(F_{x1} - F_{x2}\right) \cdot r_L. \tag{7.15}$$

To steer the wheels, a steering torque (converted to rack and pinion level) must be applied by the driver, if necessary with the assistance of a power-assistance torque, which corresponds to the amount of this restoring torque. Without power assistance, the mechanical trail is decisive for the force required at the steering wheel. With the use of power-

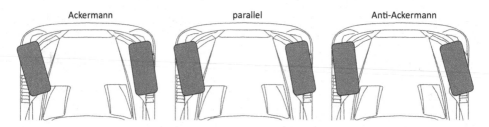

Fig. 7.68 Possibilities for generating differences in steer angles

assisted steering systems, effects that increase the steering force are no longer significant. Older Formula 1 and IndyCar cars, which do not have power steering, have caster angles of about 2°. Modern power-assisted Formula 1 cars have caster angles of 9–12° to take advantage of the compensating effects of the negative camber created at the outer wheel.

The geometry of the steering system also includes the arrangement of the kingpin axis, tie rods and steering gear in relation to each other. This arrangement determines the difference in steer angles when turning the front wheels. There are three possibilities for the course of out of the difference in steer angle, which are summarized in Fig. 7.68. The first possibility is the so-called Ackermann steering. Its design objective is not to generate slip at the front wheels when driving at low speed. To fulfil this condition, the longitudinal planes of both wheels must lie tangentially on their respective circular arcs. In this case, one also speaks of a 100% Ackermann design. Since the wheel on the inside of the curve always moves on a smaller arc than the wheel on the outside of the curve, the wheel on the inside of the curve must be turned in more than the wheel on the outside of the curve. The deviation from the ideal Ackermann angle is given as a percentage. A 90% Ackermann steering is 10% below the ideal Ackermann angle. An Ackermann design between 100 and 50% is common for production passenger cars. Among other things, this prevents increased tire wear during maneuvering.

Further design options are the realization of a parallel steering angle, in which both wheels are always turned to the same extent or the difference in steer angle is zero, or an anti-Ackermann steering system. In the latter case, the outer wheel is turned in more than the inner wheel. In addition to tire wear, the design of difference in steer angle has a noticeable influence on the maximum adhesion potential of the front axle.

The cause of this influence is explained in Fig. 7.69 using an Ackermann steering system. Figure 7.69 first shows how the steering components must be arranged in relation to each other in order to achieve an Ackermann geometry. Here, the lever betwenn wheel carrier and track rod runs along the connecting line between the king pin axis (at the height of the track rod) and the center of the rear axle. The track rod and rack are in line. The translatory movement of the track rod thus leads to a higher angular change at the wheel on the inside of the curve than at the wheel on the outside of the curve. This is explained by the fact that the greatest angular change is achieved by the track rod when the track rod is at

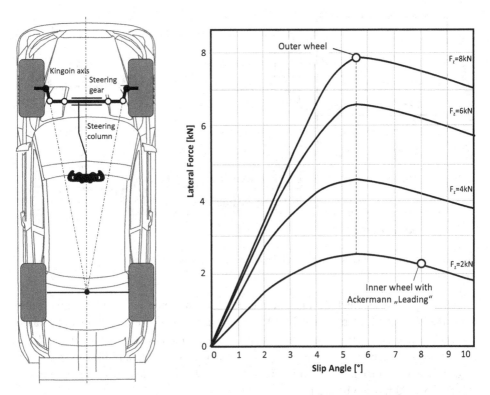

Fig. 7.69 Steering geometry of an Ackermann steering system and its influence on the adhesion potential of the front axle

right angles to the lever. In the case shown, the inner wheel is moving towards this state, while at the outer wheel it is moving away from this state.

The slip angle lateral force diagram in Fig. 7.69 illustrates the driving dynamic disadvantage of an Ackermann steering system. In this example, the outer wheel reaches the maximum transmittable lateral force at about 5.5° slip angle. The slip angle at which the maximum lateral force is reached is independent of the wheel load. This means that the slip angle would have to be identical at the inside and outside wheel to fully utilize the adhesion potential at the front axle. With Ackermann steering, however, the slip angle at the inside wheel is significantly larger due to difference in steer angle, which leads to an understeer tendency. In order to exploit the maximum adhesion potential, this tire behavior requires a parallel steering wheel of both wheels, which would also lead to a reduction in the understeer tendency. For this reason, racing vehicles have an approximately parallel steer angle or even a slight anti-Ackermann tendency.

The optimum design of the steering geometry depends on the behavior of the tire. Figure 7.70 associates the design of the steering geometry with the behavior of the optimum slip angle as a function of the wheel load.

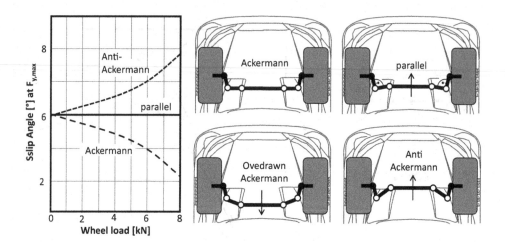

Fig. 7.70 Steering geometry designs and dependence on tire behavior

If the slip angle at which the maximum lateral force occurs is independent of the wheel load, the maximum lateral force is achieved by a parallel steering. If this slip angle increases as the wheel load increases, which is usually the case (cf. Figs. 4.9 and 4.10), anti-Ackermann steering is required to exploit the adhesion potential. If this slip angle were to decrease with increasing wheel load, again a steering system with an Ackermann tendency would have to be used.

7.8.2 Power

Power steerings are standard equipment in production vehicles. In the passenger car sector, vehicles without power steering are the absolute exception these days. In racing, however, there are still numerous racing series without power steering. These include the US IndyCar Series. In motorsport, power-assisted steering systems are used in individual high-performance racing series such as Formula 1 and the Le Mans prototypes or in near-series racing cars. In 1994, a power steering system was used for the first time in Formula One in the Williams FW16. However, power steering systems did not become established as standard equipment in Formula 1 until after the 2000 s. When using a power steering system, the following three principles are available for generating steering assistance:

- Hydraulic power steering (HPS)
- Electro-hydraulic steering assistance (EHPS, Electrohydraulic Power Steering)
- Electromechanical steering assistance (EPS, Electric Power Steering, or EPAS, Electric Power Assisted Steering)

Figure 7.71 shows the hydraulic power steering system of a Formula 1 car. The rack is mounted in the steering gear housing. Two working pistons are located on the rack, so that

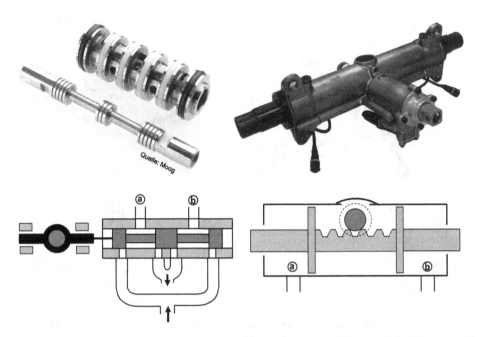

Fig. 7.71 Power steering system of an F1 car with steering gear and servo valves. (Courtesy of © MOOG INC. 2018. All Rights Reserved)

two working chambers are created between the rack and the steering gear. The pinion is guided at the end of the steering column in a translationally movable bearing, which in turn is directly connected to a hydraulic valve. This valve is connected to the central hydraulic system. The central hydraulic system is a constant pressure system which permanently provides a pressure to actuate the actuators present in the vehicle (gearshift actuation, DRS, etc.). In the middle position or when driving straight ahead, the valve is closed so that both working chambers of the steering gear are depressurized. This is referred to as "closed-center" steering.

Hydraulic steering systems in passenger cars, on the other hand, feature an "open-center" arrangement in which pressure is generated via a power steering pump driven directly by the combustion engine. The volume delivered when driving straight ahead is circulated in a closed circuit with frictional losses, resulting in relatively high power losses. The "closed-center" principle is therefore significantly more efficient. If the driver now applies a steering torque, the pinion is deflected according to the direction of the curve and the load, and the pressure and return lines are opened accordingly. The hydraulic pressure now acting on the piston supports the driver in his steering movement.

At maximum steering assistance, the valve is deflected by 0.5 mm and assistance powers of up to 2.2 kW are delivered [14]. As an alternative to the linearly moving valve described above, a rotary slide valve can also be used, the actuation of which is based on a torsionally elastic deformation of the steering column. This type of valve is also used in hydraulic and

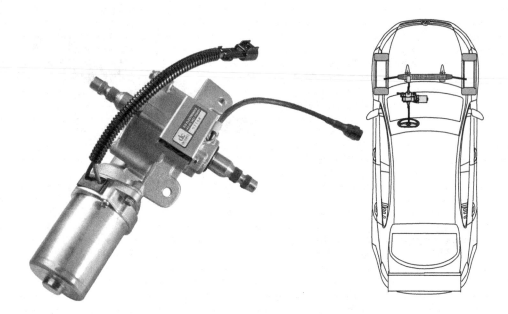

Fig. 7.72 EPS steering system of a BTCC touring car. (Courtesy © DC Electronics Motorsport Specialist Ltd. 2018. All Rights Reserved)

electrohydraulic systems in series production. An active throttle valve may be used in the control circuit to implement speed-dependent steering assistance. However, the steering system must not perform any function beyond reducing the steering forces.

In an electrohydraulic system, the servo pump is not driven directly by the combustion engine, but by an electric motor whose power is taken from the vehicle electrical system. The volume flow of the steering system can thus be controlled as required.

In the passenger car sector, however, new developments almost exclusively use electromechanical steering systems. In racing, too, the trend outside Formula 1 is towards the use of electromechanical steering systems. In Formula 1, the use of electromechanical power steering systems has been prohibited by the technical regulations since 2002.

Electromechanical steering systems have the best efficiency and thus the lowest power loss of all power steering systems. The disadvantage, however, is that the electric motor required to generate steering assistance must be mounted directly on the steering rack or column. Servo pumps, on the other hand, can be placed almost anywhere in the vehicle by using hydraulic lines. Depending on the location of the electric motor, a distinction is made between different designs of electromechanical steering systems. Figure 7.72 shows a steering column-assisted variant (C-EPS, with C for "column"), in which the electric motor applies its assistance torque directly to the steering column. This variant from the company DC Electronics is used, for example, in the touring cars of the BTTC.

Electromechanical power steering is also used in the Audi R18, as can be seen in Fig. 7.73 from the electric motor located above the stabilizer bar. Efficiency and fuel consumption play a crucial role in endurance racing, which is the decisive argument for the

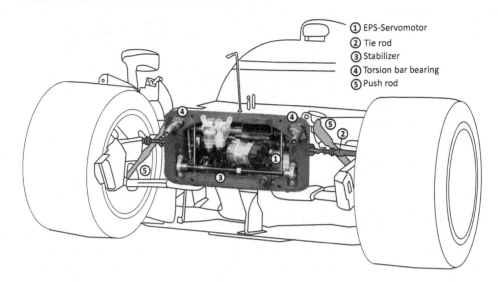

Fig. 7.73 Push rod front axle of the Audi R18 with electromechanical power steering

use of electromechanical systems. Le Mans prototypes require powerful servo motors due to the high downforce and the associated high lateral and steering forces. However, the latter require a correspondingly large amount of installation space, which is not available in every vehicle category. For this reason, too, these systems have not yet been able to establish themselves in monoposti.

The central advantages of power steering are, on the one hand, reduced driver fatigue and, on the other hand, the fact that steering ratio and steering geometry can be optimized exclusively according to driving dynamics. An example of this is the Pratt & Miller Intrepid (Fig. 7.74), which was the first GTP vehicle with power steering in 1991 [15]. By using power steering, it was possible to implement a caster angle of 6–7° at the front axle, which allowed the static camber at the front axle to be reduced to 1–1.5°. Without servo assistance, the caster angles of the GTP vehicles of the time were only in the range of 2–4°. The lack of negative camber increase during steering had to be compensated by a higher static camber, which was 2.5–3° for cars without servo assistance [16]. The caster angle of modern Formula 1 cars is about 9–12°. Even in Formula 3 cars, but they are much lighter and generate significantly less downforce, the caster angle is about 12°. In production vehicles, caster angles of 3.5–8° are common.

However, the use of power steering can also have safety-related advantages, as steering corrections can be made much more quickly. Another positive effect of power steering systems is the dampening effect on impacts ("kickbacks") that are transmitted from the wheel to the steering train and place considerable strain on the wrists and have a long-term detrimental effect on health.

The disadvantages of servo steering systems are limited to the costs, the required installation space and the energy requirements of these systems.

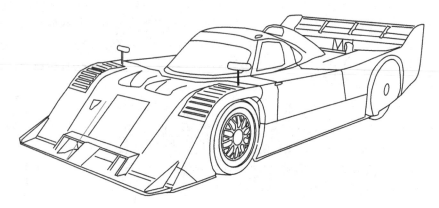

Fig. 7.74 Pratt & Miller Intrepid (1991–1993), GTP vehicle with power steering

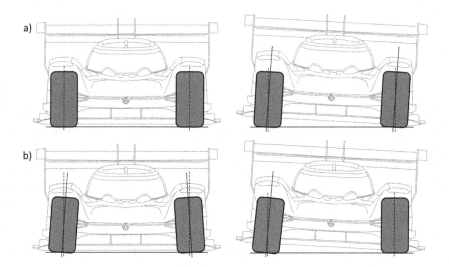

Fig. 7.75 Effect of preset camber

7.9   Toe-in and Camber Adjustment

In the previous sections, the influence of various chassis parameters on the wheel position during compression and rebound as well as during steering was described. To amplify or compensate for various kinematic effects, defined toe-in and camber values are set on the chassis already in the static equilibrium position, usually taking into account the driver and half a tank of fuel.

Figure 7.75 shows the effect of a static negative camber. Without static camber (Fig. 7.75a), when the vehicle rolls, the outside wheel on the curve goes into positive camber, which reduces the wheel's adhesion potential. This effect was referred to as camber loss in Sect. 7.7. The camber loss can be reduced if a negative camber is already

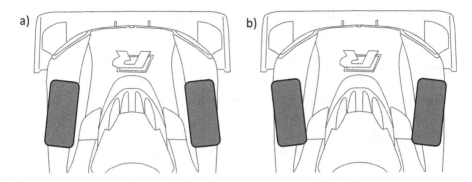

Fig. 7.76 Front axle (**a**) with toe-in, (**b**) with toe-out

set on the wheels in the static equilibrium position, as illustrated in Fig. 7.75b. This increases the wheel contact area of the outer wheel when cornering. The negative effect of the increased negative camber on the inside wheel of the curve has no significant effect on the adhesion potential of the corresponding axle due to the dynamic wheel load changes. Typical static camber values for various applications are:

- -1 to $-1.5°$ on front and rear axles of series passenger cars
- Approx. -4° on the front axle and approx. -1° on the rear axle of Formula 1 vehicles
- Approx. -5° for touring cars

Different camber settings on the front and rear axles also influence the steering tendency of the vehicle. To increase stability, a slightly higher value is often set on the rear axle of standard passenger cars than on the front axle. Since the camber loss at the rear axle is thus reduced in relation to the front axle, the understeer tendency increases. The maximum possible camber values are limited by tire wear and wheel load dependence of the camber generated lateral force. Asymmetrical wheel load distributions generated by road uneven-ness thus also generate asymmetrical camber laterla forces at cambered wheels, which imprint a yaw disturbance on the vehicle and possibly cause nervous straight-line running. Furthermore, care must be taken to ensure that the camber setting on the two wheels of an axle is as symmetrical as possible, otherwise the vehicle will pull to one side. For this reason, the specified tolerance bands must be observed in the production of series-produced vehicles.

Analogous to the camber, the toe of the vehicle is also set in the static equilibrium position. Figure 7.76 shows the definition of toe-in and toe-out. With toe-in, the wheels converge in the direction of travel, while with toe-out they diverge. This definition is identical on the front and rear axles. Moderate toe-in angles have the following driving dynamic effect on the front axle:

- Stable straight running due to self-correcting wheel position
- Improved steering response due to preload of the elasticities in the chassis
- Compensation of an Ackermann geometry, which reduces understeer in the limit range under certain circumstances.

The adjustment of a toe-out on the front axle leads to:

- Nervous straight running
- Increased agility or willingness to turn in at the entrance to the bend

Production vehicles always have toe-in on the front axle to stabilize straight-line running. On racing cars, on the other hand, a moderate toe-out is often set on the front axle.

At the rear axle, moderate toe-in is used to increase understeer tendency or vehicle stability. Setting toe-out on the rear axle usually leads to an unacceptable tendency to oversteer, which is why low toe-in values are set on the rear axle of both production and racing vehicles. In special applications, however, toe-out on the rear axle can increase a vehicle's drifting ability. The toe-in values of a 1997 F1 car were between $0.125°$ ("toe-in") and $- 0.125°$ ("toe-out") per wheel at the front axle and about $0.125°$ ("toe-in") at the rear axle.

Both toe-in and toe-out values increase the rolling resistance of the vehicle. The toe-in and toe-out values cause the tire to slip when driving straight ahead. This slip angle generates a lateral force, the proportion of which in the longitudinal direction of the vehicle causes braking of the vehicle or an increase in rolling resistance. In addition, the magnitude of these lateral forces depends on the wheel load. For this reason, excessive toe-in values can also lead to nervous straight-line running due to wheel load fluctuations caused by the road surface. Furthermore, toe-in and toe-out angles have the disadvantage of increasing wear on the inside and outside of the tire respectively. Toe-in and toe-out values can, however, also be used to increase the temperature in the tire if this makes sense due to the weather conditions. Since toe and camber values can change during compression and rebound, the static values may need to be readjusted after lowering the vehicle.

7.10 Brake System

7.10.1 Brake Force Distribution

Sect. 5.5 explains the basic relationship between brake force distribution and the dynamic behavior of a vehicle. Overbraking the front axle leads to understeering and overbraking the rear axle leads to oversteering. Brake force distribution must therefore meet two essential requirements:

- Ensuring a stable driving condition
- Achieve maximum deceleration

When a kinetic energy recovery system (KERS) is used, there are additional requirements from energy and battery management, which are discussed in the following paragraph. With the aid of Fig. 7.77, the essential variables which influence the brake force distribution are considered first. For this purpose, a simplified model is developed for a braking process in which the dynamic wheel load changes result directly from the center of gravity position and wheelbase. In reality, they are also determined in particular by the spring and damper system.

With the simplified approach, the maximum transmittable braking forces on the front and rear axles are obtained:

$$F_{xV,max} = \mu_V \cdot \left(m \cdot g \cdot \frac{l_H}{l} - \frac{h}{l} \cdot \ddot{x} - c_{zV} \cdot A \cdot \frac{\rho_L}{2} \cdot v^2 \right), \tag{7.16}$$

$$F_{xH,max} = \mu_H \cdot \left(m \cdot g \cdot \frac{l_V}{l} + \frac{h}{l} \cdot \ddot{x} - c_{zH} \cdot A \cdot \frac{\rho_L}{2} \cdot v^2 \right). \tag{7.17}$$

The braking force required for a given deceleration is:

$$F_{xV} + F_{xH} = m \cdot \ddot{x} + c_W \cdot A \cdot \frac{\rho_L}{2} \cdot v^2. \tag{7.18}$$

Brake force distribution is defined as:

$$i = \frac{F_{xV}}{F_{xV} + F_{xH}}. \tag{7.19}$$

We speak of ideal brake force distribution when the relative use of adhesion is identical on both axles. The following applies:

$$q_{ideal} = \frac{F_{xV}}{F_{xH}} = \frac{F_{xV,max}}{F_{xH,max}}. \tag{7.20}$$

From this follows for the ideal brake force distribution:

$$i_{ideal} = \frac{1}{1 + \frac{1}{q_{ideal}}} \tag{7.21}$$

The brake force distribution of high-performance racing vehicles is determined by the mechanical balance, i.e. the adhesion coefficients at the front and rear axles, the mass geometry as well as the wheel load changes and the aerodynamic balance. The theoretical

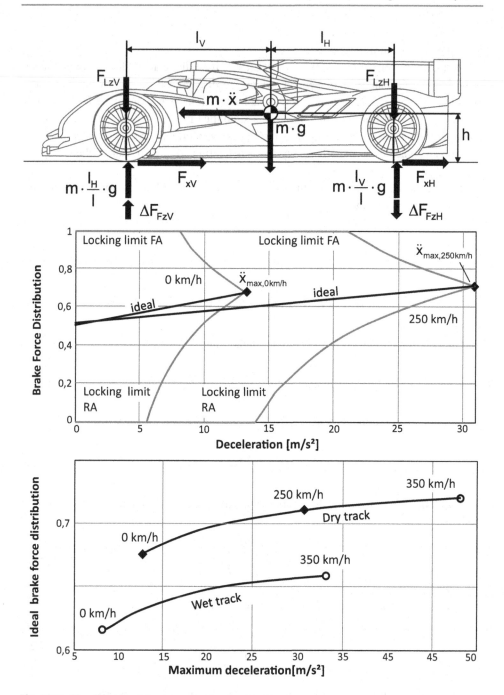

Fig. 7.77 Simplified model assumption for deceleration

maximum longitudinal deceleration can only be achieved with the ideal brake force distribution, which means that both wheels reach their adhesion or locking limit at the same time. For reasons of driving dynamics or to influence the oversteer and understeer behavior, it can be useful to deviate from this distribution. Another reason for adjusting the brake force distribution can be the temperature level in the tires. With increased tire temperature, one will try to relieve this axle by adjusting the brake force.

Due to the aerodynamic forces, the ideal brake force distribution changes above the driving speed. However, this change is relatively small, at least near the adhesion limit, as long as the aerodynamic balance is close to the mechanical balance. If the aerodynamic balance deviates strongly from the mechanical balance, the ideal braking force distribution shifts with increasing speed in favor of the more aerodynamically accentuated axle. The aerodynamic balance can also change here due to the pitch sensitivity of the vehicle.

Another factor influencing the distribution of braking force is the coefficient of friction. If, for example, the friction coefficients decrease on a wet road, the maximum longitudinal accelerations and thus the dynamic wheel load changes decrease. For this reason, the ideal brake force distribution shifts towards the rear axle. The ideal brake force distribution calculated with this simplified approach is only valid for driving straight ahead. When cornering, there is an additional influence due to the roll rate distribution between the front and rear axles, as the more stabilized axle tends to be able to generate less braking force. For these reasons, each section of the track with its individual speed characteristics has a different ideal braking force distribution, which is why an adjustment of the braking force by the driver at the steering wheel can often be observed in Formula 1 television broadcasts.

In production cars, the brake force distribution in the limit range is monitored by the brake control systems, in particular the ABS, and, if necessary, adjusted to the driving condition. Such systems are not permitted in high-performance racing vehicles. Brake force distribution must be purely mechanical and may only be adjusted manually by the driver, which is usually done by the balance bar system shown in Fig. 7.79. The absence of brake control systems is often observed when the inside cornering wheels lock up on racing cars.

7.10.2 Hydraulic Braking System with Balance Bar

The hydraulic brake system of a racing car is similar in design to that of a production car. For regulatory and safety reasons, it consists of two independent brake circuits, one acting on the front axle and the other on the rear axle. This ensures that if one brake circuit fails, it is still possible to bring the vehicle to a halt with sufficient residual deceleration. The braking system of a high-performance racing vehicle essentially consists of the following components:

- Brake pedal with balance bar
- Master brake cylinder with reservoir

- Brake lines
- Brake caliper with brake piston
- Brake disc and brake pad

The use of brake boosters and brake control systems is generally not permitted in high-performance racing. In near-series racing series and in amateur sport, the braking system may be supplemented by these components. These systems are not considered here.

Figure 7.78 shows a schematic of the brake circuit and pedal box, including clutch, brake and accelerator pedals of a high performance racing vehicle. The clutch pedal is only required for the starting process in a sequential manual transmission. On the accelerator pedal you can see the pedal travel sensor, which is needed for the electronic control of the throttle (throttle-by-wire). Next to the brake pedal, the two brake master cylinders can be seen, one each for the front and rear axle brake circuits, as well as the balance bar.

Figure 7.79 shows the operating principle of the balance bar. The driver's pedal force is initially amplified by a factor of three to four by the lever action. The balance beam divides the force acting on the main bearing between the two brake circuits and it applies:

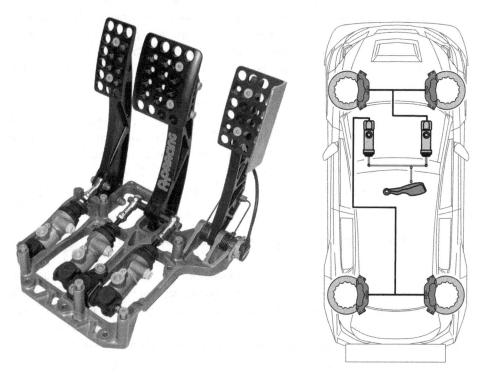

Fig. 7.78 Brake circuit and pedal box of a high performance racing car. (Photo at left courtesy of © AP Racing Ltd. 2018. All Rights Reserved)

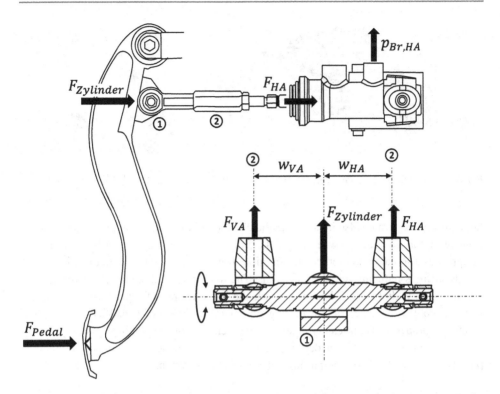

Fig. 7.79 How a balance bar works. (Courtesy of © AP Racing Ltd. 2018. All Rights Reserved)

$$F_{Cylinder} = F_{VA} + F_{HA}, \tag{7.22}$$

$$F_{VA} \cdot w_{VA} = F_{HA} \cdot w_{HA}, \tag{7.23}$$

$$F_{VA} = F_{Cylinder} \cdot \frac{1}{\frac{w_{HA}}{w_{VA}} + 1} \quad \text{and} \quad F_{HA} = F_{Cylinder} \cdot \frac{1}{\frac{w_{VA}}{w_{HA}} + 1}. \tag{7.24}$$

This corresponds to a constant brake force distribution. The brake force distribution can be adjusted by adjusting the lever arms. In high-quality systems, this adjustment can be made by the driver while driving. To do this, the balance bar is turned in a thread inside the main bearing. The adjustment can be made with a steel cable ("adjuster cable"), which is either turned directly by hand or remotely via an auxiliary motor. In both cases, however, the adjustment must be made by the driver. The use of the system as a controlled driving condition-dependent brake force distribution system is not permitted.

The forces acting on the master cylinders generate a brake pressure which is transmitted to the brake caliper and brake piston via the brake lines. Figure 7.80 shows a rear axle brake caliper made by Brembo for use in a Formula 1 vehicle. The brake pads are pressed against the brake disc by six pistons to generate the braking torque. A large number of pistons or a

Fig. 7.80 Brembo F1 brake caliper. (Courtesy of © Brembo S.p.A 2018. All Rights Reserved)

large piston area, which also increases in the direction of entry, is advantageous in order to achieve uniform contact between the brake pad and the brake disc. The brake caliper is designed as a so-called monobloc construction: It was manufactured from one piece. This avoids stiffness-reducing screw connections. The weight of an aluminum brake caliper for F1 cars is only about 1.5–2 kg. In contrast, the brake caliper of a production sports car weighs about 5 kg. The friction pairing of brake disc and brake pad as well as their geometric properties significantly determine the behavior of the braking system.

The coefficient of friction is determined by the characteristics of the brake disc and the pad. The design of the friction pairing is based on the following aspects:

- Bite
- Temperature or fading stability
- Coefficient of friction
- Lifetime

There are basically two types of friction pairings used in racing. These are cast iron brake discs in conjunction with brake pads made of organic materials and carbon brake discs in conjunction with carbon brake pads. So-called carbon-carbon brakes are reserved for use in high-performance racing. Figure 7.81 shows a compilation of brake discs from Brembo used in racing.

Two essential characteristics are the surface finish of the brake disc and the design of the internal ventilation. The surface finish primarily determines the initial bite of the brake disc and the wear behavior on the lining. The internal ventilation attempts to achieve the highest possible surface-to-volume ratio so that cooling air can flow efficiently through the brake disc. The operating temperatures of cast iron discs are within a range of 400–600 °C. Carbon brakes operate in a window of 350–800 °C in an LMP car and up to 1000 °C in an F1 car. Outside these windows, either friction coefficients drop sharply or wear increases excessively. Figure 7.82 shows the temperature-dependent behavior of the coefficient of friction of some friction pairs used in motorsport.

The aerodynamics of high performance racing cars allow the transmission of very high braking torques which result in a corresponding power input to the brake. Figure 7.83

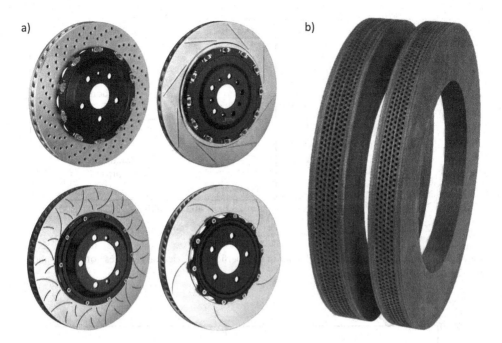

Fig. 7.81 Brembo brake discs (**a**) cast iron, (**b**) carbon. (Courtesy of © Brembo S.p.A 2018. All Rights Reserved)

shows the estimated power input to the brakes of an F1 car and a GT car. The Brembo company gives the braking power of an F1 car for a 5 g deceleration from 310 km/h as about 2500 kW, which is very close to the estimates. At this point, carbon brakes offer the advantage that they can operate at very high temperatures and have a stable as well as very high coefficient of friction over a wide temperature range. High coefficients of friction are particularly important in F1 cars as the use of brake boosters is prohibited. The braking torque is calculated according to Fig. 7.82:

$$M_{Brems} = \mu_B \cdot A_{Kolben} \cdot p_{Br,VA,HA} \cdot r_B. \tag{7.25}$$

Equation (7.25) shows that the brake pressure required for a given braking torque decreases as the coefficient of friction increases. The effective friction radius of the brake is limited by the rim diameter. Despite the high coefficients of friction of carbon-carbon brakes, which exceed 0.6 in the optimum operating window, F1 drivers have to generate a brake pedal force of more than 160 kg during emergency braking.

The temperature management of the brake is of great importance, whereby the requirements in detail can differ depending on the vehicle and track layout. In F1 cars, the aim is usually to dissipate heat as quickly as possible. Today's Formula 1 cars have

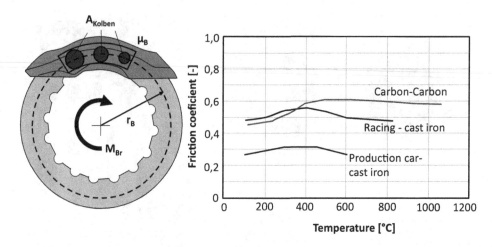

Fig. 7.82 Temperature-dependent behavior of the coefficient of friction. (Courtesy of © Brembo S. p.A 2018. All Rights Reserved)

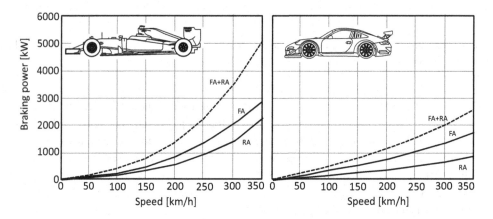

Fig. 7.83 Braking power

brake discs with up to 1400 vents. The carbon brake disc on an LMP1 car, on the other hand, has only about 600 ducts; due to the lower decelerations, the sometimes high recuperation rates, and the lower temperatures at night, the goal here is more to keep the heat in the system to reach the necessary temperature. Incidentally, it is not uncommon for LMP cars to compete in a 24-hour race without having to change the disc or pads. This circumstance also illustrates the high performance of carbon brakes.

Figure 7.84 shows the cooling behavior of an F1 brake system. The cooling rate is thus about 50 °C/s on average. In contrast, the brake system of a standard compact sports car

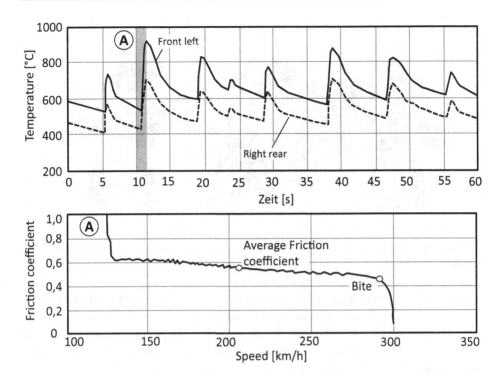

Fig. 7.84 Behavior of temperature and coefficient of friction of a F1 carbon brake system. (Courtesy of © Brembo S.p.A 2018. All Rights Reserved)

only loses about 3 °C/s. Another advantage of carbon is its low specific weight, which reduces the contribution of the brake system to the unsprung masses.

The power absorbed by the brake is converted into heat, which can be clearly seen from the outside by the glow of the brake disc, and dissipated into the ambient air. In modern production cars and now also in high-performance racing, attempts are increasingly being made to recover part of the braking energy by electrifying the drive train (Fig. 7.85). For this purpose, part of the braking torque is not generated in the friction brake but by an electric motor operated as a generator. The current generated by the generator is temporarily stored in a battery or accumulator and later used to drive the vehicle. In racing, this principle has become known as KERS (Kinetic Energy Recovery System). However, the recuperation of high braking power, which in a Porsche 919 Hybrid can theoretically amount to up to 294 kW, requires a special braking system.

7.10.3 Braking Systems for KERS Use

Hybridized powertrains (Fig. 7.85) have been used in Formula 1 since 2009 and at Le Mans since 2012. In these powertrains, at least one additional electric motor powers the vehicle in

Fig. 7.85 Hybrid powertrain of the Porsche 919. (Courtesy of © Dr. Ing. h.c. F. Porsche AG 2018. All Rights Reserved)

addition to the combustion engine. In the case of vehicles in the LMP1 hybrid category, the electric motor can be located on both the front and rear axles, or an electric motor can even be used on both axles. When braking, the electric motor acts as a generator. The braking torque generated by the generator acts in parallel with the hydraulic braking system. There are two approaches for integrating the generator brake:

- The generator braking torque is superimposed on the hydraulic braking torque without further measures.
- The sum of hydraulic and regenerative braking force is kept constant for a defined pedal pressure.

In the first case, no modification is made to the braking system. However, the braking effect of the generator depends on the state of charge of the battery. When the battery is fully charged, the generator does not brake with it. This means that the braking effect that starts at a given pedal operation can vary depending on whether or not a generator torque is added to the hydraulic pressure. In addition, the braking generator torque depends on the driving speed. As long as the generator power is relatively small, the effects on braking torque and braking balance remain relatively small, especially at higher speeds. This solution was used in Formula 1 during the first phase of the KERS system between 2009 and 2013. At high generator outputs, however, the use of a conventional braking system is no longer expedient.

With increasing generator power, the influence of the regenerative braking torque becomes continuously greater (Fig. 7.86). Without measures on the braking system, the braking force achieved at a certain pedal travel would then no longer remain constant, possibly leading to unexpected driving behavior. A continuous pedal feel could only be achieved by limiting the generator braking torque to its maximum braking torque in the high-speed range. However, the recuperation potential is then not optimally utilized.

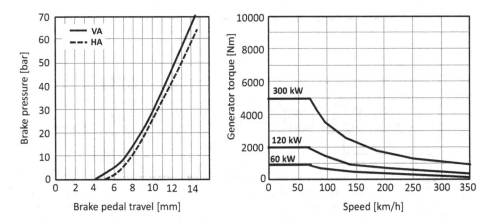

Fig. 7.86 Brake pedal behavior and speed-dependent generator torque

The requirements for the braking system of a hybrid or electric vehicle are therefore:

- Clear assignment of pedal travel to vehicle deceleration
- Feedback pedal feel
- Optimum utilization of the generator braking torque

The recuperation potential is used optimally if the friction brake is only applied as soon as the generator braking torque is no longer sufficient to achieve the desired deceleration. Such a coordination of friction brake and generator is called "brake blending". This process requires a decoupling of the brake pedal from the hydraulic brake circuit, which is achieved by using a brake-by-wire system. Brake-by-wire systems are used in today's Formula 1 and in the LMP1 hybrid category.

Figure 7.87 shows the functional diagram of the braking system used in Formula 1 since 2014. The front axle brake circuit remains unchanged and is actuated by pedal pressure alone. The rear axle brake circuit is split. The pedal pressure generated by the driver runs into a pedal travel simulator. This has two tasks. Firstly, it must provide the driver with the familiar and feedback pedal feel, and secondly, the pedal pressure is measured and communicated to a control unit. The control unit determines the required braking force and the optimum distribution between the friction and generator brakes. The friction braking torque is generated by an electronically controlled actuator, which builds up the required braking pressure. In addition, the control unit requests the required generator torque. The brake-by-wire system must not allow any deviation between the braking request made by the driver and the brake force distribution set by the driver. The braking force associated with a given pedal travel must remain constant. Only the distribution of the total braking torque between the friction brake and the generator torque may be varied

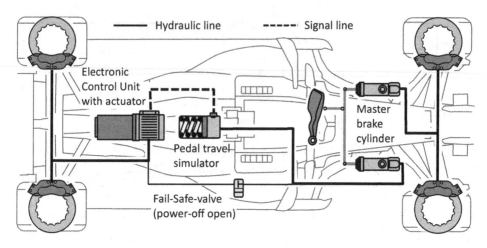

Fig. 7.87 Brake-by-wire system of a Formula 1 car

depending on the driving speed and the battery charge level. This regulation prevents the use of the brake-by-wire systems as an ABS system or for speed-dependent brake force distribution. In LMP-1 hybrid vehicles, depending on the configuration of the powertrain, a brake-by-wire system is used which acts on both axles.

References

1. Trzesniowski, M.: Handbuch Rennwagentechnik – Fahrwerk. Springer Vieweg, Wiesbaden (2017)
2. Heißing, B., Ersoy, M., Gies, S. (eds.): Fahrwerkhandbuch. Springer Vieweg, Wiesbaden (2013)
3. Wright, P.: Formula 1 Technology. Society of Automotive Engineers, Warrendale (2001)
4. Weber, W.: Fahrdynamik in Perfektion – Der Weg zum optimalen Fahrwerk-Setup. Motorbuch, Stuttgart (2011)
5. Reimpell, J.: Fahrwerktechnik: Stoßdämpfer. Vogel, Würzburg (1983)
6. Dixon, J.C.: The Shock Absorber Handbook. Wiley, West Sussex (2007)
7. Rottenberger, T.: Dämpferkonzepte für Formel 1 Rennfahrzeuge. Fahrwerk-Vertikaldynamik: Systeme und Komponenten, Haus der Technik Essen (2007)
8. https://scarbsf1.wordpress.com/2011/11/25/analysis-rotary-dampers/.   Zugegriffen   am 06.06.2017
9. Rendle, S.: Red Bull Racing F1 Car 2010 (RB69 – Owner's Workshop Manual). Haynes Publishing, Sparkford (2011)
10. http://theracingline.net/2018/race-car-tech/race-tech-explained/porsche-919-front-suspension-part-1/. Zugegriffen am 20.08.2018
11. Smith, C.: Tune to Win. Aero Publishers, Inc., Fallbrook (1978)
12. Newey, A.: How to Build a Car. HarperCollinsPublisher, London (2017)
13. Milliken, W.F., Miliken, D.L.: Race Car Vehicle Dynamics. Society of Automotive Engineers, Warrendale (1995)
14. McBeath, S.: More power to your elbows. Racecar Eng. **21**(1), 61–67 (2011)

15. Martin, J.A., Fuller, M.J.: Inside IMSA's Legendary GTP Race Cars – the Prototype Experience. Motorbooks, Minneapolis (2008)
16. Binks, D., DeWitt, N.: Making it Faster – Tales from the Endless Search for Speed. CreateSpace, North Charleston (2013)

Limited Slip Differentials

8.1   Design of Differential Gears

Limited slip differentials are a component of the drive train and are used in the final drive or, in the case of mechanical all-wheel drive systems, in the transfer case. Since mechanical all-wheel drive systems are not permitted in high-performance racing with the exception of rally or rally-raid vehicles, the following explanations are limited to the use of limited slip differentials in final drives.

Figure 8.1 shows the drive train of a Formula 1 vehicle. The drive torque delivered by the combustion engine and, if applicable, by the engine-generator units of a kinetic energy recovery system (KERS or MGU-K) flows through the clutch to the transmission input shaft of the semi-automatic shifting transmission. From the transmission output shaft, the transmission output torque is transmitted to the final drive. The final drive has the task of amplifying the drive torque and then distributing it to the two driven rear wheels in a defined ratio.

The function of the final drive, which in Formula 1 cars is designed as a spur differential with an electrohydraulically operated limited slip differential, is discussed in detail in this chapter. In the first step, the function of open or conventional differentials, which have no or only a negligible locking effect, is explained. Subsequently, the properties of limited slip differentials or differential locks are considered.

Figure 8.2 shows a summary of various differential types used as final drives. The main distinguishing feature is the design of the differential and output gears, which are either bevel, spur, worm or crown gear pairs. Bevel gear differentials are predominantly used in both production and racing vehicles. Spur gear differentials, however, have the advantage of being much more compact and about 15% lighter than bevel gear differentials due to lower gearing forces, which is a major argument for their use in Formula 1 cars. In series

© The Author(s), under exclusive license to Springer Fachmedien Wiesbaden GmbH,   343
part of Springer Nature 2023
L. Frömmig, *Basic Course in Race Car Technology*,
https://doi.org/10.1007/978-3-658-38470-8_8

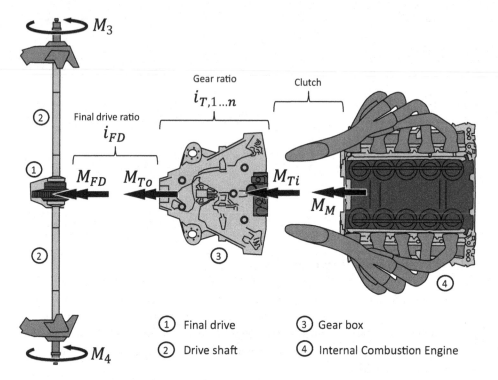

Fig. 8.1 Drive train of a Formula 1 car

production vehicles, they are only used in exceptional cases due to the higher production costs. Worm gears are used, for example, in torque-sensing limited slip differentials (Sect. 8.6).

The basic functional characteristics of a differential are independent of its design and are derived in the following using the bevel gear differential as an example. The resulting basic formulas for describing these functional properties also apply without restriction to the remaining differential designs.

Figure 8.3 shows the section through the bevel gear differential of a vehicle with front-transverse engine installation. The differential cage carries the drive wheel (often called a ring gear), which transmits and amplifies the torque delivered by the manual transmission. The drive gear is usually bolted onto the differential cage. It is designed as a spur gear when the axis of rotation is the same as the axis of the gearbox (typical for transverse engines), and as a bevel gear when rotated 90° (typical for longitudinal engines). Within the differential cage are the differential pinion or planetary gears and the side or sun gears. The differential pinion gears are rotatably mounted in the differential cage via the differential pinion. They mesh with the side gears, which are connected to the axle shafts via splines. At the opposite end of the axle shafts are the receptacles for the tripod joints of the

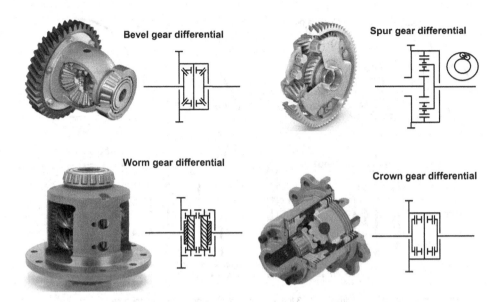

Fig. 8.2 Types of differential gears

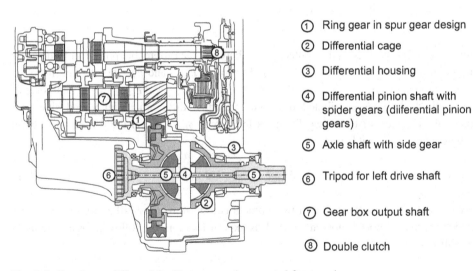

① Ring gear in spur gear design

② Differential cage

③ Differential housing

④ Differential pinion shaft with spider gears (diiferential pinion gears)

⑤ Axle shaft with side gear

⑥ Tripod for left drive shaft

⑦ Gear box output shaft

⑧ Double clutch

Fig. 8.3 Bevel gear differential with transversely mounted front engine

drive shafts. The combination of axle shaft and side gear is also rotatably mounted in the differential cage by means of a friction bearing.

The mathematical laws for the distribution of the drive torque to the two wheels and the speed ratios can be derived from the free section in Fig. 8.4. For simplicity, complete freedom from friction is assumed at all contact points. The axle drive torque or ring gear

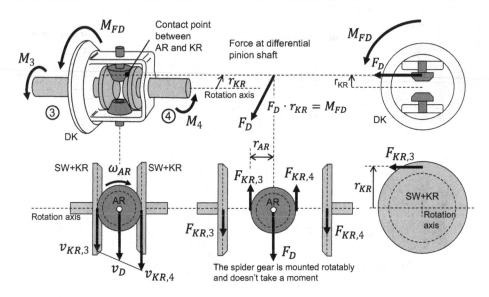

DK: Differential cage/ SW:drive shaft, 3 left, 4 right / KR: Side gear / AR: spider gear

Fig. 8.4 Free cut of a bevel gear differential (DK differential cage; SW, axle shaft; 3, left; 4, right; KR, side gear; AR, differential pinion gear)

torque M_{FD} is supported at the contact points between the differential pinion and side gears in the form of the forces $F_{KR,3}$ and $F_{KR,4}$. The forces $F_{KR,3}$ and $F_{KR,4}$ are derived from the free section in Fig. 8.4. In this consideration, the forces $F_{KR,3}$ and $F_{KR,4}$ are the total force of the individual forces distributed over the individual differential pinion and side gear pairings. The sum of these two forces is the force F_D acting on the differential pinion (abbreviated D). It holds true:

$$F_{KR,3} + F_{KR,4} = F_D. \tag{8.1}$$

By the additional consideration of the moment equilibrium at the differential pinion gear, which cannot support a moment around its axis of rotation, the following relationship is established:

$$F_{KR,3} \cdot r_{AR} - F_{KR,4} \cdot r_{AR} = 0. \tag{8.2}$$

The forces $F_{KR,\,3}$ and $F_{KR,\,4}$ have the effective lever arm r_{AR}, which corresponds to the pitch circle radius of the differential pinion gears. By transforming and inserting the two equations, the following is obtained:

$$F_{KR,3} = F_{KR,4} = \frac{F_D}{2}. \tag{8.3}$$

The resulting total force on the differential pinion depends on the final drive torque and is:

$$F_D \cdot r_{KR} = M_{FD}. \tag{8.4}$$

In this equation, r_{KR} is the pitch circle radius of the side gears (abbreviated KR), which also represents the lever arm around the rotational axis of the differential. The input torques transmitted from the differential or side gears to the axle shafts and thus to the wheels are obtained as:

$$M_3 = F_{KR,3} \cdot r_{KR} = \frac{F_D}{2} \cdot r_{KR}, \tag{8.5}$$

$$M_4 = F_{KR,4} \cdot r_{KR} = \frac{F_D}{2} \cdot r_{KR}. \tag{8.6}$$

The result of the calculations is the first basic equation of an open differential:

$$M_3 = M_4 = \frac{M_{FD}}{2}. \tag{8.7}$$

▶ This basic equation states that the input torque arriving at the drive wheel (with identically sized side gears) is always distributed symmetrically to both axle and drive shafts. This statement has general validity and is in particular independent of the road conditions.

The second basic equation describes the kinematic behavior of the differential, by which is meant the relationship between the speeds of the side gears or the axle shafts and the differential cage. To derive this relationship, the speed conditions at the differential pinion gear are considered. For this purpose, the circumferential speed of the differential pinion gear at its bearing point is determined at the level of the contact point. The following applies:

$$v_D = \omega_{AG} \cdot r_{KR}. \tag{8.8}$$

In the second step, the circumferential speeds of the side gears $v_{KR,\,3}$ and $v_{KR,\,4}$ are calculated as a function of the speed of the differential pinion gear ω_{AR}. The case is shown that the right rear wheel (4) rotates faster than the left rear wheel (3). It follows:

$$v_{KR,3} = v_D - \omega_{AR} \cdot r_{AR} = \omega_3 \cdot r_{KR}, \tag{8.9}$$

$$v_{KR,4} = v_D + \omega_{AR} \cdot r_{AR} = \omega_4 \cdot r_{KR}. \tag{8.10}$$

Adding Eqs. (8.9) and (8.10), with previous substitution of Eq. (8.9), leads to the second basic equation of the open differential:

$$\omega_{AG} = \frac{\omega_3 + \omega_4}{2}. \tag{8.11}$$

▶   This equation expresses that the speed of the differential cage is just equal to the average speed of the two axle shafts. This means that a different rotational speed of the two axle shafts leads to a corresponding change in the rotational speed of the differential cage. Viewed in this way, the two speeds at the sideshafts are independent of each other. This is also referred to as speed compensation by the differential.

The significance of the basic properties of the differential described by Eqs. (8.7) and (8.11) for the vehicle dynamic behavior is explained in the following section.

8.2   Driving Dynamics Function of Conventional Axle Differentials

Conventional axle differentials are used to distribute the torque delivered by the transmission symmetrically to the two driven wheels while allowing for the kinematically induced wheel speed differences. Symmetrical drive force distribution is necessary to avoid drive force-induced skewing(torque steer) when driving straight ahead. Allowing differential speeds between the wheels serves to avoid severe understeer when cornering and to reduce component loads and tire wear. This effect of the differential is explained with reference to Figs. 8.5 and 8.6. Figure 8.5 shows the kinematic conditions at free rolling rear wheels for an unaccelerated left turn. The vehicle turns with the yaw rate $\dot{\psi}$ around the instantaneous center, which lies on the extension of the rear axle. The points K3 and K4 form two points fixed to the body, which lie at the height of the wheel centers. The rigid body speeds of these points are:

$$v_{K3} = \left(\rho_H + \frac{S_H}{2}\right) \cdot \dot{\psi}, \tag{8.12}$$

$$v_{K4} = \left(\rho_H - \frac{S_H}{2}\right) \cdot \dot{\psi}. \tag{8.13}$$

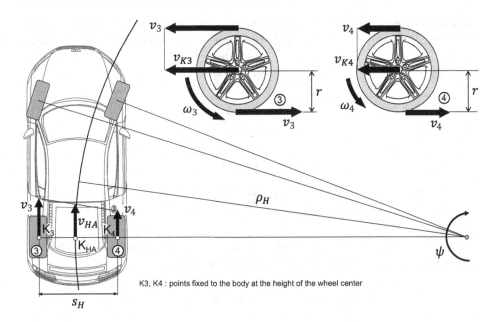

Fig. 8.5 Kinematic wheel speed ratios on free rolling wheels

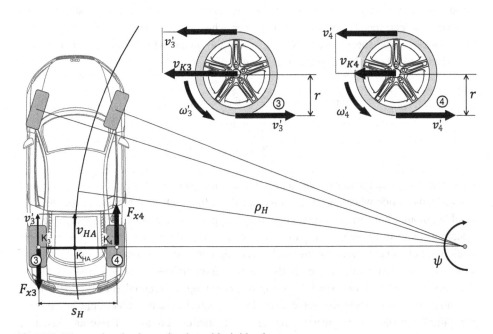

Fig. 8.6 Kinematic wheel speed ratios with rigid axle

These speeds correspond to the movement of the wheel center or wheel hub above ground. In the case of the slip-free and slip angle-free cornering assumed here, the wheel circumferential speeds v_3 and v_4 are identical in magnitude to the rigid body speeds. Therefore, the following wheel speeds are obtained for free rolling:

$$\omega_3 = \frac{v_3}{r} = \frac{v_{K3}}{r}, \tag{8.14}$$

$$\omega_4 = \frac{v_4}{r} = \frac{v_{K4}}{r}. \tag{8.15}$$

To enable slip-free cornering, the outer wheels must turn faster than the inner wheels. This is ensured by the speed compensation in the differential, which is described by the second basic equation (Eq. (8.11)).

In contrast, in Fig. 8.6 the rear wheels are forced by a rigid connection to rotate at the same wheel speed. The following condition is thus introduced:

$$v_3' = v_4' = \rho_H \cdot \dot{\psi}. \tag{8.16}$$

Both rear wheels adopt the rigid-body speed of the axle center K_{HA} as their circumferential speed. The rigid body speeds v_{K3} and v_{K4} remain unchanged. A comparison with the associated wheel speeds shows that the outer wheel now rotates more slowly than it corresponds to its speed over ground, while the inner wheel rotates faster than it corresponds to its speed over ground. It is valid:

$$v_{K3} = \left(\rho_H + \frac{s_H}{2}\right) \cdot \dot{\psi} > v_3' = \rho_H \cdot \dot{\psi}, \tag{8.17}$$

$$v_{K4} = \left(\rho_H - \frac{s_H}{2}\right) \cdot \dot{\psi} < v_4' = \rho_H \cdot \dot{\psi}. \tag{8.18}$$

Thus, brake slip occurs at the outer wheel and a corresponding braking force is generated. Similarly, the inside wheel has drive slip and a drive force. Drive force and brake force are equal in amount, but form a yaw moment that is opposite to the direction of the curve. The vehicle tends to understeer. The yaw moment is greater the higher the differential speed would be if the wheels were rolling freely. The slip that occurs increases tire wear, and the forces that occur twist the drive shafts and therefore increase component stress. These effects are avoided by differential gearing and allowing differential speeds. Speed compensation is mandatory if the vehicle has brake control systems (electronic brake force distribution, anti-lock braking system, electronic stability control). These are based on wheel-specific braking interventions which can only be carried out without interference if the speed compensation is not affected.

However, the combination of symmetrical drive force distribution and speed compensation has decisive disadvantages when it comes to achieving the highest possible performance. The disadvantages are explained by comparing accelerated straight-line driving on

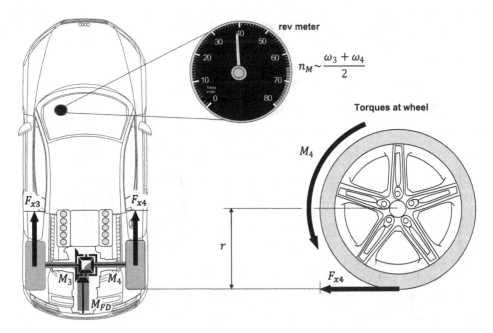

Fig. 8.7 Force and moment ratios on a homogeneous road surface

a homogeneous road surface and on a μ-split. Figure 8.7 shows the force and torque ratios for acceleration on homogeneous high friction conditions, which means that both wheels have the same adhesion potential.

In this case, the drive shaft torques on both wheels can be fully transmitted to the road surface. The drive forces transferred to the road surface are in equilibrium with the drive shaft torques, so that the following applies:

$$F_{x3} = \frac{M_3}{r} = F_{x4} = \frac{M_4}{r}. \tag{8.19}$$

If one of the two wheels is at low friction, as shown in Fig. 8.8, the incoming drive shaft torque at the low friction wheel can no longer be fully transmitted to the road and it no longer forms a moment equilibrium with the driving force. The difference that is not transferred is used to accelerate the low friction wheel. In this case, the following applies to wheel 4:

$$M_4 = F'_{x4} \cdot r + J_R \cdot \dot{\omega}_4 \Rightarrow \dot{\omega}_4 = \frac{M_4 - F'_{x4} \cdot r}{J_R}. \tag{8.20}$$

The inertia of the wheel is relatively small compared to the incoming drive torques, which results in the low friction wheel spinning up very strongly and in a very short time. Along with the wheel, the entire drivetrain and motor spin up, symbolized at this point by running

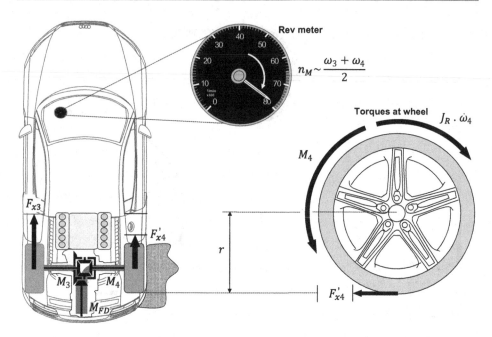

Fig. 8.8 Unstable force and moment ratios on μ-split

into the speed limit. As the drivetrain spins up, a significant portion of the engine's power is lost in the inertias of the drivetrain and can no longer be converted into propulsion. This situation is further specified in connection with cornering. For a stable acceleration process, the drive shaft torque at the low friction wheel must be dosed in such a way that a state of equilibrium is re-established (Fig. 8.9). The following condition must be fulfilled:

$$M_4 = F'_{x4} \cdot r = \mu_{low} \cdot F_{z4} \cdot r. \tag{8.21}$$

Thus, from the first basic equation of the differential (Eq. (8.7)), the moment transmissible at the high friction coefficient is given by:

$$M_3 = M_4 = F'_{x3} \cdot r = \mu_{low} \cdot F_{z4} \cdot r. \tag{8.22}$$

Thus, the frictional potential of the low friction wheel also determines the level of the high friction wheel. If, in an extreme case, the low-friction wheel is on polished ice ($\mu = 0$), it is no longer possible to start with an open differential. The actually available frictional potential of the high friction value wheel is not optimally used with a conventional axle differential. Different adhesion potentials on the wheel side result not only from different friction values on the wheel side, but also from the dynamic wheel load changes that occur during cornering.

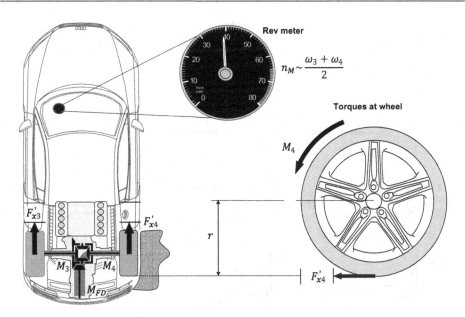

Fig. 8.9 Stable force and moment ratios on µ-split

Therefore, as Fig. 8.10 shows, a direct analogy to accelerated cornering can be established via the friction circles. The behavior of a differential – divided into three phases – during an acceleration process from the apex of the curve is shown. Until the apex is reached, acceleration is not yet occurring and the outside rear wheel is turning faster than the inside rear wheel. The dynamic wheel load changes lead to an additional load on the outside rear wheel and to a relief of the inside rear wheel, so that friction circles with different diameters are formed at these wheels. Phase I marks the beginning of the acceleration phase. Both wheel speeds increase. The slip builds up and is higher at the inside wheel than at the outside wheel – due to the dynamic wheel load distribution. The increase in speed is therefore higher on the inside wheel than on the outside wheel.

In phase II of the acceleration phase, the inside rear wheel approaches its adhesion limit. Due to the significantly higher wheel load, the frictional potential at the outside wheel is not yet exhausted at the same time. In addition, the slip requirement for transmitting the drive torque is significantly higher at the inside wheel, which is why the differential speed between the two wheels decreases progressively. As the duration of the acceleration phase increases, the inside wheel finally overtakes the outside wheel. The increased drive slip development leads to a decrease in the lateral force potential at the wheel on the inside of the curve. The friction circle deforms into an oval. For this reason, the theoretical sum force that the two friction circles exhibit in phase I cannot be fully utilized under high drive load.

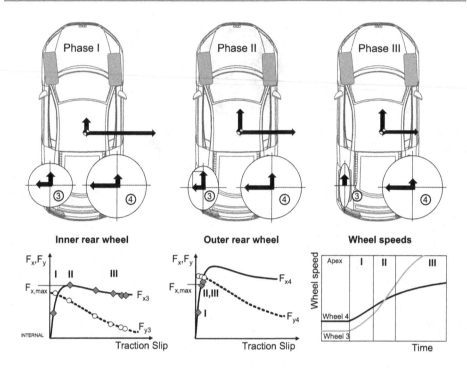

Fig. 8.10 Accelerated cornering with differential open

During Phase III, the inside front wheel then spins, causing it to virtually lose its ability to transmit lateral forces. Due to the relatively low rotational inertias in the driveline, the spinning up of the front wheel occurs very quickly, changing the flow of power and torque through the driveline and over the drive shafts. As Fig. 8.11 shows, during Phase I, with the inside wheel spinning stably, virtually all of the engine torque – translated through the transmission and final drive – is still fully transmitted to both wheels. In phase III, the spinning of the inside wheel leads to the fact that a large part of the torque delivered by the engine or a large part of its power is used to accelerate the driveline components, i.e. the crankshaft drive, the clutch, the gearbox and the axle drive with drive shafts and wheels. The individual moments of inertia of these components reduce the drive torque arriving at the final drive. The torque or power arriving at the outer wheel is not high enough to cause the outer wheel to spin. Thus, in conjunction with a conventional axle differential, the ability to transmit lateral forces to the outside wheel is always maintained. This is a kind of protective mechanism against a complete breakdown of the lateral force at a driven axle. This effect means that the understeer or oversteer reactions caused by the spinning of the inner wheel remain predictable and controllable, at least for experienced drivers.

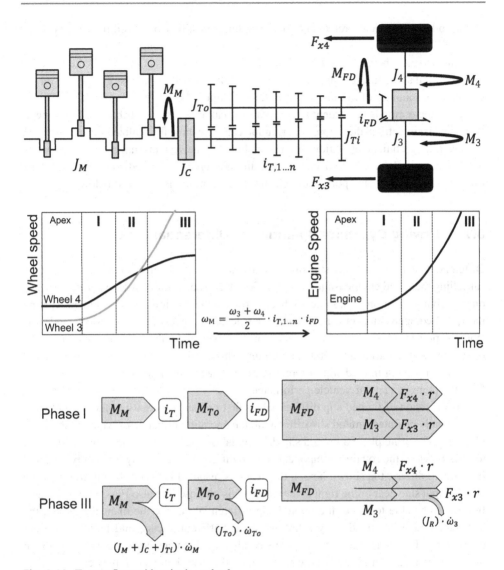

Fig. 8.11 Torque flow with spinning wheel

▶   A spinning of the outer wheel would result in the loss of lateral control not only of the
    wheel on the inside of the curve, but also of the entire axle.

This would result in abrupt and difficult to control oversteer ("snap oversteer") in
rear-wheel drive vehicles and abrupt understeer ("snap understeer") in front-wheel
drive vehicles.

In production vehicles, relatively good handling can still be achieved by using conventional differentials even without or with deactivated traction control. However, a loss of performance must be accepted for this, as the adhesion reserves at the outer wheel are not yet exhausted at the time of spinning. In order to fully exploit the adhesion potential on the driven axle, spinning of the inside wheel must therefore be prevented. The prevention of speed compensation is equivalent to an asymmetrical drive force distribution. The generation of an asymmetrical drive force distribution can be technically realized in various ways. However, the technical regulations set narrow limits to the systems that can be used, which is made clear in Sect. 8.12. For this reason, different types of limited slip differentials are used as standard in motorsport, and their mode of operation is analyzed below.

8.3   Driving Dynamics Operation of Differential Locks

Differential locks or limited slip differentials have the task of partially or completely cancelling the speed compensation in the differential. The optimum locking effect depends on the driving situation. It has already been shown in the previous sections that preventing the speed compensation in a rigid connection leads to undesirable understeer tendency. Without preventing the speed compensation, however, the adhesion potential cannot be optimally utilized during accelerated cornering. This conflict of objectives can be partially eliminated by using limited slip differentilas, or at least a better compromise can be found from the point of view of vehicle performance.

The operation of limited slip differentials is first explained here using the example of a hydraulically operated limited slip dfferential, the operating principle of which is shown in Fig. 8.12. The principle of a limited slip differential is based on the generation of frictional torques between the rotating components of the differential. In the hydraulically operated limited slip differentila shown here, a disk pack is installed between the differential cage and the right side gear or the right axle shaft for this purpose (see also Figs. 8.13 and 8.21). In order to be able to install the limited slip differential outside the manual gearbox, the differential cage is extended by a hollow shaft. Differential cage and hollow shaft are brought together by a spline. The discs are axially movable via splines, but are mounted in the differential cage or in the hollow shaft and on the axle shaft so that they cannot rotate. The outer disks engage in the hollow shaft, and the inner disks engage in the axle shaft, which can be seen from the teeth in opposite directions. The disk pack is compressed by a working piston which transmits the contact pressure via an axial needle bearing. The hydraulic pressure required for this is generated by a pump and adjusted as required by a pressure control valve. This generates a frictional torque which is proportional to the hydraulic pressure. This frictional torque brakes the faster rotating component and accelerates the slower rotating component. The differential speed is reduced and the speed compensation is partially or completely cancelled.

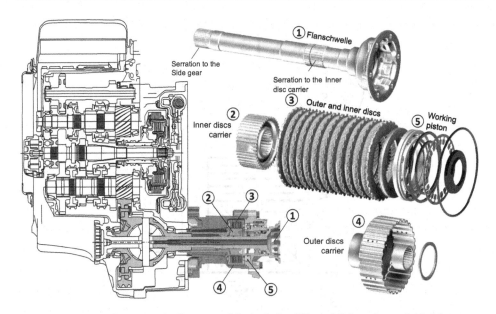

Fig. 8.12 Design of a hydraulically actuated limited slip differential for a front-wheel-drive transverse engine vehicle. (Photo at right courtesy © BorgWarner Inc. 2018. All Rights Reserved)

▶ In other words, a differential lock behaves in such a way that it transfers a torque from the faster rotating component to the slower rotating component. This means that the direction of action depends on the differential speed applied to the wheels.

Figure 8.13 illustrates the differential speed dependent torque flow. In case a, the right axle shaft rotates faster than the left axle shaft. Therefore, the right axle shaft also rotates faster than the differential cage (see Eq. (8.11)). The outer disks rotate at the speed of the differential cage, and the inner disks rotate at the speed of the right axle shaft. The locking torque is transmitted through the disks between the differential cage and the right axle shaft. The generated friction or locking torque brakes the right-hand axle shaft and accelerates the differential cage. The locking torque is thus withdrawn from the right-hand side shaft and transmitted to the differential cage. The sum of the locking torque and the axle gear torque is distributed to both sideshafts in the ratio 50:50 according to the first basic equation of the differential (Eq. (8.7)). In case b the ratios are reversed. This behavior is described by the following equations:

$$M_3 = \frac{M_{FD} + M_{Sperr} \cdot \text{sgn}(\omega_4 - \omega_3)}{2}, \tag{8.23}$$

$$M_4 = \frac{M_{FD} - M_{Sperr} \cdot \text{sgn}(\omega_4 - \omega_3)}{2}. \tag{8.24}$$

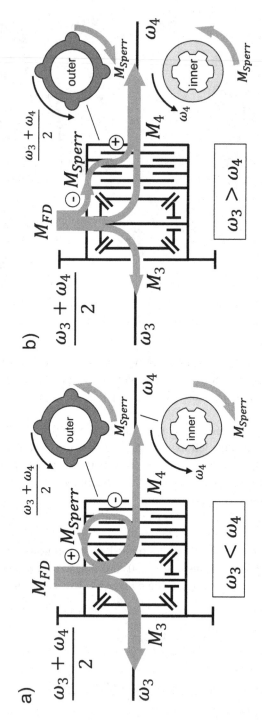

Fig. 8.13 Torque flow in the limited slip differential at different wheel speeds

Due to this torque transfer, the drive torque is now no longer distributed to the two side shafts in a 50:50 ratio, but an asymmetrical drive torque or drive force distribution occurs. The differential torque between the two drive shafts corresponds to the locking torque:

$$M_3 - M_4 = M_{Sperr} \cdot \mathrm{sgn}(\omega_4 - \omega_3). \tag{8.25}$$

If the differential speed is zero in the open state, then the differential is initially loaded by the closing of the disks, but there is no torque transfer between the drive shafts. The dependence of the torque transfer on the differential speeds and its influence on the driving dynamics are discussed below for various driving situations.

8.3.1   Locking Behavior During Steady-State Cornering

A characteristic feature of steady-state cornering is that the vehicle speed and the radius of curvature travelled remain constant. The longitudinal acceleration or deceleration of the vehicle is therefore zero, and there is no significant traction or brake slip. The tires therefore transmit virtually only lateral forces. In this case, the wheel speeds are mainly determined by the kinematic conditions described in Sect. 8.2. In this driving situation, the outer wheels turn faster than the inner wheels. The torque transfer by a limited slip differential can therefore only take place from the outside wheel to the inside wheel. Figure 8.14

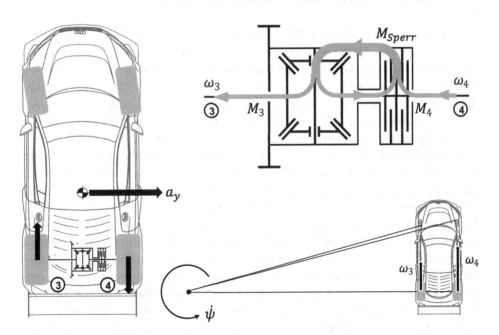

Fig. 8.14 Torque transfer during stationary steady-state cornering ($M_{FD} = 0$, $a_x = 0$)

illustrates, using the example of a rear axle driven vehicle, that a braking force is generated at the outside rear wheel and a driving force at the inside rear wheel. These braking and driving forces generate a turning out yaw moment which increases the understeer tendency of the vehicle. This effect is more pronounced the higher the kinematic speed difference would be in an open differential. Higher driving speeds and tighter radii of curvature therefore increase the understeer effect. The increase in understeer tendency due to a turning out locking torque can have both negative and positive influences on the driving condition. In slow to medium speed corners, where agility is usually the primary concern, one wants to avoid understeer effects. In high-speed curves, where there may also be vertical excitation from road bumps, such a locking effect can increase the stability of the vehicle because the yaw damping is increased by the turning-out moment. With tirning-out locking torques, it is possible to counteract an unwanted oversteer reaction of the vehicle. A spinning-out moment can also be an advantage when braking (Sect. 8.3.3).

These examples already show that the optimum locking effect depends on the driving condition. The ideal situation would therefore be a variable locking effect that can be set depending on the driving condition, but this is usually not feasible due to the restrictive specifications in today's technical regulations. The configuration of an uncontrolled limited slip differential must therefore result from a compromise that takes into account all relevant driving situations.

8.3.2 Locking Behavior During Accelerated Cornering

The typical example of accelerated cornering is the acceleration process from the apex to the exit of the corner. At the apex, the vehicle reaches approximately its maximum lateral acceleration. Due to the dynamic wheel load changes, the wheel on the inside of the curve has a significantly lower wheel load than the wheel on the outside of the curve. Therefore, significantly more slip is generated at the inside wheel than at the outside wheel in order to transmit the drive torque. These conditions are illustrated in Fig. 8.15. The increased traction slip causes the wheel speed of the inside rear wheel to increase sharply and it eventually overtakes the outside rear wheel. The inside rear wheel turns faster than the outside rear wheel. The direction of the torque transfer via the multi-plate clutch of the limited slip differential is reversed. The torque transfer now takes place from the inside rear wheel to the outside rear wheel. The outside rear wheel thus transmits a higher drive torque than the inside rear wheel. The relative reduction of the drive torque at the inside rear wheel prevents its tendency to spin. This results in an asymmetrical drive force distribution, which corresponds to the dynamic wheel load distribution and thus makes optimum use of the adhesion potential of the driven axle. At the same time, the asymmetrical drive force distribution creates a yaw moment that pulls the vehicle into the curve. This turning-in yaw moment reduces the vehicle's tendency to understeer, and the steering wheel angle requirement is reduced. Depending on the characteristics of the limited-slip differential,

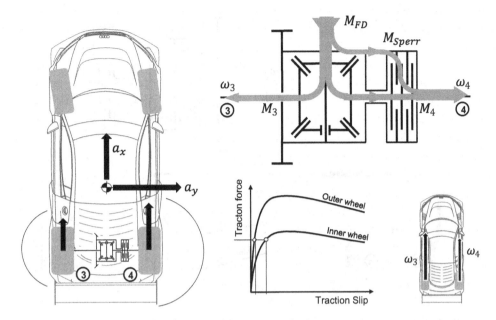

Fig. 8.15 Torque transfer during accelerated cornering ($M_{FD} \ll 0$, $a_x \ll 0$)

this effect is more spontaneous or gentle. The driver must adapt his driving style to this effect.

The maximum locking effect is achieved when both drive wheels rotate at the same speed. This condition is also referred to as synchronous operation. In synchronous operation, no slip or differential speed occurs between the clutch plates. In this state, the limited slip differential does not generate any frictional losses, which minimizes component wear and thermal stress.

8.3.3   Locking Behavior During Braking

In racing vehicles, which generally do not have any brake control systems, the stabilizing effect of a limited slip differential is often also used to optimize braking behavior. During deceleration, the rear axle is relieved of wheel load, which, depending on the brake balance setting, can lead to an oversteering effect or at least to an nervous rear end of the vehicle. In this case, the locking effect counteracts any yawing movement of the vehicle, and the vehicle behaves much more smoothly when braking. In addition, a limited slip differential reduces the tendency of a braked wheel to lock. This case is illustrated in Fig. 8.16 for braking in a curve ("trail braking"). In addition to the relief of the rear axle, there is a wheel load transfer from the inside to the outside rear wheel. The inside rear wheel therefore has an increased tendency to lock. The locking rear wheel turns significantly slower than the

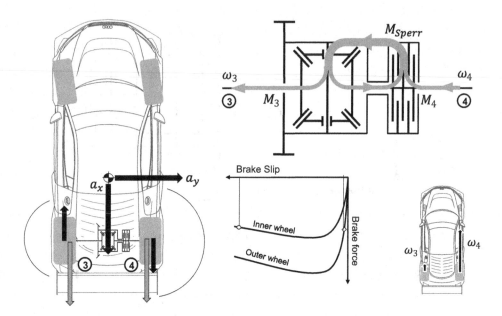

Fig. 8.16 Torque transfer during braking processes ($M_{FD} = 0$, $a_x \gg 0$)

stable rear wheel on the outside due to the higher brake slip. The torque transfer of the differential lock therefore takes place from the outside to the inside rear wheel. At the locking inside rear wheel, the differential lock generates a drive torque which counteracts the hydraulic braking torque. The effective braking effect is reduced at the locking rear wheel, which stabilizes the wheel. The effective braking effect is increased at the opposite rear wheel. This results in a braking force distribution that corresponds to the dynamic wheel loads. This behavior is similar to the effect of an ABS system, whereby a limited slip differential can noticeably stabilize the braking behavior.

On vehicles equipped with the ABS and ESC brake control systems, the use of a limited-slip differential during braking is unnecessary and in some cases even impedes the effect of these systems. If brake control systems are used, it must therefore be ensured that the locking effect is completely withdrawn during braking.

The positive effect of this ABS effect is not limited to the driven axle, but can also be achieved on non-driven axles. For this reason, on the 1999 Benetton B199 and the 2004 BAR 006 (see also [1]), the two wheels of the front axle were coupled to a hydraulic clutch via two drive shafts, as shown in Fig. 8.17. These systems became known as the Front Torque Transfer System (FTT). However, the use of FTT systems was banned at the end of the 2004 season. Since the 2005 season, any form of torque transfer between the front wheels has been banned in Formula One (see Sect. 8.12). In the all-wheel-drive hybrid cars of the LMP1 class, however, the use of limited slip differentials is also permitted on the front axle, with the front axle being driven by a powerful electric motor. According to factory information, the 2016 Audi R18 e-tron quattro transmits more than 350 kW at the

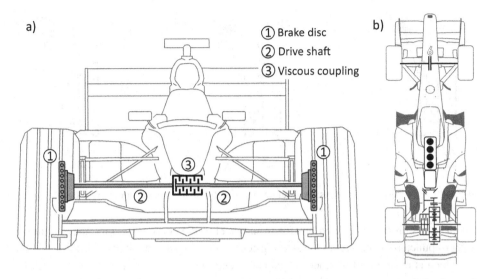

Fig. 8.17 FTT systems on the front axle: (**a**) Benetton B199 with Visco clutch (1999), (**b**) BAR 006 Honda of the 2004 season

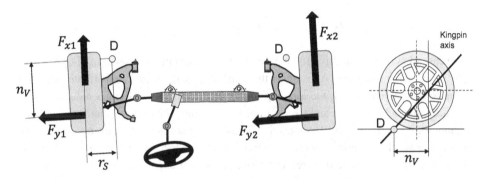

Fig. 8.18 Influencing the steering return by differential locks

front axle with a total system output of more than 728 kW, which is why a limited-slip differential is integrated into the motor-generator unit.

8.3.4 Interaction with the Steering System with the Front Axle Driven

The driving dynamics effects explained above are independent of the type of drive. However, in vehicles with a driven front axle, there is an interaction with the steering system resulting from the moment effect of the lateral and longitudinal forces around the kingpin axis. Figure 8.18 shows the effect of the tire forces on the steering return. The

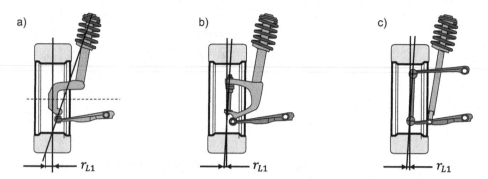

Fig. 8.19 Front axle suspensions: (**a**) McPherson, (**b**) RevoKnuckle, (**c**) double wishbone

lateral forces on the inside and outside front wheel of the curve form a moment about the point of puncture that resets to the steering center. The associated lever arm is the total trail n_V, which is made up of the mechanical caster and the pneumatic caster (see Sect. 7.8.1). The driving forces act around the disturbance force lever arm r_S, whereby the driving forces of the two front wheels generate a moment in opposite directions. Neglecting other effects, the following thus applies to the total aligning torque of the steering system:

$$M_S = \left(F_{y1} + F_{y2}\right) \cdot n_V + \left(F_{x1} - F_{x2}\right) \cdot r_S. \tag{8.26}$$

In the case of symmetrical drive force distribution, the aligning torques caused by the drive force cancel each other out. As soon as the limited slip differential redistributes the drive torque to the outside wheel, the aligning torque of the steering system is reduced. When designing the locking effect and steering geometry, it must be ensured that the sign of the aligning torque is not reversed as a result of the drive force-induced restoring. In this case, the steering would no longer automatically return to the steering center when the steering wheel is released, but would increase the steering wheel angle. The drive force-induced return is significantly influenced by the disturbance force lever arm. In order to achieve low interference effects due to drive forces, this must be designed as small as possible. Various front axle suspensions and their influence on the lever arms are shown in Fig. 8.19.

One of the standard designs of driven front axles is the McPherson axle. It essentially consists of the lower wishbone and a unit consisting of a suspension strut and wheel carrier, which is rotatably mounted between the lower wishbone and the strut bearing and thus represents the steering function. Due to its principle, the McPherson axle has a relatively large interference force lever. A solution presented by Ford to reduce the interference lever arm is the so-called RevoKnuckle axle. The basic principle of the McPherson axle is retained, but the wheel carrier is now mounted separately in the suspension strut. The changed position of the kingpin axis leads to the desired reduction of the disturbing force lever. This axle principle is now used in many powerful front-wheel-drive vehicles. A complex possibility for reducing the disturbing force lever arm is the use of a double wishbone axle. In the case of front-wheel-drive series-produced vehicles, however, this is often opposed by economic reasons.

8.4   Classification of Limited Slip Differentials

There is a wide range of different designs and functional principles for the use of limited slip differentials. Which form is used is often determined by the requirements of the technical regulations. The characteristic distinguishing features of limited slip differentials are compiled in Fig. 8.20. At the top level, a distinction can first be made between positive-locking, self-regulating and controlled limited slip differentials. This provides a higher-level statement on the dosing capability of the locking effect.

The category of positive-locking connections includes dog clutches as well as the rigid connection between both wheels. With a dog clutch, only digital switching between an open differential and a rigid connection is possible. Such constructions are mainly used in off-road vehicles. The locking effect cannot be further dosed.

In the case of self-regulating differentials, the locking effect cannot be controlled externally, but depends on the driving condition. Torque and speed-sensing differentials fall into this category. In the first case, the locking effect depends on the drive torque and in the second case on the differential speed between the driven wheels. Both types are considered in detail in the following sections.

Electronically controlled limited slip differentials are versions in which the locking effect can be adjusted externally in several stages or continuously. Depending on how the locking torque is generated, a distinction is made in this category between semi-active and active systems. In the case of semi-active systems, the locking torque is generated without

Positive locking	self-regulating (passive)		controlled	
rigid and switchable	Speed sensing	Torque sensing	semi-active	active
Rigid shaft without differential (spool) Dog clutch, manually operated	Viscous coupling with fluid friction Multi-plate clutch with internal Feed pump (ViscoLok)	Multi-plate clutch actuated by axial gear forces Multi-plate clutch operated by profiled thrust rings Friction generation by axially acting gear forces *Thrust cone actuated by axially acting gear forces*	*Multi-plate clutch with pressure build-up by internal Feed pump, combined with Pressure control valves (Haldex clutch of the 2nd generation)*	Multi-plate clutch with electro-hydraulic Actuation *Multi-plate clutch with electromechanical Actuation* *Multi-plate clutch with electro-magnetic Actuation*
Weismann-Differential Detroitlocker				

Fig. 8.20 Classification of differential locks

auxiliary energy, but can be controlled. An example of this is the second generation Haldex clutch. Pressure is generated by a passive feed pump as soon as a differential speed is established. A control unit energizes electromagnetic pressure control valves for a driving condition-dependent pressure build-up at the multi-plate clutch. In active systems, the locking torque is generated by an actuator that is operated with auxiliary power. This means that any locking torque can be set at any time, irrespective of the driving condition. The plates are actuated electrohydraulically, electromechanically or electromagnetically. Active and controlled limited slip differentials, in conjunction with appropriate detection and determination of the driving condition, permit an optimum locking effect. The potential of an active and controlled limited slip differential in terms of driving dynamics is shown in the following section. Based on this, the following sections explain the compromises that must be made when using self-regulating differentials.

8.5   Active and Controlled Limited Slip Differentials

In the hydraulic limited slip differential shown at the beginning (Fig. 8.12), external energy is added to an actuator – in this case the hydraulic pump – for the actuation of the disks, which is why we speak of an active system. This becomes a controlled system when driving state variables are fed back in a control loop to generate the locking torque. These state variables, which are of particular relevance for the control of a limited slip differential, include wheel speeds, yaw rate, lateral acceleration and steering wheel angle. The use of closed control loops is prohibited in most professional racing series. This also applies to Formula 1, where the use of an active hydraulic system is, however, permitted. The main advantage of an active differential is that its characteristics can be adjusted during driving for different driving situations and track sections, but only the driver is allowed to do this. In controlled systems, this task is performed by a control unit which calculates the optimum locking torque in a few milliseconds.

Controlled limited slip differentials are (apart from off-road or all-wheel-drive vehicles) mainly used in sporty production vehicles or in near-production racing vehicles for amateur sport. An example of such a system is the electrohydraulic front axle differential lock from the Golf VII GTI Performance (see also [2]), which was already tested in the Scirocco GT24 during the 24-h race at the Nürburgring before it went into series production. In the meantime, this differential lock is also used in various other near-production racing vehicles of the Volkswagen Group. Figure 8.21 shows the installation situation and a section through the front axle differential lock.

The controlled front axle differential lock of the Golf VII GTI is flange-mounted to the manual gearbox as a separate module with its own housing, as shown in Fig. 8.21. The module also includes the right-hand axle shaft, which functions as an inner disk carrier and accommodates the right-hand drive shaft via a tripod. The outer disk carrier is designed as a hollow shaft and is pushed onto the differential cage via splines. The disks are actuated by a working piston. The hydraulic pressure required for actuation is generated by a

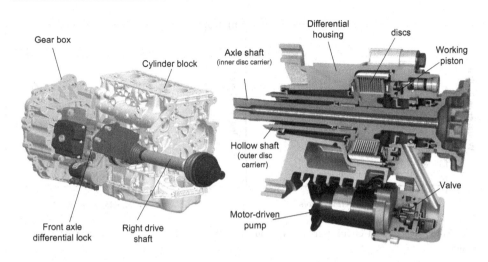

Fig. 8.21 Front axle differential lock in the Golf VII GTI Performance. (See also [2]; courtesy of © Volkswagen AG 2018. All Rights Reserved)

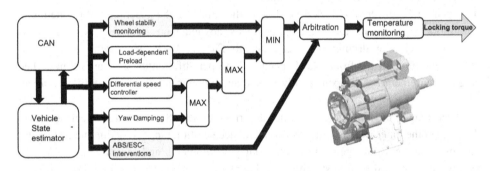

Fig. 8.22 Functional structure of the controlled front axle differential lock in the Volkswagen Golf VII GTI [2]

reciprocating pump driven by an electric motor. A centrifugal valve regulates the level of pressure so that the hydraulic pressure is directly proportional to the pump speed. The maximum locking torque is 1600 Nm. This is sufficient to fully lock the differential in any situation and to release the drive torque via just one wheel if required. The locking torque is adjusted depending on the driving situation by controlling the pump speed. Figure 8.22 shows the corresponding functional structure.

In the low-speed range, there is a throttle-dependent pre-load. This means that when the accelerator pedal is pressed, a low locking torque is applied, irrespective of the slip occurring at the drive wheels. This ensures optimum traction when starting off on all surfaces.

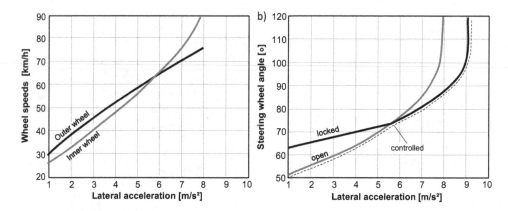

Fig. 8.23 Accelerated circular drive: (**a**) Course of the wheel speeds, (**b**) Steering wheel angle course

The central module of the function software is the differential speed controller. It monitors the wheel speeds and prevents one drive wheel from spinning up on one side. By opening the differential lock according to the situation, the generation of unintentional turning-out yaw moments is prevented. This advantageous feature is explained for accelerated circular driving with reference to Fig. 8.23a, b. The lateral acceleration-dependent curve of the front wheel speeds of an open differential and the steering wheel angle requirement for an open, a fully locked and a controlled limited slip differential are shown.

According to the kinematic properties described in Sect. 8.2, the outer wheels turn even faster than the inner wheels in the lower lateral acceleration range. Locking the differential in this situation would generate turning-out yaw moments and increase the understeer tendency of the vehicle. The steering wheel angle requirement of the fully locked differential in this range is significantly higher than that of the open differential. As lateral acceleration increases, the dynamic wheel load transfer increase. The inside front wheel is unloaded and develops increased drive slip. This slip requirement eventually causes the inside front wheel to overtake the outside front wheel. As the limit is further approached, the inside front wheel eventually spins, resulting in the characteristic power-on understeer of front-wheel drive vehicles. This behavior is reflected in the progressive increase in steering wheel angle demand. After overtaking the cornering outside front wheel, locking generates turning-in yaw moment that reduce the understeer tendency and redistribute the drive torque to the outer front wheel according to the dynamic wheel loads. The steering wheel angle requirement of the locked differential decreases and the maximum achievable lateral acceleration increases. By monitoring the wheel speeds, a controlled limited slip differential allows the disks to be actuated according to the situation. In this case, actuation only takes place after the inside front wheel has overtaken the outside front wheel. In this way, the potential of a limited slip differential can be optimally exploited, whereas negative effects cannot be completely avoided with speed- and torque-sensing differentials.

Figure 8.24 underlines the effectiveness of the controlled front axle differential lock during the passage of the Klostertal curve with a Golf VII GTI on the Nordschleife. The behavior of different variables for a vehicle with and without differential lock is compared. The turn-in process starts at 293 s, at 295 s the apex is reached. In this range, both vehicles still behave identically. At the apex, the acceleration process begins. Due to the high dynamic wheel load changes, the front wheel on the inside of the curve tends to spin very quickly. The controlled front differential lock counteracts this with a rapid build-up of locking torque. A turning-in yaw moment reduces the vehicle's tendency to understeer, which can be seen impressively in the steering wheel angle curves of the two vehicle variants. The steering wheel angle requirement of the vehicle without front axle differential lock is significantly higher on corner exit. The g-g diagram shows how with the front axle differential lock the lateral accelerations increase for the same longitudinal acceleration. The speed curve makes it clear that the vehicle with the controlled front axle differential lock exits the Klostertal curve with a considerable speed advantage. As a simple rule of thumb, a time saving of approx. 0.2 s per bend can be assumed, which adds up to a time advantage of around 8 s on the Nordschleife [2].

Another key advantage of a controlled differential lock is that yawing moments can be used specifically to stabilize the vehicle, e.g. by increasing yaw damping, and to reduce load change reactions. To this end, a yaw damping module has been integrated into the function software of the Golf VII GTI Performance. This module uses a vehicle model to calculate a target yaw rate, which is constantly compared with the measured actual yaw rate of the vehicle. If the control unit detects an oversteer situation, an outward turning yaw

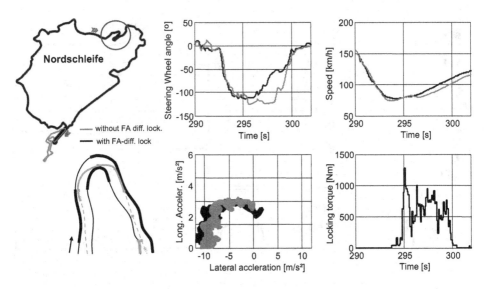

Fig. 8.24 Driving through the Klostertal curve [2]

moment is generated by actuating the disks, which reduces the yaw rate. The effectiveness of this module is shown in Fig. 8.25 using a single lane change simulating a sudden evasive maneuver. The test shown was conducted on a wet road with a normal driver and with ESC deactivated. The diagrams show the progression over time of the steering wheel angle, yaw rate and locking torque for a vehicle with the front axle differential lock activated and deactivated. It can be seen that with the start of the reverse steering movement (at approx. 0.45 s) after leaving the first lane, a turning-out locking torque is applied, which leads to a faster yaw rate reduction compared to the vehicle with the conventional differential. Without the controlled front differential lock, the violent reverse steering movement leads to an oversteering vehicle reaction when entering the second lane (after approx. 1 s). The yaw rate does not reduce in line with the driver's steering movement. In this phase, the front axle differential lock stabilizes with about 1000 Nm of locking torque. This supports the driver and the yaw rate is reduced much faster compared to the second vehicle. The steering angle requirement for the final countersteering movement is reduced, and the vehicle returns to straight-ahead driving sooner. The strain on the driver is reduced and confidence in the vehicle increases.

The controlled front axle differential lock is also used in the Golf VII GTI for active damping of oversteering load change reactions. Figure 8.26 shows the implementation of this function for a power-off load change from an accelerated circular drive. Vehicles with limited slip differentials – even with non-controlled systems – generally tend to have a more neutral load change response (dashed line) than vehicles with conventional differentials. This is due to the fact that the effects that cause an oversteer reaction, such as the wheel load transfer to the front axle and the reduction of drive slip, are superimposed by the shortfall of the turning-in locking torque. The turning-in yaw moment cannot be

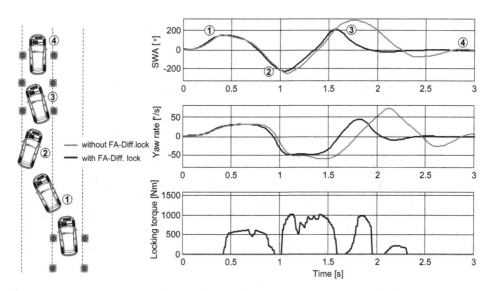

Fig. 8.25 Stabilization during lane change [2]

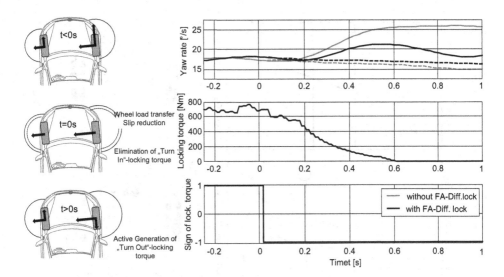

Fig. 8.26 Stabilization during power-off load change [2]

maintained because the wheel speeds want to return to their kinematic speeds after the load change. The sign of the locking torque is reversed at this point, and turning-out yaw moments are generated. This effect can be used specifically for vehicle stabilization. In the application of the front axle differential lock, the dynamics of the locking torque reduction are modeled in a defined way. The following statement applies: The slower the locking torque is reduced after the load change, the lower the load change oversteer.

In the Golf VII GTI, the overall locking torque is controlled by a wheel stability monitoring system. This monitoring function ensures that both front wheels do not spin at the same time when traction control is switched off and the front axle thus abruptly loses its capability of transmitting lateral forces. Figure 8.27 explains the function for different vehicle variants accelerating from the apex. In Fig. 8.27a the variant with open differential is shown first. As already explained, the dynamic wheel load changes cause the inside front wheel to spin. This loses its ability to transmit lateral forces. On the other hand, the outer front wheel retains this ability, since – as shown in Sect. 8.2 – the power of the engine is used up for the acceleration of the drive train.

This property is cancelled out by locking the differential so that, as Fig. 8.27b shows, both front wheels spin simultaneously as soon as the adhesion potential on the outer front wheel is exhausted. This results in a loss of lateral forces at both front wheels. The result is a sudden massive understeer ("snap understeer"). In the case of a rear-wheel drive vehicle, the result is a sudden oversteer ("snap oversteer"). It is this snap-off of the front axle that is prevented by the wheel stability monitoring system. For this purpose, the wheel speed of the outside front wheel is compared with the speed of the outside rear wheel. If the deviation is too great, the differential lock is deliberately opened so that the inside front wheel can turn up again independently of the outside wheel. In this way, abrupt understeer

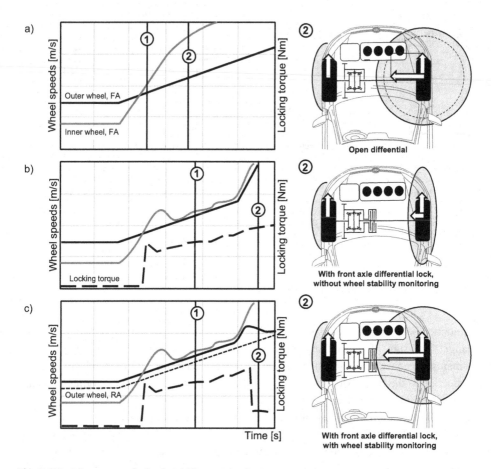

Fig. 8.27 Advantages of wheel stability monitoring

is prevented and the grip level is at least equal to that of a vehicle without differential lock. Such a function can only be implemented with a controllable system.

In the Golf VII GTI, the differential lock opens when braking is detected, as driving stability is ensured by the slip control systems. However, for a variant used in racing, functions have also been developed which exploit the stabilization effects described in Sect. 8.3.3.

In summary, it is concluded from this section that the optimum locking torque is strongly dependent on the momentary driving situation. The essential requirements for the locking torque are summarized in Fig. 8.28. In addition, the load condition on the axle drive is also listed here, which is important in connection with the characteristics of torque and speed-sensing limited slip differentials.

Driving situation		Requirements for the Final drive	Load condition, Torque at the final drive
braking for a cornerup	ENTRY	Locking torque corresponding to the locking tendency of the wheels Preload if necessary, significant oversteering with reduction of the brake pressure decreasing Locking torque (approach to apex)	S ▦ Z Engine in overrun mode S: Overrun Mode, Z=Traction
Turn In	ENTRY	a) No locking torque, if vehicle is stable b) moderate locking torque, if vehicle is nervous	S ▦ Z Engine in overrun mode
Turn in Hairpin		little to no locking torque	S ▦ Z
Apex	MID	No locking torque, free rolling of the wheels	S ▦ Z
Accelerate from apex	EXIT	Maxmimum locking torque Preload, if necessary, when accelerating very hard	S ▦ Z
Power-Off Load Change	EXIT	High blocking locking torque and if necessary with preload if vehicle shows heavy power-off oversteer load change oversteer, depending on driver preferences	S ▦ Z
Highspeed		If necessary, defined preload to increase stability for Passage of unevenness	S ▦ Z

Fig. 8.28 Optimum locking torque depending on the driving situation

With controlled limited slip differentials, the optimum locking torque can be achieved by processing the driving condition signals in every driving situation. A Formula 1 driver is not allowed to use control loops, but he can adjust the locking behavior via the steering wheel in different curve sections and thus solve at least some target conflicts. Figure 8.29 shows the corresponding dials on the steering wheel of a Formula 1 car.

For self-regulating limited slip differentials, it is the race engineer's job to adjust the differential to achieve the best possible compromise of driving dynamics potentials and negative interactions.

8.6   Torque-Sensing Limited Slip Differentials

Torque-sensing limited slip differentials have the property that the generated locking torque is proportional to the input torque at the axle drive. This property is described by the locking value S. The following applies:

$$S = \frac{M_{Sperr}}{M_{FD}} = \frac{|M_3 - M_4|}{|M_3 + M_4|}. \tag{8.27}$$

The most commonly used torque-sensing limited-slip differential in racing is the Salisbury or multi-plate limited-slip differential, as shown in Fig. 8.30. Because of its high

Fig. 8.29 Steering wheel of the BMW Sauber F1.09 with the (white circled) setting options for the limited slip differential. (Courtesy of © BMW AG 2018. All Rights Reserved)

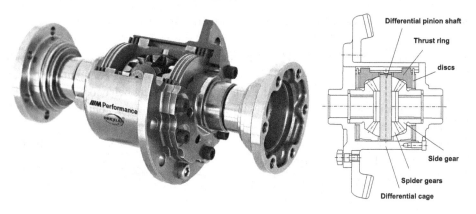

Fig. 8.30 Salisbury or multi-plate limited slip differential

variability, its range of applications extends from sporty production cars to high-performance race cars. The characteristic design feature of the Salisbury differential is that the differential pinions are not mounted directly in the differential cage, but in two pressure rings (Figs. 8.31 and 8.32). These pressure rings are guided in the differential cage via a groove so that they can move axially but are fixed against rotation. A ramp-shaped profile is worked into the pressure rings. The differential pinion is supported on this ramp. Behind the pressure rings is a package of inner and outer disks. The outer disks engage in the differential cage, the inner plates in the axle shafts. As soon as the differential has to transmit an input torque, the differential pinion and pressure rings press the disk packs together, and a frictional or locking torque is generated. Depending on whether driving or

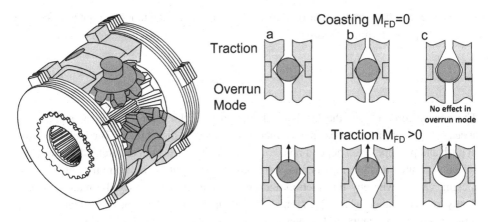

Fig. 8.31 Bearing arrangement of the pinion gears in two profiled pressure rings

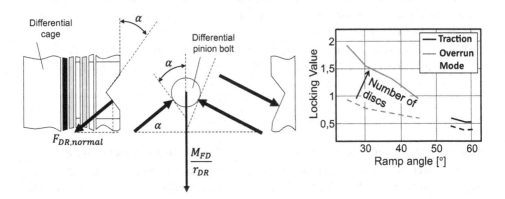

Fig. 8.32 Frictional torque generation in pressure ring profiling

engine braking torques are received at the differential input, support is provided on the upper or lower side of the profiling. The ramp angles for traction and engine braking are usually designed differently. The pulling direction is designed for optimum utilization of the adhesion potential during acceleration, the pushing direction for increased load change and braking stability. If necessary, the locking effect can be completely eliminated, e.g. if braking stability is already ensured by slip control systems, as is the case with series-production vehicles.

Figure 8.32 shows the geometric relationship between the locking effect, ramp angles and input torque. A force proportional to the final drive torque M_{FD} is generated at the differential pinion (cf. Fig. 8.4). This force is supported perpendicularly on the contact surface between ramp and differential pinion and can be divided into an axial component (in the direction of movement of the pressure rings and disks) and a tangential component.

The axial component creates a normal force that compresses the inner and outer disks. The following applies to this component:

$$F_{DR,axial} = \frac{M_{FD}}{r_{Dr}} \cdot \frac{1}{\tan(\alpha)} \cdot \qquad (8.28)$$

The normal force causes the mechanical friction or locking torque, which impedes the speed compensation or causes an asymmetrical drive force distribution. The locking values in push and pull operation depend on the ramp angle in the form of the arctan function. The selection of the pressure rings is one of the adjustment options for optimum setting of the locking values, whereby the traction and engine braking settings can be made independently of each other using the ramps. A very large spread of locking values can be represented via the ramp angles, whereby locking values significantly greater than 1 can also be realized due to the wedge effect. However, such large locking values are only used on the engine braking side in order to compensate for the relatively low braking torques of the engine. Figure 8.32 shows the measured locking values of some setup variants for the application in a GT vehicle.

If the locking value and ramp angle of a Salisbury differential are known, the locking value resulting from an exchange of the pressure rings can be predicted as follows:

$$S_{neu} = S_{alt} \cdot \frac{\tan(\alpha_{neu})}{\tan(\alpha_{alt})} \cdot \qquad (8.29)$$

In addition, the locking value can also be influenced via the disks. A frictional torque only arises between the surfaces of an outer and an inner disk. If one outer and one inner disk are stacked next to each other, the highest possible locking effect is obtained. The locking value can be reduced by placing identical disks next to each other. This is proportional to the effective number of contact surfaces z. Figure 8.33 shows an example of the halving of the locking value by rearranging the inner and outer disks. A rearrangement – in contrast to the ramp angles – always affects the traction and engine braking operation simultaneously. For a left and right symmetrical layering of the disks, the following applies in the case of a rearrangement:

$$S_{neu} = S_{alt} \cdot \frac{z_{neu}}{z_{alt}} \cdot \qquad (8.30)$$

It should be noted that Eq. (8.30) is only valid for symmetrical layering. In the case of asymmetric stratification, different locking effects occur in left-hand and right-hand curves in both traction and engine braking modes. In general, a Salisbury differential can have four different locking values depending on whether it is in traction or engine overrun mode, or whether the left or right wheel is turning faster. The locking action is described by a four-square diagram as shown in Fig. 8.34. There, the progression of the signed locking torque

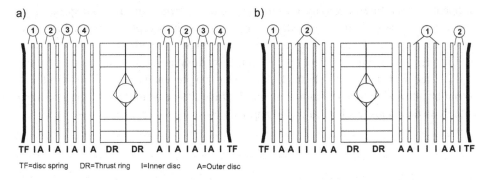

Fig. 8.33 Halving of the locking value from (**a**) to (**b**) by regrouping (TF, disk spring; DR, pressuere ring; I, inner disk; A, outer disk)

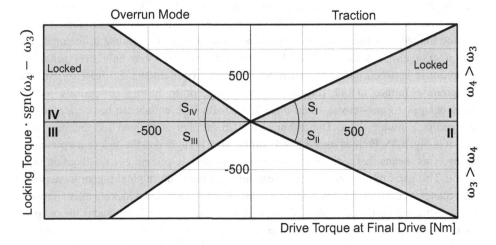

Fig. 8.34 Four-square diagram of a torque-sensing limited slip differential

as a function of the axle torque is plotted in the four quadrants. The slopes of the four straight lines correspond to the locking value effective in the respective quadrant.

The main advantage of torque-sensing limited slip differentials is that their locking effect starts immediately with the drive torque. This means that the locking torque builds up even before a wheel starts to spin. However, the proportionality to the axle torque also has two noticeable disadvantages: When the engine is in braking mode, only a fraction of the engine torque is available compared to traction mode. This can be partially compensated for by very shallow ramp angles. But even this measure is usually not sufficient to prevent a wheel from locking when braking, especially at high vehicle speeds. There, the available axle torques drop further due to the required gear ratio. Torque-sensing limited slip differentials therefore offer only limited stabilization potential in the high-speed range. Furthermore, locking torques that reduce agility must also be accepted within certain limits. In addition, it should be noted that a torque-sensing limited slip differential in the case of a completely unloaded drive wheel (e.g. μ-split, curb crossing) cannot prevent the spinning of this wheel with a sensibly selected locking value. To prevent wheel spin, no drive torque would be allowed to be directed to that wheel. Drive torque would have to be transferred entirely to the opposite wheel. It would apply:

$$M_3 = 0, \tag{8.31}$$

$$M_4 = M_{FD}. \tag{8.32}$$

According to the definition of the locking value according to Eq. (8.27), the locking value in traction operation would then have to be at least 1. The vehicle would then behave under traction practically like a vehicle with a rigid wheel connection. This is not always sensible under traction. In extreme situations with very low adhesion potential on one side, a torque-sensing limited slip differential behaves in principle like a conventional differential. The design of a torque-sensing limitedslip differential can therefore only ever achieve a compromise between the various driving dynamics requirements. In order to shift the compromise further in one direction or the other, or to further compensate for the disadvantages, torque-sensing multi-plate limited-slip differentials can be operated with a pre-load or a counter-load. The technical implementation of preload and counterload is shown in Fig. 8.35. Preload means that a normal force is applied to the disks via a preloaded spring. This means that a defined locking effect is already present even without an input torque. This can improve the response behavior, especially during braking, or increase the load change stability. Even when a wheel is lifted, at least a residual locking effect remains. The disadvantage is the permanent loss of agility. The counterload has exactly the opposite effect. A preloaded spring pushes the pressure rings apart. Thus, the disks are only pressed together as soon as the axle drive torque compensates for the counteracting force, which reduces the agility-reducing locking effect of these differentials, especially in tight curves.

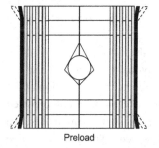
Preload

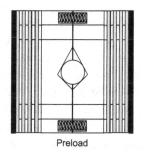
Preload

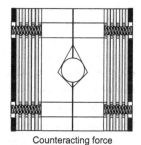
Counteracting force

Fig. 8.35 Generation of preload and counteracting forces

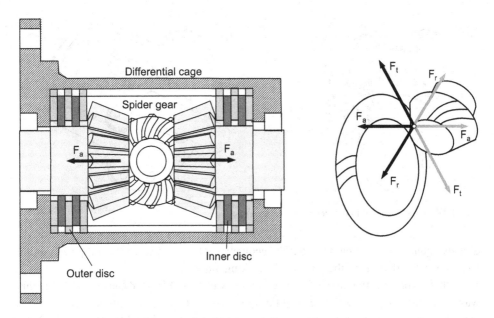

Fig. 8.36 Dana-Trac-Loc multi-disc limited slip differential

Due to the great flexibility in the design of the locking value, the principle of the Salisbury differential has established itself as a standard solution, particularly in high-performance racing.

Another type of torque-sensing multi-plate limited slip differential is the Dana-Trac-Loc differential shown in Fig. 8.36, which operates on a similar principle to the Salisbury differential. A set of disks is located between the differential cage and the left axle shaft. The axial force required to actuate the disk pack is generated here by the gearing forces occurring in the tooth flanks. A contact surface is created in the three-dimensionally shaped helical teeth of the differential pinion and side gears, on which the resulting gearing force vector is supported vertically. This force vector can be decomposed into an axial, tangential and radial component. The tangential component F_t generates the torque, the axial component F_a the normal force acting on the disks. The radial component essentially generates friction in the bearing.

Probably the best-known type of torque-sensing limited-slip differential, however, is the Torsen differential, whose function is already included in its name (TorSen = "torque sensing"). Torsen differentials have become popular mainly due to their use in Audi's all-wheel drive vehicles, which are marketed under the term "quattro". They exist in a variety of designs. Figure 8.37 shows the type A and B Torsen differentials, which are suitable for use as final drives. Torsen differentials also use the force components of helical-toothed worm and gear wheels to generate frictional torques. However, the normal forces generated in the gears do not act on disk packs; instead, the output gears are pressed

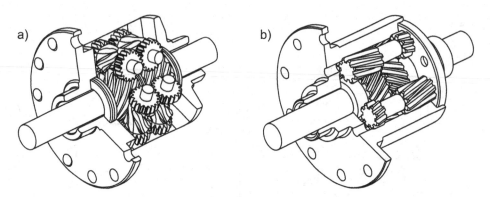

Fig. 8.37 Torsen differentials: (**a**) Type A, (**b**) Type B

directly against the differential cage. The geometry of the gearing and the properties of the friction surface determine the level of the locking value.

With the Torsen differential of type A, locking values of 0.1 to 0.8 can be realized in this way. Since the two differential gears of this type move in the same direction under load, but have different directions of movement in engine braking operation, different tuning for the two operating modes can be carried out via the condition of the two friction surfaces between the left and right side gears and differential cage. In principle, the Torsen differential type A also allows distributions that deviate from a 50:50 basic distribution, but this is only important for use as a central differential in all-wheel drive vehicles. With the type B Torsen differential, locking values of 0.2 to 0.5 can be achieved. It is used, for example, in the Alfa Romeo 147 Q2 and the Ford Focus RS (up to model year 2016).

As these two examples already indicate, Torsen differentials are used today as final drives essentially in sporty production vehicles. Compared to multi-plate limited-slip differentials, however, Torsen differentials only have less flexibility when it comes to tuning the locking value. They are therefore of rather less interest for modern high-performance racing. At the end of the 1980 s, however, they were used in the McLaren-Honda MP4/4 in Formula 1 [3].

Further advantages of torque-sensing differential locks, which they offer above all in comparison to viscous clutches, are their locking properties, which are largely independent of service life and ambient temperature.

8.7 Speed-Sensing Limited Slip Differentials

The best-known representative of speed-sensing limited slip differentials is the Visco clutch. A characteristic feature of Visco clutches is that they transmit torque by fluid friction and not by mechanical friction. The basic operation is explained by the thought experiment in Fig. 8.38. The figure shows on the left two discs or plates enclosed by a

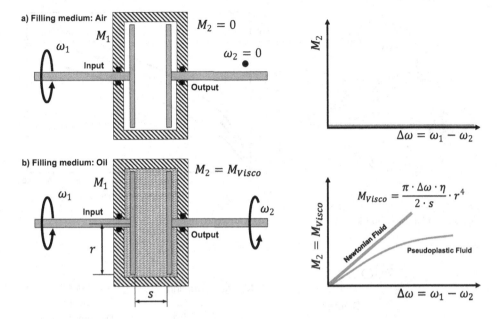

Fig. 8.38 Thought experiment on the Visco coupling

sealed housing, each connected to a driving and a driven shaft. In case a the housing is filled with air and in case b with oil. If the housing is filled with air, the output shaft remains stationary when the input shaft rotates. No torque is transmitted to the output side. If, on the other hand, the housing is filled with oil, a speed is also set on the output side as soon as the input shaft is driven. The fluid between the plates therefore transmits a torque M_{Visco} from the input to the output side. Ideally, Newton's law of shear stress applies to the torque transmission, which is as follows:

$$M_{Visco} = \frac{\pi \cdot \Delta \varpi \cdot \eta \cdot r^4}{2 \cdot s}. \tag{8.33}$$

This linear dependence applies to the so-called Newtonian fluids for which the viscosity η is not dependent on the differential speed $\Delta \omega$. However, silicone oils, as used in Visco couplings, belong to the pseudoplastic viscous fluids, i.e. their viscosity decreases with increasing differential speed. Thus, the degressive course of the viscous or blocking torque over the differential speed, which is typical for the viscous coupling, is obtained and is also shown in Fig. 8.38. According to Eq. (8.33), the properties of the Visco coupling can be tuned for the particular application by the geometry, the number of disks as well as the properties of the silicone oil. A comprehensive description of the properties of viscous couplings and their design criteria can be found in [4].

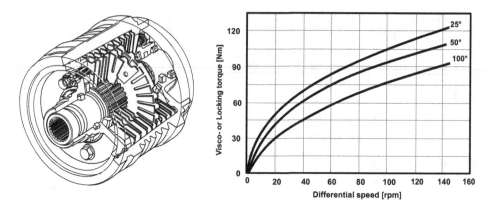

Fig. 8.39 Cut-away and locking torque curve of a Visco coupling

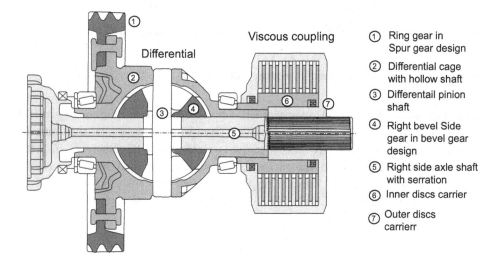

Fig. 8.40 Arrangement of the viscous coupling next to the differential

Figure 8.39 shows the typical structure of a Visco coupling for use in a drive train. The housing of the Visco coupling acts as an outer disk carrier and is connected to a shaft via splines. A short shaft is rotatably mounted inside the housing, which acts as an inner disk carrier and is sealed against the housing by a sealing ring. A serration on this side also allows a shaft to be fitted. The outer and inner disks are arranged at a defined gap width *s to* each other. Figure 8.40 shows a possible installation scenario for the use of a Visco coupling in a final drive. The housing with the outer disks is connected to the right-hand side shaft via splines. The inner shaft of the Visco coupling is designed as a hollow shaft and is placed on the differential cage via splines.

Figure 8.39 also shows the characteristic degressive locking torque curve and its dependence on the oil temperature. The transmittable locking torques decrease with increasing temperature, since the viscosity of the oil decreases with increasing temperature. The degressive course of the locking torque has the fundamental disadvantage that when high locking torques are generated, a relatively high power loss is also always produced, which heats up the silicone oil in the coupling. This creates a kind of vicious circle, since the rising temperatures cause the differential speeds to rise further due to the falling blocking torque. At the same time, this process leads to increasing heating of the silicone oil. The heating of the silicone oil causes an increase in the internal pressure of the coupling. When designing the Visco clutch, it must therefore be ensured that the permissible operating temperature of the silicone oil is not exceeded and that the internal pressures do not cause the differential to burst or the seal to fail.

One way to protect the Visco clutch is to use the hump effect. Here, the increasing internal pressure mechanically presses the disks against each other so strongly that they overlock the differential. As a result, no more differential speeds occur, allowing the oil to cool down. After the oil has cooled down, normal operation is resumed. For an application in racing vehicles, however, the unsteady overlocking is not desirable, which is why Visco clutches can also be designed without hump effect for this application. In normal operation of the viscous coupling, no synchronism is established between the shafts. The locking torque is transmitted under slip, which is why the theoretical lateral force potential of a driven axle is not fully achieved. The viscous coupling also operates more slowly than a torque-sensing coupling, since the buildup of locking torque always requires a differential speed first. However, as a rule, its locking effect is also more gentle as a result.

However, the independence of the locking torque from the drive or engine brake torque is of particular importance for the driving dynamics of racing vehicles. In this way, the stabilizing effect of the Visco clutch can be used during braking processes if, for example, only low locking torques can be generated with torque-sensing locks when braking from high speed due to the high gear and the relatively low engine brake torque. A fundamental disadvantage is that turning-out torques are also generated at the apex of the curve. This understeer effect is all the stronger the tighter the curve. The setting of a viscous coupling is therefore also subject to compromise.

8.8   Torque and Speed Sensing Limited Slip Differentials

In high-performance racing cars, limitedslip differentials are often used that have both a torque-sensing and a speed-sensing component. This is the most efficient compromise when the use of active differentials is prohibited by the regulations. Figure 8.41 shows the structure of such a limited slip differential. The torque sensing portion is integrated into the differential cage as a Salisbury differential. The viscous coupling is connected to the two axle shafts by a complex construction. The outer disk carrier of the Visco clutch sits on the right-hand axle shaft and is connected there to the tripod mount for the constant velocity

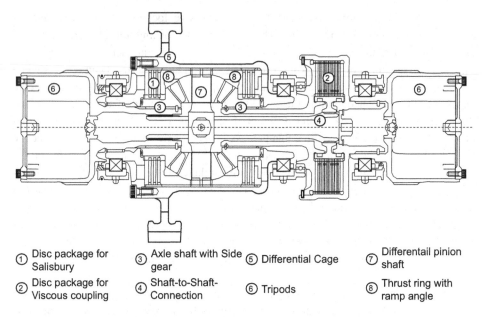

①	Disc package for Salisbury	③	Axle shaft with Side gear	⑤	Differential Cage	⑦	Differentail pinion shaft
②	Disc package for Viscous coupling	④	Shaft-to-Shaft-Connection	⑥	Tripods	⑧	Thrust ring with ramp angle

Fig. 8.41 Combination of Salisbury differential and Visco coupling

joints. The right-hand side shaft is designed as a hollow shaft and is inserted into the right-hand side gear of the bevel gear differential. An auxiliary shaft is pushed into the left axle shaft, which extends the axle shaft and passes through the right axle shaft. This auxiliary shaft acts as the inner disk carrier of the Visco coupling. This design combines the advantages of the torque-sensing limited slip differential with those of the Visco clutch. In this case, the Visco clutch can be designed without hump effect without any problems, since the torque-sensing multi-plate limited slip differential ensures that the wheel differential speeds are minimized under load, thus sufficiently reducing the thermal load on the Visco clutch. A significant power input into the Visco clutch then only takes place if one wheel tends to lock in the event of a braking operation.

Such limited-slip differentials are or were used, for example, in the Porsche RS Spyder (LMP2), in ChampCar or GT vehicles.

8.9   Spool

In the meantime, torque and speed-sensing limited slip differentials have become part of the standard equipment of even very powerful racing vehicles. Until late in the 1980s, the reliability of these components still left a lot to be desired or required an extreme development effort. Many high-performance racing vehicles therefore used a rigid connection between the wheels instead of a limited slip differential, which is known as a "spool". With a spool (Fig. 8.42) the differential speed between the two driven wheels is always zero,

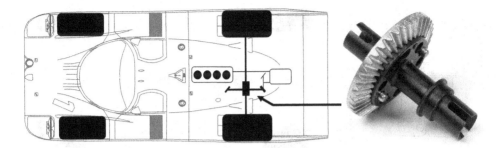

Fig. 8.42 Transaxle drive train with spool in the Porsche 956

which leads to the disadvantages explained in Sect. 8.2. The main disadvantage of the rigid connection is the strong understeer at the turn-in ("turn-in understeer"). However, the resulting loss of time is more than compensated for by powerful vehicles during acceleration on the exit of the bend. A spool was used, for example, in the Porsche 917, the Porsche 956 and the Sauber Mercedes C9 (1989), all of which won the 24 Hours of Le Mans. Spools operate without wear and allow a return to the pits even if a drive shaft breaks, which can be of great importance in endurance racing. However, due to the disadvantages in terms of driving dynamics, spools are hardly used today. Basically, they are only an alternative to the conventional differential if the technical regulations do not allow any other option. Spools cannot be used in conjunction with brake control systems.

8.10   One-Sided Brake Application

During the 1997 Formula 1 World Championship, a small additional foot pedal on the McLaren-Mercedes MP4/12 caught the eye. With the introduction of semi-automatic gearboxes, the clutch pedal had actually become needless. The clutch was now only required to be operated for the starting process, which was, however, carried out by hand. Nevertheless, as Fig. 8.43 illustrates, a third foot pedal was located to the left of the brake pedal. This foot pedal could be used to apply brake pressure to one of the two rear wheels independently of the main brake system. For this purpose, the foot pedal was coupled to its own brake cylinder, which was connected to the brake line to the rear wheel via a T-piece. With this one-sided brake actuation, the driver was able to impose a yaw moment on the vehicle, which primarily counteracted the vehicle's tendency to understeer. This system allowed the vehicle to tend to have a more stable set-up, which was achieved mainly by reducing the rear stabilizer stiffness. This measure additionally helped the vehicle to improve traction. The advantage of the system was put at around 0.3 s per lap. The presence of the system could be detected by the fact that at the exit of the corner – despite clear vehicle acceleration – one rear brake disc was partially glowing. In order not to get into a grey area of the regulations, McLaren decided to let the system work on only

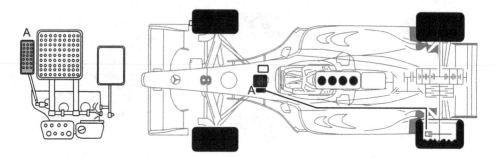

Fig. 8.43 One-sided brake application on McLaren-Mercedes MP4/13

one rear wheel [5, 6]. So the system only worked in one cornering direction. However, a switchable version was to be added for the following season. In the first race of the 1998 season, the McLaren-Mercedes MP4/13 took a superior double victory, which intensified the protests against the one-sided brake intervention. By the second race of the season, the system had already been banned as unacceptable all-wheel steering. This argumentation was quite controversial. The technical regulations were subsequently supplemented by a clear provision requiring identical brake pressure on both wheels of an axle. The fundamental disadvantage of using the brake to generate yaw moments is that high power losses occur, as these are directly proportional to braking torque and wheel speed. When the wheel is not locked, the peripheral wheel speed is approximately equal to the vehicle speed. The following therefore applies approximately to the power loss in a brake:

$$P_{Brake} = M_{Brake} \cdot \frac{v}{r}. \tag{8.34}$$

Due to these high power losses, unrestricted use of the brake to influence the driving condition is neither possible nor sensible. However, it can be used for a short time to simulate a differential lock or to stabilize critical driving conditions, as is done as standard in conjunction with modern brake control systems in production vehicles. The losses that occur are significantly lower when the brake intervention is used as a supplement to an existing limited-slip differential.

8.11 Superposition Differentials and "Torque Splitters"

Superposition differentials are used in very sporty variants of today's production vehicles and also in some SUVs. These have their origins in rallying, where they were first used by Mitsubishi in the 1990s. Superposition differentials can create asymmetrical drive force distributions virtually independent of the driving condition, thus influencing the driving condition regardless of the load condition and differential speed of the wheels. Influencing

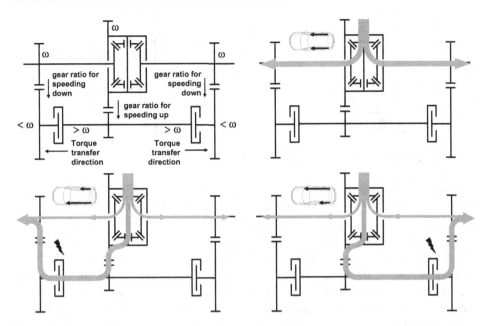

Fig. 8.44 Functional principle of a superposition differential

the driving condition by asymmetrical drive force distributions is also commonly referred to as "torque vectoring". A mechanically operating superposition differential requires two independent clutches to represent this characteristic. The principle of mechanical superposition differentials consists of a specific generation of relative speeds between two differential components, which are coupled via a clutch. Figure 8.44 shows the functional diagram of superposition differentials using the example of a relatively simple transmission structure. The differential cage drives a transmission shaft, which is translated to high speed via a spur gear stage. At each end of the transmission shaft there is a coupling half. The second coupling half is driven by the corresponding axle shaft via a spur gear stage that is translated into low speed. This results in a defined differential speed and transfer direction in the clutches, so that in this case actuation of the right-hand clutch causes a drive torque transfer to the right-hand wheel, while actuation of the left-hand clutch transfers the drive torque to the left-hand wheel. In a left turn, the first case corresponds to an in-turning yaw moment, while the second case corresponds to an out-turning yaw moment. When the clutches are not actuated, the system behaves like a conventional differential.

In the known series applications, controlled clutch systems are used exclusively, whereby electromechanical, electrohydraulic or electromagnetic actuators can be used just as in the active limited slip differentials. In series-production vehicles, much more compact transmission structures are used, which are compiled, for example, in [7]. Audi uses such a principle under the term "sport differential", at BMW it is used under the term "Dynamic Performance Control".

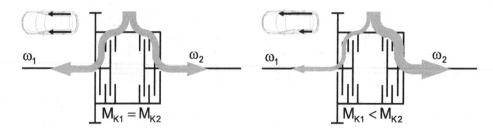

Fig. 8.45 Functional principle of a torque splitter

So-called torque splitters distribute the drive torque, as Fig. 8.45 shows, without basic distribution through a differential directly via two controlled clutches. With this arrangement, just as with superposition differentials, it is in principle possible to distribute incoming drive andengine brake torques between the wheels of an axle regardless of the sign of the wheel differential speed. However, inward yaw torques are limited to the level of the input torque, which results in a dependency on the load condition of the engine, similar to limited slip differentials. Torque splitters in single-axle drive vehicles have so far only been implemented in technology carriers or demonstrator vehicles. In current production vehicles, torque splitters are always used as "hang-on" four-wheel drive clutches. One well-known application is the Ford Focus RS from the 2016 model year, which uses a torque splitter on the rear axle.

8.12   Technical Regulations for the Final Drive

The technical regulations in force today only permit the use of limited slip differentials for drive torque distribution. This is achieved by the following formulation:

▶   Any system or device capable of transferring or diverting torque from a slower rotating wheel to a faster rotating wheel is prohibited.

The use of torque splitters and superposition differentials is already no longer permitted with this simple regulation. If the use of controlled and active limited slip differentials is also to be excluded, the following supplement is added.

▶   Only mechanical limited slip differentials that operate without the aid of a hydraulic or electric system are permitted. Visco-clutches are not considered to be a hydraulic system provided that their behavior cannot be modified while the vehicle is in motion.

In this case, therefore, only torque-sensing limited slip differentials and Visco clutches or a combination of both can be used. It has already been mentioned in Sect. 8.3.3 that the use of couplings on non-driven axles is not permitted. By two to three sentences, this provides all the essential information on the use of final drives in high performance racing.

For the use of limited-slip differentials in near-series racing vehicles, there are usually a number of supplementary regulations which state, for example, that the limitedslip differential must be housed in the series-production casing and what materials its components must be made of.

8.13   Weismann Differential and "Detroit Locker"

In connection with spools, it has already been mentioned that a central challenge for the use of limited-slip differentials in high-performance racing vehicles was reliability (Sect. 8.9). One of the first axle drives that was able to solve the conflict between an open and a spool was the Weismann differential [8] (Fig. 8.46). It was used, among others, in the 527 hp. McLaren M6A, which won the North American CAN-AM title in 1967.

In a Weismann differential, the axle shafts are not connected to the differential cage via gears, but via cylindrical rollers. Under load, the cylindrical rollers are driven from position A to position B [7] because the differential cage is accelerated by the axle gear torque and tries to overtake the axle shafts. During this process, the differential components tighten and become a rigid connection. Without load, the cylindrical rollers loosen again. It should be noted here that when the vehicle transitions into a load-free cornering maneuver, initially only the outside wheel will disengage, as the differential cage rotates faster than the inside wheel. Actuating the accelerator pedal then initially leads to an understeer impulse, as the drive force on the inside wheel of the curve generates a turning-out yaw

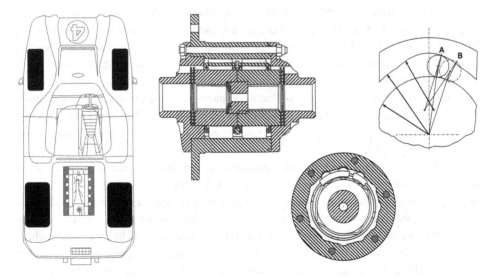

Fig. 8.46 Operating principle of a Weismann differential [7, 8]

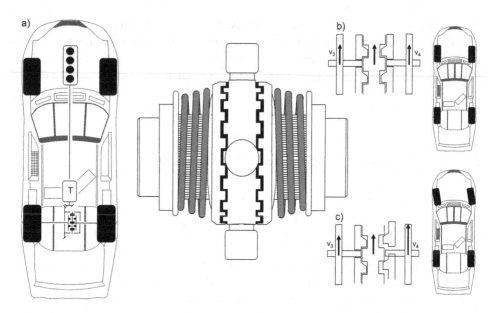

Fig. 8.47 Functional principle of the Detroit Locker [7]

moment. To avoid this effect as far as possible, the driver has to step on the accelerator strongly so that the differential cage takes the outside wheel along as quickly as possible. Only a few drivers have mastered this demanding driving technique to perfection. This is another reason why the Weismann differential is no longer used in modern racing cars, but it is still produced today as a spare part for historic racing cars.

Figure 8.47 shows a so-called Detroit Locker, which is a type of axle gearbox that is largely unknown in Europe. Detroit Lockers are used in the NASCAR vehicles of the Monster Energy Cup and in drag racing. Torque is transmitted from the drive wheel to each of the two driveshafts by a dog clutch. The axle shafts are connected to the two output shafts of the Detroit Locker. The coupling half, which is seated on the output shaft, is mounted on the shaft with a keyway so that it can move axially but is fixed against rotation. A return spring presses the coupling halves of the output shafts and the drive wheel against each other. This positive locking creates a rigid connection in straight travel. This condition is shown in Fig. 8.47a, b.

During load-free cornering, the release clutch arranged on a second pitch circle of the clutch halves becomes active. This operating principle is illustrated in Fig. 8.47c. Since the outside wheel turns faster than the inside wheel, it overtakes the differential cage. During this process, the coupling half runs over a ramp profile, whereby this coupling half is axially disengaged. The outside wheel can therefore rotate freely, while the inside wheel remains in positive engagement. This disengagement process takes place cyclically and is also noticeable acoustically. As with the Weismann differential, only the inside wheel is driven initially when accelerating into a corner, which is why NASCAR drivers enter the corner with a strong blast of the throttle. However, the Detroit Locker generally does not meet the high driving dynamic requirements of modern race cars.

In addition, the sliding block differential ("Cam and Pawl") from ZF should be mentioned here, which was one of the very first limited slip differentials and was used, for example, in the Mercedes W25 from 1935. Since it is no longer important today, no explanation is given here. However, its function is described in [9], for example.

References

1. Piola, G.: Formula 1 Technical Analysis 2004/2005. Giorgio Nada Editore, Vimodrone (2005)
2. Frömmig, L., Apel, A., Gieße, S.: Golf VII GTI – Fahrdynamik mit geregelter Differenzialsperre. 22. Aachener Kolloquium Fahrzeug- und Motorentechnik, S. 1493–1509. RWTH Aaachen (2013)
3. Bamsey, I.: McLaren Honda Turbo – a Technical Appraisal. Haynes Publishing, Sparkford (1990)
4. Pesch, W.: Die Wirkungsweise einer Visco-Kupplung und ihr Einfluss auf die Traktion eines Allradfahrzeuges. Dissertation, Universität Hannover (1989)
5. Collins, S.: Left foot forward. Racecar Engineering, Bd 19 No 9 (2009)
6. Piola, G.: Formula 1 Technical Analysis 1998. Giorgio Nada Editore, Vimodrone (1999)
7. Frömmig, L.: Simulation und fahrdynamische Analyse querverteilender Antriebssysteme. Dissertation, TU Braunschweig. Shaker, Aachen (2012)
8. Weismann, A.A., Weismann, P.H.: Positive Drive Differential. US-amerikanische Offenlegungsschrift – 3283611 (1966)
9. Trzesniowski, M.: Rennwagentechnik. Vieweg+Teubner, Wiesbaden (2008)

9.1 Overall Vehicle Concept

The term "overall vehicle concept" covers all the basic technical solutions or components that are required to perform the tasks of a vehicle. The vehicle structure or body-in-white acts as the carrier for the components. Figure 9.1 summarizes the main influences on the overall vehicle concept and its structure.

First of all, the occupants must be accommodated in the vehicle and the driver must be given access to the central controls, i.e. the steering wheel and foot pedals. In the case of racing vehicles, it should be noted that the driver's seating position has a major influence on the center of gravity. However, ergonomic aspects must also be taken into account to ensure that physical and mental performance is maintained for as long as possible. In the cockpit area, the central task of the structure is to provide occupants or pilots with a survival space in the event of an accident. Survival space means that the occupants in this area are protected from intrusion by their own vehicle parts and foreign objects, from fire and direct contact with obstacles. This zone must have minimal deformation in the event of an accident and is also known as the safety cell. In order to keep the loads on the human body within a range of high survival probability during a collision, the kinetic energy of the vehicle must be dissipated by defined deformations of the remaining vehicle structure. Compliance with these criteria is monitored for both series-production and racing vehicles by mandatory crash tests conducted by the approval or sports authorities. A further criterion for the design of the structure is the connection of the chassis, which is discussed in Sect. 9.4. The provision of usable space does not play a role in the case of racing vehicles.

The higher-level powertrain concept includes the definition of the driven axles and the positioning of the engine (or engines in the case of hybrid drive), manual transmission and differential. The positioning of the energy storage systems, i.e. the fuel tank and the

L. Frömmig, *Basic Course in Race Car Technology*,
https://doi.org/10.1007/978-3-658-38470-8_9

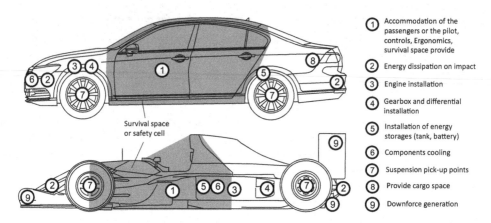

Fig. 9.1 Influences on vehicle concept and structure

batteries, and the arrangement of the cooling circuits are also part of the powertrain concept. This determines to a large extent the mass geometry and defines important basic driving dynamics properties of the vehicle. While the powertrain concept for production vehicles is essentially determined by market positioning and economic aspects, the focus for high-performance racing vehicles is exclusively on driving dynamics performance or, if applicable, compliance with the regulations.

Particularly in high-performance racing vehicles, the individual technical solutions of the aspects mentioned above are significantly determined by the interaction with aerodynamics, since optimum downforce generation often imposes a compromise solution on the other areas. Powertrain concept and structural design are key features of a racing vehicle. For this reason, both features are discussed in separate sections. In the case of near-production racing vehicles, the concept is already determined by the homologation vehicle, which is why concepts that are not optimal in terms of driving dynamics often have to be used. For this reason, in many racing series there is a "Balance of Performance (BoP)", by which the performance level of the different concepts is aligned. Typical measures to achieve a "Balance of Performance" are the specification of concept-dependent minimum weights, combustion air volume limits, tank capacities or comparable characteristics.

9.2   Powertrain Concept

With the engine, the transmission and the differential, with the tank and the batteries, the powertrain comprises the essential mass points of the vehicle. The arrangement of these components in the vehicle determines the basic mass geometry of the vehicle. The mass geometry primarily includes the position of the center of gravity in the longitudinal direction as well as the center of gravity height and the moment of inertia about the vertical axis. These parameters have a significant influence on the steering tendency, agility, traction, and braking and yaw stability of a vehicle. The dependence of traction and braking stability on mass geometry has already been shown in Fig. 5.47.

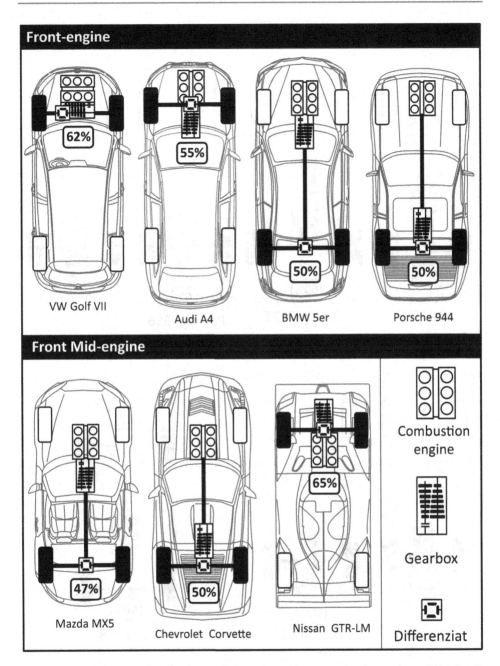

Fig. 9.2 Powertrain concepts with front and front-mid engine as combustion engine and single-axle drive

Figures 9.2 and 9.3 show the main drive concepts, divided into front-engine and front-mid-engine as well as mid-engine and rear-engine concepts, for vehicles with an internal

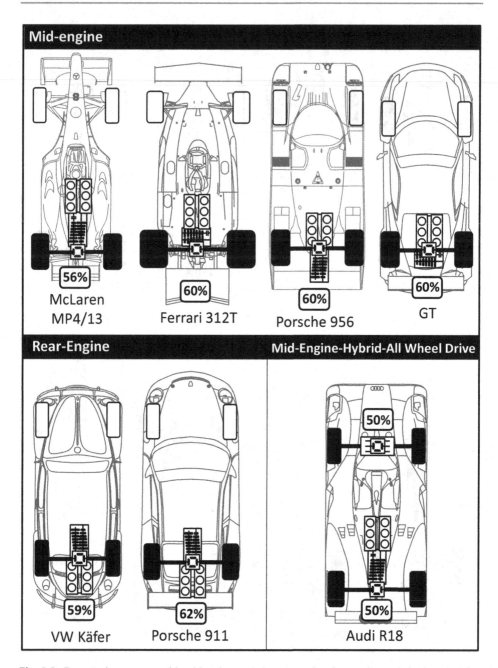

Fig. 9.3 Powertrain concepts with mid and rear engine as combustion engine and single-axle drive

combustion engine and single-axle drive. The hybridized powertrain of an LMP1 vehicle with electric all-wheel drive is also shown there. Mechanical all-wheel drive trains are not discussed here because, with the exception of rally cars, they do not play a role in high-performance racing (due to regulations).

The powertrain concepts are primarily characterized by the positions of the combustion engine, transmission and differential. The absolute weight distribution is further determined by the driver's position as well as the safety cell, the fuel tank, the cooling circuit and the other auxiliary units.

In addition to the static center of gravity position, traction and braking stability also interact with the dynamic wheel load transfer. The wheel load transfer to the rear axle that occurs during acceleration is compensated for in front-wheel drive vehicles by a higher static wheel load on the front axle. The majority of front-wheel-drive production vehicles in the A and B segments have a transverse engine which, together with the transmission, is located in front of the front axle. The weight share on the front axle is on average about 62%. The high weight on the front axle leads to a high slip angle there and thus to a relatively strong tendency to understeer.

An alternative concept for front-wheel drive is to install the engine longitudinally with a transmission located behind the front axle. This allows the weight share of the driven front axle to be reduced to around 55%, which reduces the tendency to understeer. This concept is used primarily by Audi in the B-, C- and D-Class vehicles. The longer wheelbase of these vehicles reduces the influence of dynamic wheel load transfer on traction capacity.

In the standard drive, which is typical of many BMW and Mercedes-Benz model series, among others, the longitudinally mounted front engine is coupled to the transmission, and the rear axle is driven via a cardan shaft. By shifting the engine to the front axle, a very balanced weight distribution is achieved. The weight share of the driven rear axle is about 50% for this concept. The wheel load transfer takes place to the driven rear axle. In a transaxle design, the manual transmission is separated from the engine and located on the driven rear axle. The Porsche 944 also achieved a roughly balanced weight distribution in conjunction with a transaxle concept and an engine located in front of the front axle.

With the front mid-engine, the engine moves behind the front axle. The Mazda MX-5, with a transmission flanged to the engine, has 47% of the weight of the driven rear axle. The Chevrolet Corvette's combination of front-mid engine and transaxle design increases the weight percentage of the driven rear axle to 50%. With the combination of front-mid engine and driven front axle, the Nissan GTR-LM features a very exotic drive concept. The front axle, with the transmission in front of it, carries a weight share of 65%, which is necessary to transfer the high performance potential of the vehicle to the road despite dynamic unloading of the front axle.

The standard configuration for high-performance racing vehicles is the mid-engine design, in which the engine is located in front of the driven rear axle. Variations of the mid-engine design result from the location of the transmission and differential. In today's Formula One cars, the engine and gearbox are mounted longitudinally, and the differential is at the gearbox end. Since 2013, the weight distribution for F1 cars has been fixed so that the rear axle accounts for 56% of the weight. A variant of this concept is the transverse installation of the gearbox, such as the Ferrari 312T, which has 60% of its static weight on the driven rear axle. However, this configuration was still found on many F1 cars in the early 1990s. In addition to the more favorable weight distribution resulting from the lower distance between the engine and gearbox centers of gravity, this gearbox position allowed the diffuser to rise earlier in the center of the car. However, the restrictions on the permissible length of the central diffuser duct (Fig. 6.79) introduced for the 1994 Formula One season cancelled out the aerodynamic advantages of this concept. The longitudinal installation of the gearbox allows a much narrower run in the "coke-bottle" section, so that longitudinal installation of the engine and gearbox is now standard in all F1 cars. In the Porsche 956, the gearbox is located behind the rear axle. This design is also sometimes used in GT cars with a transversely mounted gearbox.

The main advantages of the high weight on a driven rear axle are good traction and high braking stability as well as significantly more neutral handling. Another feature of the mid-engine concept is that the overall vehicle center of gravity and the engine center of gravity are very close together. This reduces the overall moment of inertia of the vehicle by minimizing the Steiner components. The vehicle responds spontaneously to steering inputs. However, the high agility also reduces stability at the limits, as high side slip angles can build up very quickly when oversteering. In principle, the mid-engine design allows a very wide spread in weight distribution. It ranges from the almost balanced weight distribution (approx. 50:50) of a modern LMP1 car to a significantly rear-heavy layout (approx. 35:65) for the ground-effect Formula 1 cars used in the early 1980s. Shifting weight to a driven rear axle increases traction, but also the lateral force required to support lateral acceleration. Figure 9.4 shows the mid-engine concepts of an Audi R18 e-tron quattro and a Sauber-BMW F1.08, underscoring the importance of this design for modern high-performance racing cars. The combination of high drive power and a high proportion of weight on the rear axle usually requires mixed tires. Likewise, the aerodynamic balance must be designed according to the mass and force distribution. In a rear-engine design, the engine is located behind the rear axle and the transmission is usually located in front of the rear axle. This design is typical above all for Porsche's 911 series. The weight share on the driven rear axle is about 62%.

For the reasons mentioned in Chap. 5, the reduction of the center of gravity height is of enormous importance in the design of racing vehicles. The engine design has a significant influence on the overall center of gravity. Figure 9.5 shows the V6 monoturbo of the Audi R18.

a)

b)

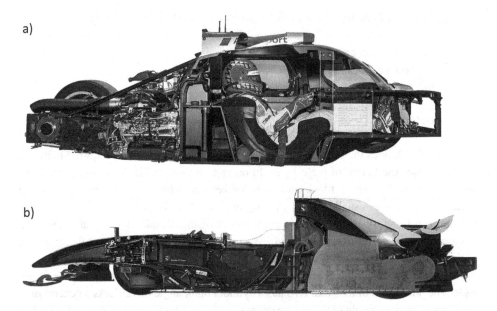

Fig. 9.4 Mid-engine concepts: (**a**) Audi R18 e-tron quattro. (2012; courtesy of © Audi AG 2018. All Rights Reserved), (**b**) Sauber-BMW F1.08. (2008; courtesy of © Sauber Motorsport AG 2018. All Rights Reserved)

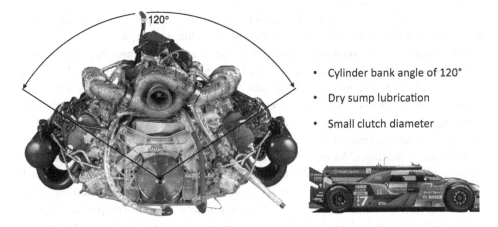

- Cylinder bank angle of 120°
- Dry sump lubrication
- Small clutch diameter

Fig. 9.5 4.0-liter monoturbo V6 TDI of the Audi R18. (Courtesy © Audi AG 2018. All Rights Reserved)

▶   The center of gravity height of the motor is essentially determined by

- the lubrication concept,
- the coupling diameter and
- the cylinder bank angle

Determined.

The oil pan defines the lowest point of the engine. The volume of oil required in the oil pan determines the height of the oil pan. In order to lower the heavy components of the engine such as the engine block and cylinder head as much as possible, a dry sump lubrication system is usually used in racing engines. In this case, the crankshaft drive does not run in an oil bath, but oil is pumped out of the oil pan or sump and delivered specifically to the points to be lubricated. This method has the advantage that the supply of lubricant to the engine is ensured even at high lateral accelerations.

Another factor that influences the engine's center of gravity height is the cylinder bank angle. The V6 engine of the Audi R18 has a cylinder bank angle of 120°. As a general rule, the center of gravity height of the engine decreases as the cylinder bank angle increases. For Formula 1 engines, the cylinder bank angle has been fixed at 90° since 2006. Before that, concepts between 72 and 111° existed. The reasonable height of the cylinder bank angle is often limited by the fact that the engine components should not penetrate the permissible volume for the rear diffuser.

Due to the low oil sump of the dry sump lubrication, the coupling diameter determines the contact point of the engine with the underbody in this concept, which can be seen in Fig. 9.6. The larger the coupling diameter, the higher the crankshaft must be, which in turn dictates the distance from the engine block and cylinder head to the underbody. The center of gravity of the engine moves with the axis of rotation of the crankshaft. A low clutch diameter therefore favors a low engine center of gravity. Not least for this reason, clutches for racing cars are made of extremely high-performance materials. The center of gravity of the gearbox can be lowered if the differential is as low as possible, which also offers aerodynamic advantages. An extremely shallow gearbox was used on the Williams FW33 in 2011. However, lowering the differential requires the drive shafts to be more inclined, which increases the deflection angles in the constant velocity joints. The deflection angles in the constant velocity joints degrade the efficiency of the drivetrain. This provides another example of why classical mechanical design goals are deviated from due to aerodynamic requirements.

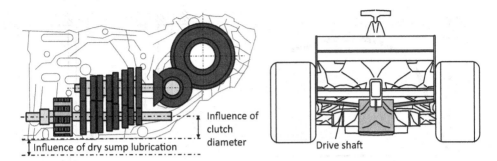

Fig. 9.6 Position of coupling, axle drive and drive shafts

9.3   Structural Designs

9.3.1   Basic Concepts of Lightweight Construction

The design of a load-bearing structure for automotive applications requires comprehensive competencies in the fields of mechanics, materials science and manufacturing technology, which are summarized under the term lightweight construction. Mechanics provides the relationship between the geometric properties of a component, the external forces and moments, and the loads in the internal structure. In conjunction with the material properties provided by materials science, the behavior of a load-bearing structure can be predicted. Finally, manufacturing engineering provides the processes required to produce the desired shape and, where appropriate, defines the trade-offs to be made in the execution of a design. The primary goal of lightweight design is to construct a structure that is load-optimized and weight-optimized. According to [1], the following lightweight design strategies exist to reduce the overall weight:

- Lightweight fabric
- Molded lightweight construction
- Concept Lightweight
- Conditional lightweight construction
- Lightweight manufacturing

Lightweight material design is concerned with the selection of materials that can perform their task with the lowest possible weight. Some of the essential characteristic properties of materials are determined, for example, by tensile tests. The result of such a tensile test is the stress-strain diagram shown in Fig. 9.7. In this diagram, the stress applied in the tensile specimen is plotted against the strain that occurs. The following applies for stress and strain:

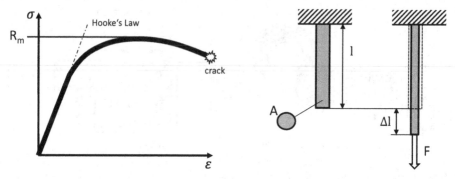

Material	$\rho \left[\dfrac{kg}{dm^3} \right]$	E[GPa]	G[GPa]	R_m [GPa]	$\dfrac{E}{g \cdot \rho}$ [km]	$\dfrac{G}{g \cdot \rho}$ [km]	$\dfrac{R_m}{g \cdot \rho}$ [km]
Steel	7,85	210	79,3	0,7	2675	1010	8,92
Aluminium	2,70	70	25,5	0,4	2592	944	14,80
Magnesium	1,74	45	17,0	0,3	2586	977	17,24
Titanium	4,50	110	41,4	1,0	2444	920	22,22

Fig. 9.7 Stress-strain diagram of a tensile specimen

$$\sigma = \frac{F}{A}, \tag{9.1}$$

$$\epsilon = \frac{\Delta l}{l}. \tag{9.2}$$

In the elastic range, the workpiece returns to its original shape after a load. This range is described by the hooked straight line. The modulus of elasticity E is derived from the slope of this straight line. It is defined as:

$$E = \frac{\sigma}{\epsilon}. \tag{9.3}$$

The higher the modulus of elasticity, the smaller the deformations on the workpiece. In addition to the change in length, the tensile specimen also experiences an elastic transverse strain, the so-called transverse contraction, as a result of which the diameter of the tensile specimen decreases. The shear modulus G results from this deformation. For isentropic materials, the modulus of elasticity and shear modulus are interdependent via the Poisson's ratio.

$$G = \frac{E}{2 \cdot (1 + \nu)} \tag{9.4}$$

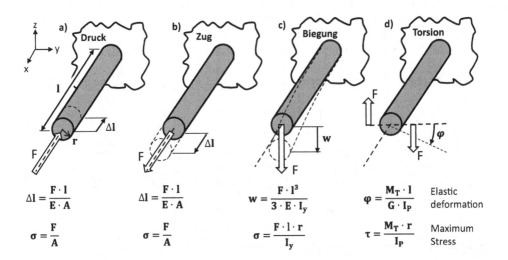

Fig. 9.8 Basic load types of a simple structure in (**a**) compression, (**b**) tension, (**c**) bending and (**d**) torsion

As the stress σ increases, the tensile specimen undergoes permanent plastic deformation and eventually fails. The maximum stress occurring until the tensile specimen fails is referred to as the tensile strength R_m. In simple terms, the design of a high strength and stiff structure requires the use of materials that have high elastic moduli and strength values at low density. Figure 9.7 summarizes in tabular form the above properties of some metallic materials.

For further explanation of the lightweight structure of material and form, the four basic load conditions according to Fig. 9.8 are considered first. Accordingly, a distinction is made between compressive, tensile, bending and torsional loads, which cause corresponding stresses and deformations in the loaded workpiece. The behavior in compression and tension is almost identical and corresponds to the behavior described by the stress-strain diagram. However, it should be noted that components subjected to compressive stress can buckle, which must be taken into account when dimensioning the workpiece. Components subjected to compressive stress must therefore generally be dimensioned more strongly than corresponds to the actual compressive stress. The bending and torsional deformations that occur and the corresponding stresses depend not only on the material properties but also on the polar moments of inertia. Figure 9.9 shows the formulas for determining the moments of inertia of some simple basic shapes.

The polar moments of inertia of solid sections and shell-shaped hollow sections are a good example of the importance of lightweight design. Figure 9.10 shows a comparison of a solid and a hollow section with identical torsional stiffness. The polar moment of inertia I_P of a circular cross-section increases quadratically with its radius. In other words, the core of the cross-section contributes only slightly to the polar moment of inertia. Accordingly, it is more advantageous to use thin-walled hollow sections, since they have a significantly

		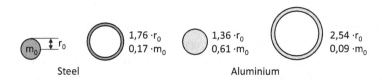		
I_y	$\dfrac{\pi \cdot r^4}{4}$	$\pi \cdot t \cdot \left(r - \dfrac{t}{2}\right)^3$	$\dfrac{b \cdot h^3}{12}$	$\dfrac{b \cdot h^3}{12} - \dfrac{(b - 2t)(h - 2t)^3}{12}$
I_z	$\dfrac{\pi \cdot r^4}{4}$	$\pi \cdot t \cdot \left(r - \dfrac{t}{2}\right)^3$	$\dfrac{b^3 \cdot h}{12}$	$\dfrac{b^3 \cdot h}{12} - \dfrac{(b - 2t)^3(h - 2t)}{12}$
I_p	$\dfrac{\pi \cdot r^4}{2}$	$2 \cdot \pi \cdot t \cdot \left(r - \dfrac{t}{2}\right)^3$	$\dfrac{b \cdot h}{12} \cdot (h^2 + b^2)$	$\dfrac{t \cdot \left((b \cdot h) + (b - 2t)(h - 2t)\right)^2}{2 \cdot (b + h - 2t)}$
A	$\pi \cdot r^2$	$\pi \cdot (r^2 - (r - t)^2)$	$b \cdot h$	$b \cdot h - (h - 2t)(b - 2t)$

Fig. 9.9 Polar moments of inertia of some basic shapes

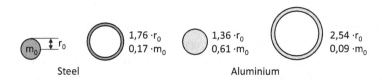

Fig. 9.10 Cross-sections of steel and aluminum with identical torsional rigidity

smaller cross-sectional area and thus a lower weight for an identical polar moment of inertia. In the example shown, the weight of the thin-walled hollow section is less than 20% of the weight of the solid section. However, the hollow section requires a significantly larger installation space. The weight of the structure can be further reduced by using aluminum instead of steel. When using thin-walled profiles, it should be noted that they tend to buckle, which usually requires additional stiffeners. The design of structures with thin-walled hollow sections is known as shell construction and is a widely used lightweight construction method, particularly in the field of automotive structures and in aerospace engineering. Figure 9.11 shows three superordinate structural designs using the example of aircraft fuselages.

For a long time, the classic concept for the construction of aircraft fuselages was the frame construction method. The load-bearing structure is made up of hollow sections that form a three-dimensional framework. All loads acting on the structure are absorbed by the frame. The body fairing is not load-bearing. In vehicle construction, this is also referred to as a stubular frame. With the integrated construction, the normal stresses are usually absorbed by beams, but the outer skin is integrated into the structure in a load-bearing manner to absorb shear stresses. In addition to shaping, the outer skin thus now also has a load-bearing function, which saves weight. In vehicle construction, this design is most comparable to the box frame design. In the case of the shell construction method, the

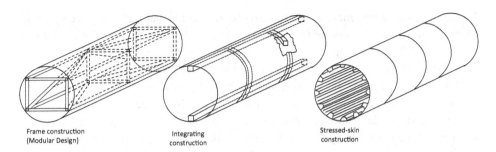

Fig. 9.11 Structural designs of aircraft fuselages [2]. (Courtesy of © Springer Fachmedien Wiesbaden 2013, All Rights Reserved)

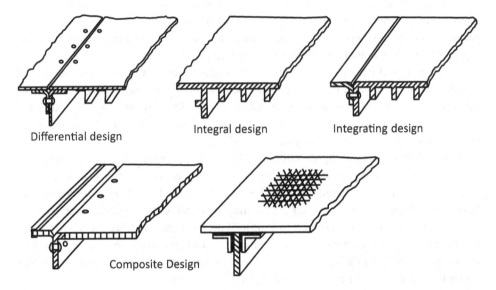

Fig. 9.12 Lightweight construction methods [2]. (Courtesy of © Springer Fachmedien Wiesbaden 2013. All Rights Reserved)

cladding absorbs all the forces, for which it must, however, be reinforced by stringers and ribs in order to maintain dimensional stability. The design shown here is called monocoque construction, which means that the construction consists of only one (largely) closed shell. Shell and monocoque construction methods are used in various forms in vehicle construction, whereby the term monocoque in particular is often not used faithfully in the construction of racing cars.

Conceptual lightweight construction is understood to mean the definition of a lightweight construction method which fulfils the set requirements in terms of application and costs, safety and reparability as well as manufacturability. In addition to the construction methods for aircraft fuselages discussed above, the lightweight construction methods shown in Fig. 9.12 also belong to the basic lightweight construction concepts. In the differential construction method, the function-bearing individual parts are joined together additively using various joining techniques, such as bolting, riveting, gluing or welding. The weight of the connections stands in the way of achieving a minimum weight. However, advantages may be the feasibility of repairs and better "fail-safe" properties.

In integral design, the aim is to achieve an absolute minimization of the individual components, resulting in a reduction of the total weight. Since the component to be designed must then fulfil several functions, this is also referred to as lightweight construction through functional integration. The application of this principle usually requires a higher effort in manufacturing technology. In the event of damage, the entire component often has to be replaced.

A compromise solution is the integrating construction method, which attempts to combine the advantages of both of the aforementioned lightweight construction methods. Composite construction methods are understood to mean the substitution of metallic materials with fiber composites. Hybrid construction methods combine different materials in order to make optimum use of their specific material properties.

A racing-related example of conditional lightweight design is the limited number of engines and transmissions that may be used per season in Formula 1. This increases the mileage per unit and has a massive influence on the dimensioning of individual components. The racing teams then define for themselves how extremely they use these power units or to what extent they calculate with safety factors.

Lightweight manufacturing describes the fact that a theoretically load-optimized but geometrically very complex structure cannot necessarily be manufactured. It is therefore often only advances in manufacturing technology that make it possible to further exploit lightweight design potential. In the conventional machining of metallic materials, the component geometry is often subject to narrow limits due to accessibility and the necessary machining time. The production of very complex and filigree structures is made possible, for example, by 3D printing processes. These include metal laser sintering, in which powdered metal is heated by a laser and welded into a solid.

The intensity with which lightweight construction is pursued is reflected in the costs. In high-performance racing, manufacturing costs play a subordinate role, which is why one also speaks of ultra-lightweight construction. However, today's regulations set limits to the pursuit of ultra-lightweight construction, for example by clearly defining the permissible material properties for certain areas.

9.3.2 Integral Body Structure

The most widely used construction method for the manufacture of a body structure in automotive engineering today is the integral body strucutre construction method, which forms a spot welded, pressed steel sheet metal body. The main reason for the use of this construction method is its cost-effectiveness in the production of large quantities, which results primarily from the favorable steel prices compared to other materials. Figure 9.13a

a)

b)

Fig. 9.13 (**a**) Body of the VW Passat B8 as a integral body structure. (Courtesy of © Volkswagen AG 2018. All Rights Reserved), (**b**) Roof rail profile made of welded pressed sheets

shows the bodyshell of a Volkswagen Passat B8. A characteristic feature of this construction method is the joining of shell-shaped steel sheets to form hollow sections, which is illustrated in Fig. 9.13b using the example of a roof bar. The sheet metal is formed by various folding and pressing processes, for which powerful, building-sized presses are required. The investment costs for these machines can only be amortized by producing very high quantities and a high degree of automation.

In order to achieve the lightest and most economical body-in-white possible, different sheet thicknesses and steel grades are used. Sheets which have locally optimized sheet thicknesses and material grades are referred to as "tailored blanks". The weight of the body-in-white of a typical mid-size car, excluding add-on parts such as flaps and doors, is about 250 kg. The engine and transmission are supported by the longitudinal members. For economic reasons in particular, this structural design does not exploit the full lightweight potential that would result from material substitution. As a result, this design leads to a relatively high total weight compared to other structural designs, which is still the case even if the add-on parts are made of fiber composites. However, due to their high prevalence among production vehicles, many near-production race cars feature a steel-sheet shell design. This has the advantage that the basic body of the vehicle can be taken over cost-effectively from series production. For use in motorsport, the main requirement is then the retrofitting of a roll cage.

9.3.3 Space Frame Design

Another structure used in series production is the so-called space frame. The basic concept is very similar to the integral body structure. The characteristic differences lie in the materials used and the forming processes. The space frame construction mainly uses aluminum, which is transformed into completely closed hollow sections by extrusion in combination with various high-pressure forming processes. Figure 9.14a shows the space frame structure of the 2015 model year Audi R8. The closed profile of a roof bar is shown

a)

b)

Fig. 9.14 (**a**) Space frame of the Audi R8 (2015) in multi-material construction. (Courtesy of © Audi AG 2018. All Rights Reserved), (**b**) Extruded aluminum profile on the roof spar

in Fig. 9.14b. The profiles are joined by cast aluminum nodes. A multi-material construction method is already used in the space frame of the second generation Audi R8. The exhaust tunnel, rear wall and B-pillar are made of fiber-reinforced plastics. The investment costs for the necessary tools are significantly lower than for the integral body structure construction method, which means that the higher material costs of aluminum can be compensated for in smaller quantities. The space frame construction method is therefore used primarily for high-priced sports cars or luxury class sedans, such as the Audi A8. For this reason, in racing it is mainly found in near-series GT vehicles. For use in motorsport, a roll cage must also be retrofitted here. The standard add-on parts can also be replaced by lightweight parts, if permitted by the regulations.

9.3.4 Roll Cage

For racing series in which production-based vehicles are used, the fitting of a roll cage certified by the sports authority is usually required. Figure 9.15 shows the bodyshell of the Audi R8 LMS (2015) extended by a roll cage in accordance with GT3 regulations. The roll cage consists of hollow steel sections welded together to form a spatial framework (see Sect. 9.3.5). The central task of a roll cage is to increase the stability of the safety cell. The cross-shaped flank protection behind the door frame, which reduces body deformation in lateral impact situations, is clearly visible. In the Audi R8 LMS, the roll cage is integrated into the structure in a fully load-bearing manner and increases the torsional stiffness of the car by 39%. However, significant increases are only achieved if the roll cage is a welded construction and the foot points are also welded to the body. In addition, the roll cage should be connected to the chassis domes of the front and rear axles in order to create a closed force path between the two axles. In older vehicle models, whose bodies have

Fig. 9.15 Space frame with roll cage in the Audi R8 LMS. (2015; courtesy © Audi AG 2018. All Rights Reserved)

relatively low torsional stiffnesses in contrast to today's production vehicles, the torsional stiffness can be increased by up to 200% by a fully supporting roll cage [3]. Roll cages retrofitted using bolted connections increase stability in the event of an accident, but they have a significantly smaller effect on structural stiffness than welded constructions.

9.3.5 Tubular Frame Construction

The classic structural design for the construction of thoroughbred racing vehicles is the steel tubular frame design. Characteristic of this construction method is the assembly of hollow steel sections to form a framework. Figure 9.16 shows the steel lattice tube frame of a stock car as used in the various series of NASCAR. In this concept, the engine and transmission are integrated into the structure in a non-load-bearing manner and are supported by various beams. The exterior design is achieved by non-load-bearing fairing parts.

Many historic racing vehicles, such as F1 cars up to the end of the 1960s, the Porsche 917 and the vehicles of the German Touring Masters up to 2011, were built according to this principle. Today, the steel tubular frame design is still used primarily in racing series in which high-performance racing is to be pursued with a limited budget. These include, for example, the US Sprint Cup series, rally raid cars such as those used in the Dakar Rally or the Australian V8 Supercars. However, a steel tubular frame construction can also be combined with other construction methods so that, for example, only some sections such as the rear or front of the car are built according to this principle. This is the case, for example, with today's DTM cars (Fig. 9.29) or the British Touring Cars.

An essential objective in the design of the steel tubular frame is to avoid bending stresses in the individual profiles, which is achieved by so-called triangulation. Figure 9.17

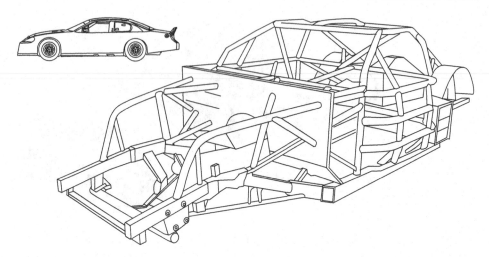

Fig. 9.16 Tubular frame of a US stock car

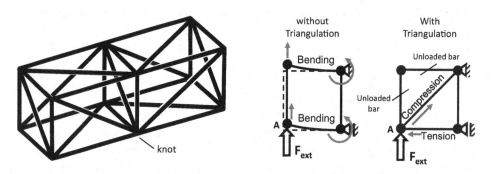

Fig. 9.17 Principle of triangulation

shows the principle of triangulation. In triangulation, different profiles converge at an account point, resulting in multiple triangles in space. If we consider a single non-triangulated frame, on which an external force F_{ext} acts at point A, we can see that this force causes a bending moment at the bearing points, resulting in a corresponding bending stress on the lower and upper cross members. Triangulation allows the external force to be fully supported in the additional strut. In the ideal case shown here, the absorbed force is coaxial with the strut, so the upper bearing does not absorb any bending moment. This strut then is loaded only by pressure. The horizontal force component of the additional strut leads to tensile stress in the lower cross-member. The upper cross member and the vertical member can be omitted. They do not absorb any forces. They are so-called zero bars. For a favorable weight/stiffness ratio, struts that do not make a significant contribution to load absorption must be avoided.

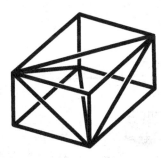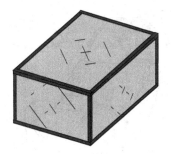

Fig. 9.18 Transition from s frame to stressed-skin box frame

Complete triangulation is not possible due to the space required for the driver and various modules such as the engine and transmission. The introduction of the forces generated by the chassis should, also to avoid bending stresses, take place in the nodes of the frame. Bending stresses lead to relatively large deformations which have a negative influence on the response behavior of the chassis (see Sect. 9.4).

The steel tubular frame allows the implementation of a package that is optimal for the application within the conceptual constraints. One is not bound to the specifications of a basic body from the series. Racing vehicles with this structural design are usually sports car prototypes. The production of a steel tubular frame requires only relatively low investment costs, since practically no special production facilities or machine tools are needed, so that even high-performance racing cars can be produced economically. However, due to unused potential through lightweight material construction and functional integration, the total weight of such vehicles still remains relatively high.

9.3.6   Stressed-Skin Design

An alternative to building the structure as a frame is to use a stressed-skin design. Both concepts are shown in Fig. 9.18. The stressed-skin design in form of rectanglular box or box frame is constructed from thin plates that are joined together, e.g. by rivets, to form a closed hollow section. Such a structure is ideal for resisting bending torsional moments and has a very good stiffness-to-weight ratio. The stressed-skin box can be stiffened by stringers and transverse frames to accommodate stresses, which enables the box to also absorb tensile and compressive stresses. The now load-bearing outer skin can simultaneously assume the shape of the outer contour. This doubling of functions leads to further weight savings.

One of the first racing cars in which the structure was consistently built from stressed-skin beams was the Lotus 25 from 1962 (Fig. 9.19). The basic structure of the Lotus 25 consisted of two parallel longitudinal beams, which were designed in a D-shape as large-surface stressed-skin beams and extended over the entire wheelbase of the vehicle.

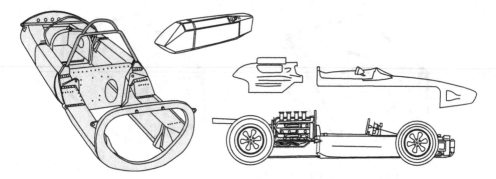

Fig. 9.19 Stressed-skin design of the Lotus 25

The outer beams were connected to each other by four transverse bulkheads and the floor panel. The rear part of the longitudinal beams formed the engine bay and served to accommodate the engine-transmission unit. The forward D-sections also acted as fuel tanks, providing an illustrative example of weight saving through functional integration. However, due to the unprotected tanks, the driver was exposed to enormous danger from leaking fuel in the event of an accident, which is why such a design is no longer permitted today. The D-sections were part of the outer contour, so additional fairing was only used for the top half of the Lotus 25. The majority of the structure was made of aluminum. The inner surfaces of the rear beam section were made of steel for better temperature resistance. The front cross member was also made of steel to support the loads from the front axle suspension [4]. The main advantage of this structural design was that the Lotus 25 had a much smaller frontal area than its competitors, while maintaining the same or higher torsional stiffness. The associated reduction in drag was a major advantage, as it at least partially compensated for the relatively low power of the 1.5-liter engine, which had been mandatory since 1961.

With the use of the engine and transmission as stressed members of the structure, the concept of an enclosed safety cell became established. This began with the front bulkhead, which served to connect the front axle suspension, and ended with the rear wall, which also served as a firewall and for flanging the engine. This design became popular with the introduction of the Ford DFV engine, which was first used in the Lotus 49 in 1967 (Sect. 2. 1.7). Figure 9.20 shows the safety cell for combination with a load-bearing engine. As a further feature, it can be seen that in this case the sheet metal of the outer skin completely forms the outer contour. This increase in the cross-section of the hollow section led to a further increase in torsional stiffness and made the use of additional fairing unnecessary. This design laid the foundation for the monocoque construction of modern high-performance racing cars.

The safety cell of the Porsche 956 used in the 1982 24 Hours of Le Mans, shown in Fig. 9.21, is another example of the box-frame design of the load-bearing structure. When designing the vehicle, the question arose as to whether to use a tubular frame or a stressed-

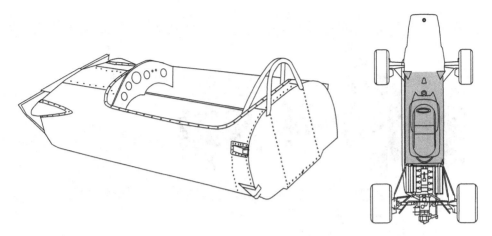

Fig. 9.20 Safety cell of the Lotus 49 with supporting engine

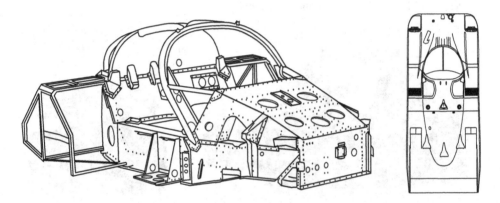

Fig. 9.21 Safety cell of the Porsche 956 (1982)

skin box frame. In the end, the decision was made to use the stressed-skin variant, as it was thought to be the only way to meet the strict safety regulations. The box frame was assembled from thin aluminum sheets. Sandwich or composite materials were not used. This was due to a lack of experience with the fatigue behavior of these materials. The torsional stiffness of the safety cell was about twice as high as that of the framework construction previously used by Porsche. Only the non-load-bearing engine, transmission and rear axle suspension were supported by a frame attached to the rear wall. The total weight achieved was only a few kilograms above the prescribed minimum weight, so that a possible weight reduction through the use of sandwich construction would not have brought any significant advantage [5]. Nevertheless, this concept with its self-contained safety cell also pointed the way for sports car prototypes to the monocoque construction with load-bearing engines and transmissions used today.

Fig. 9.22 Combination of mid-engine concept and monocoque safety cell of a BMW F1.07. (2007; courtesy © BMW AG 2018. All Rights Reserved)

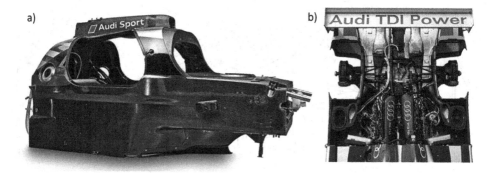

Fig. 9.23 Combination of mid-engine concept and monocoque safety cell on sports car prototypes: (a) safety cell of the Audi R18, (b) engine and transmission of the Audi R10. (Courtesy of © Audi AG 2018. All Rights Reserved)

9.3.7   Monocoque Construction and Fiber Composites

For formula vehicles and sports car prototypes, the combination of a mid-engine with a safety cell in monocoque construction has become established as the overall vehicle concept, unless otherwise prescribed by the regulations. Figures 9.22 and 9.23 show this using the example of the BMW F1.07 and the Audi R18 and Audi R10 Le Mans prototypes. It should be noted that the term "monocoque" merely describes a lightweight construction, although it is often used in common parlance as a synonym for the safety cell. It also makes no statement about the materials used, yet it is just as frequently associated (incorrectly) with the use of carbon fibers.

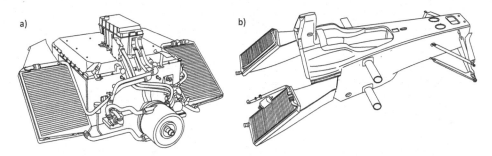

Fig. 9.24 Structure of a Formula E vehicle: (**a**) supporting battery box, (**b**) safety cell

It is characteristic of the structural design of high-performance racing vehicles that the engine and transmission are fully load-bearing components (stressed members) of the structure, which can be seen particularly well in Fig. 9.23b. The rear axle suspension components are directly attached to the gearbox housing and support for the downforce generated at the rear wing is also provided at the gearbox housing. The transmission is bolted to the engine block via threaded bolts. The safety cell and engine block are also connected by threaded bolts without any other auxiliary structures. The front axle is mounted at the front end of the safety cell. The transmission, engine and safety cell (or the monocoque) form a closed load path between the front and rear axles. If this design is used consistently, additional structural elements, such as longitudinal members, can be omitted.

The gearbox and motor are stressed by the drive forces and torques, the aerodynamic forces and the forces introduced in the suspensions. This high degree of functional integration allows a compact and very lightweight design. To meet the tight tolerances for the moving parts of the powertrain, such as gear stages and pistons, materials are required that have both high strength and high thermal stability. Monocoque construction is particularly efficient if it has a minimum of openings, which is very difficult to implement in production vehicles because of the windows, flaps and doors required. A full monocoque construction is therefore used primarily in the safety cells of formula vehicles and sports car prototypes.

The monocoque construction is also used in Formula E vehicles. The battery box, shown in Fig. 9.24, is flanged behind the monocoque instead of an internal combustion engine. This is a fully supporting element of the structure. The gearbox and electric motor are each housed in a separate casing, which in turn are also designed as fully load-bearing elements. The rear axle suspension components are accommodated by the engine and transmission housings.

Innovations in the structural design of racing cars in the late 1970s and early 1980s were driven primarily by aerodynamics and the power potential of turbo engines, which led to a significant increase in structural loading. At the same time, the optimal exploitation of the ground effect required the design of the narrowest possible safety cell. However, such a reduction in cross-section has a negative effect on the torsional stiffness of a vehicle

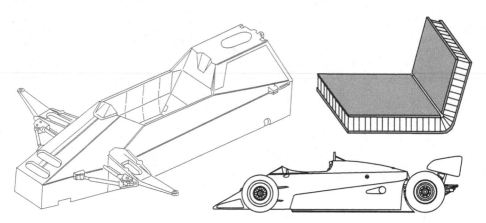

Fig. 9.25 Aluminum safety cell in cut-'n'-fold monocoque construction of the Ferrari C126 C2 (1982)

without further measures. In Formula 1, the commercialization of racing that took place during this period led to a sharp increase in the development budgets of racing teams, and corresponding resources were now available for working with cost-intensive materials.

In 1979, the Wolf WR7 was the first Formula 1 car of this generation to feature a monocoque safety cell in sandwich construction. To produce the safety cell, aluminum sandwich panels were bent around transverse ribs so that, with the exception of the opening for the driver, a closed shell was created. This construction method is also referred to as a "cut 'n' fold" monocoque, which was also used for the safety cell of the Ferrari C126 C in Fig. 9.25, among others.

The main material for this construction method is the sandwich structure shown in Fig. 9.26. It consists of two aluminum sheets, between which there is a support core made of aluminum, which in turn has a honeycomb structure and is bonded to the outer skins. This form of support core is also known as "Nomex Honeycomb". At the bending edges on the inner aluminum outer skin, a recess is incorporated for shaping purposes. Since only very simple shapes could be produced with this construction method, the aerodynamic shaping of the vehicle had to be achieved by additional fairing parts. The main advantage of a sandwich structure is that the insertion of a relatively light support core significantly increases the overall stiffness. The support core prevents the thin outer skin from buckling, which enables the outer skin to absorb compressive and bending loads. Loads acting perpendicular to the outer skin are supported by the core. Only through the connection with the support core can the outer skin, which actually determines the strength, work efficiently.

In the next evolutionary stage of monocoque construction, the aluminum outer skins were replaced by carbon fibers embedded in a plastic matrix, but the aluminum Nomex support core was retained. This design for the safety cell is now colloquially referred to as a "carbon fiber monocoque", although the structure remains a significant percentage

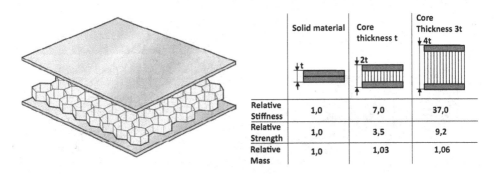

	Solid material	Core thickness t	Core Thickness 3t
Relative Stiffness	1,0	7,0	37,0
Relative Strength	1,0	3,5	9,2
Relative Mass	1,0	1,03	1,06

Fig. 9.26 Properties of a sandwich structure

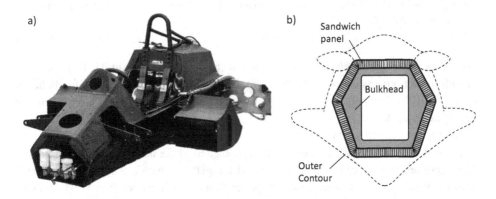

Fig. 9.27 (**a**) Monocoque of McLaren MP4/1. (1981; courtesy © McLaren Technology Group 2018. All Rights Reserved), (**b**) "Cut-'n'-fold" CFRP monocoque of Lotus 87 (1981)

aluminum. The first cars to have a safety cell with outer skins made of carbon-fiber-reinforced plastic (CFRP) and a support core made of aluminum honeycomb were the Lotus 87 and the McLaren MP4/1 in 1981 (Fig. 9.27a). This combination of materials is also used for the safety cell of modern high performance racing cars, and manufacturing processes have been continually developed. For the Lotus 87, as can be seen in Fig. 9.27b, the "cut-'n'-fold" construction method familiar from aluminum monocoques was adopted. Basically, only the material of the outer skins was substituted.

The CFRP safety cell of the McLaren MP4/1 shown in Fig. 9.27a was a joint development of the McLaren Racing Team and the company Hercules Aerospace, which at that time was already manufacturing CFRP components for the aerospace industry. A positive mold was used in the manufacture of the safety cell. A three-stage curing process in an autoclave was then used to complete the safety cell. In the first stage, the unidirectionally aligned prepreg layers (cf. Figure 9.33) of the inner surface were first cured. The adhesive film between the inner skin and the aluminum core was cured in the second stage. The third stage was required for curing the adhesive layer between the core and the outer skin and for

a) b)

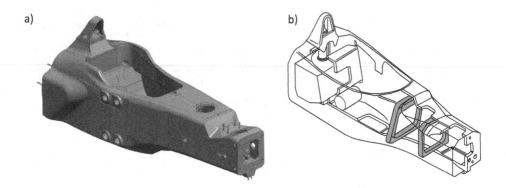

Fig. 9.28 Safety cells of modern monoposti: (**a**) Formula 1. (Courtesy of © Dallara 2018. All Rights Reserved), (**b**) Formula Renault

curing the outer skin itself [6]. As this manufacturing method is similar to the current manufacturing process of a safety cell, the McLaren MP4/1 is considered the forefather of today's CFRP safety cells. However, a separate fairing was still required to represent the aerodynamic contours.

In 1983, the CFRP safety cell of the ATS D6 was produced in a negative mold, which meant that the shape could already be largely adapted to the aerodynamic requirements. The safety cell is made of two half-shells bonded together, which is why it is also referred to as a semi-monocoque construction. The production of the safety cell from two half-shells produced in a negative mold is the standard concept of today's safety cells, as exemplarily shown in Fig. 9.28. Today's machining processes allow even complex geometries to be manufactured directly in the negative mold, which is why fairing parts are no longer required. The outer contour of the safety cell and the nose with front wing are traced by painting. In Fig. 9.28b, the reinforcing transverse bulkheads can be seen, which absorb the stresses introduced by the suspension and prevent buckling of the thin-walled structure. The transverse bulkheads are either made of a carbon fiber reinforced sandwich structure or are made of aluminum.

▶ The CFRP monocoque, or safety cell, of a modern formula car weighs about 45 kg and performs almost all the same tasks as the approximately 250 kg bodyshell of sheet steel in a mid-range production car.

In this context, the safety cell of today's DTM vehicles shown in Fig. 9.29 is also worth mentioning. This is designed in a hybrid construction, in which the lower part consists of carbon-fiber reinforced sandwich structures and the upper part is implemented as a steel tubular frame like a roll cage. The front engine and transaxle transmission are each supported as non-load-bearing components by a subframe, which is flanged to the safety

Fig. 9.29 DTM safety cell. (Courtesy of © Audi AG 2018. All Rights Reserved)

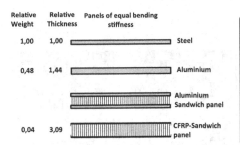

Relative Weight	Relative Thickness	Panels of equal bending stiffness	
1,00	1,00		Steel
0,48	1,44		Aluminium
			Aluminium Sandwich panel
0,04	3,09		CFRP-Sandwich panel

Material	Density [kg/m³]	Tensile Strength [N/mm²]	Elastic Modulus [N/mm²]	Relative strength
Steel	7,8	1300	200000	167
Aluminium	2,8	350	73000	124
Intermediate Modulus Carbon Fibers	1,51	2500	151000	1656
High Modulus Carbon Fibers	1,54	1550	212000	1006

Fig. 9.30 Properties of carbon compared to other materials

cell. As a cost-cutting measure, the safety cell is a standard component prescribed by the technical regulations.

The use of carbon fiber reinforced materials offers various advantages. The central advantage of carbon fibers is that although they have comparable properties to metallic materials in terms of stiffness and strength, they have a significantly lower density. According to Fig. 9.30, this means that fiber-reinforced components can be made much lighter with the same stiffness and strength properties.

The high strength of carbon fibers or carbon fiber-reinforced materials results, among other things, from the increasing strength with decreasing fiber cross-section, because the number of strength-reducing defects in the microstructure is reduced. This relationship is illustrated in Fig. 9.31. Furthermore, a cracked cross-section of a solid material is compared with a cross-section composed of fibers. The crack in the solid material will propagate with further loading and lead to complete failure of the component relatively quickly. In contrast, the failure of individual fibers will only lead to a proportional weakening of the material, while the damage remains isolated. The higher the number of fibers in the cross-section, the lower the influence of a failing fiber on the strength.

In their raw state, the fibers themselves are only capable of transmitting tensile forces. The fibers only acquire the ability to withstand compressive and bending loads while

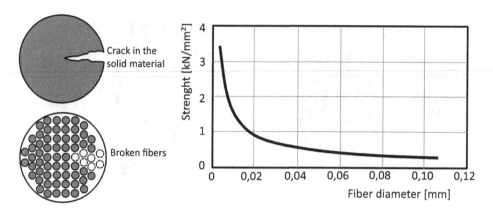

Fig. 9.31 Relationship between strength and fiber thickness

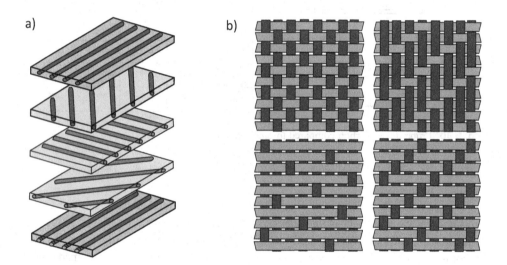

Fig. 9.32 (**a**) Matrix and fiber, (**b**) Fibrous tissue

retaining their shape when embedded in a matrix. Figure 9.32 shows this combination. The fibers are not placed individually in the matrix, but in the form of a fabric. If the fibers are oriented in only one direction, this is called unidirectional distribution. By superimposing fibers that are twisted towards each other, a multi-directional mesh is created. The variation of fiber orientation and number of layers allows a load-appropriate distribution of the material, enabling its efficient use and avoidingunnecessary weight.

Epoxy resins are used as matrix materials. To achieve maximum strength, the fiber content in the matrix must be as high as possible. This cannot be achieved with manual

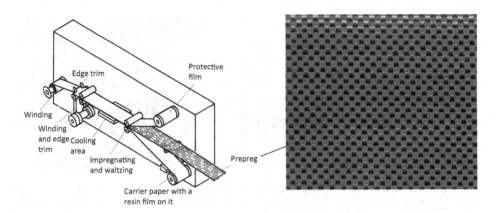

Fig. 9.33 Prepreg production [7]. (Courtesy of © Springer Verlag Berlin Heidelberg 2007, All Rights Reserved)

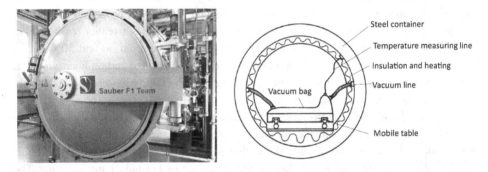

Fig. 9.34 (**a**) Autoclave in Formula 1. (Courtesy of © Sauber Motorsport AG 2018. All Rights Reserved), (**b**) Section through an autoclave [7]. (Courtesy of © Springer Verlag Berlin Heidelberg 2007, All Rights Reserved)

lamination. For this reason, the carbon fiber mats are now impregnated by machine and prepared as so-called prepregs for further processing. This process is illustrated in Fig. 9.33. After weaving, the mats are soaked in epoxy resin and a protective film is applied to both sides. The material is then cooled and rolled up.

At low temperatures, the combination of fibers and epoxy resin remains malleable, which is why the prepregs are stored at about −18 °C. In the next step, they are rolled out and cut to size. The cut prepregs are placed in the required orientation and number of layers over a positive mold or in a negative mold. Additional aluminum parts or a support core are incorporated, if necessary, using an additional adhesive film. Then the prepregs for the second outer skin are applied, also using an adhesive film. Up to 12 layers are used to manufacture an F1 safety cell. The support core thickness of a safety cell is about 3.5 mm. Epoxy resin and adhesive are then cured in a vacuum under heating and pressure in an autoclave (Fig. 9.34). The temperature in the autoclave is about 100 °C and there is a pressure of 100 PSI. This process can also be used to produce very complex molds.

a) b)

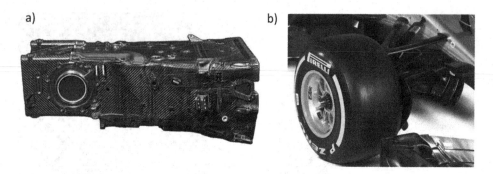

Fig. 9.35 (**a**) CFRP gearbox housing of the Audi R18. (Courtesy of © Audi AG 2018. All Rights Reserved), (**b**) CFRP suspension components of the Mercedes-AMG F1 W08 EQ Power+, 2017 (courtesy of © Daimler AG 2018. All Rights Reserved)

The weight savings resulting theoretically from Fig. 9.30 in particular are not achieved in reality for various reasons. These reasons include the anisotropic properties of the fiber composites as well as significantly more complex stress types. In reality, carbon fiber reinforced components can achieve a weight advantage of between 30 and 50% over a metallic component of comparable properties [6]. The production of carbon fibers and their further processing is a cost-intensive process, which is why their widespread use has only been economical in aviation to date.

Due to their high stiffness-to-weight ratio, carbon fiber-reinforced materials are not only used in the safety cell area, but also for various other highly stressed components. In 1998, for example, Team Arrows produced the first gearbox housing made of carbon-fiber-reinforced plastic. This design was later adopted by various Formula 1 racing teams. The Audi R18 also has a CFRP gearbox housing, as Fig. 9.35 shows. Other components of modern high-performance racing vehicles which are now made from carbon fiber-reinforced plastics include, for example, the wishbones of the wheel suspension.

9.4 Torsional Stiffness

The torsional stiffness of a structure describes the relationship between the torsional angle and an external torsional moment. It applies:

$$c_T = \frac{M_T}{\varepsilon} = \frac{\frac{G \cdot I_P}{l}}{\varepsilon}. \tag{9.5}$$

The torsion angle ε is an elastic deformation. The torsional stiffness of a vehicle structure is an important parameter for driving dynamics. To determine the torsional stiffness of a vehicle structure, one end of the vehicle is clamped firmly and a torsional moment is introduced at the opposite end. This procedure is outlined in Fig. 9.36. There are various

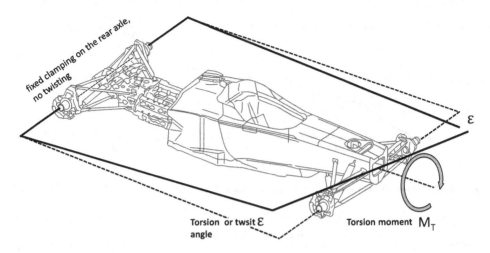

Fig. 9.36 Definition of torsional stiffness

ways of introducing the torsional moment, which is why values given in the literature are not always directly comparable. In some cases, the torsional stiffness is also specified in N/m rather than in the unit Nm/°. This is due to the fact that, in order to measure torsional stiffness, the moving bearings of the chassis are often replaced by rigid mounts. Three articulation points of the vehicle are then firmly clamped. A vertical force is applied to the remaining free pivot point and the associated vertical movement is measured.

The figures can be converted into each other if the lever arm, which corresponds approximately to the track width, is known. The torsional stiffness of a vehicle structure is determined by three factors according to Eq. (9.5):

- Construction and cross-section geometry, symbolized by parameter I_P
- Materials, symbolized by parameter G
- Vehicle concept, symbolized by parameters l

In contrast to the hollow sections shown in Figs. 9.8, 9.9 and 9.10, the stiffness of a vehicle structure is not constant along its length. This can be seen in Fig. 9.37 from the torsion angle curve along the longitudinal axis of the vehicle: relatively high deformations occur in the front area of the monocoque or the nose of the vehicle, which is due to the fact that the smallest possible cross-sections are aimed for in this area for aerodynamic reasons. In the wider area of the cockpit, only extremely small deformations occur. Only in the area of the connection to the engine and at the engine itself do the deformations increase significantly. The narrow gearbox also deforms relatively strongly. The overall stiffness of F1 cars can nevertheless be around 40,000 Nm/°, with the maximum values being reached in the 1990s, when the cross-sections of the vehicle nose were significantly higher than in today's

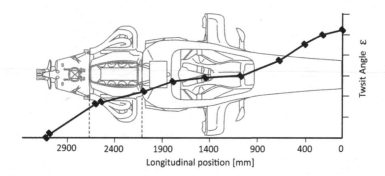

Fig. 9.37 Course of the torsion angle over the vehicle length [8]

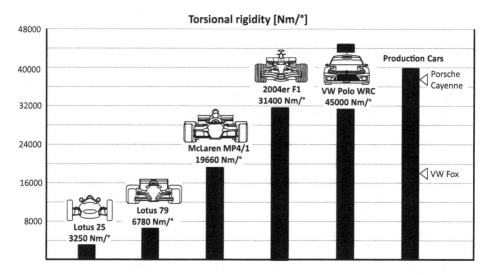

Fig. 9.38 Torsional stiffness values of different vehicles

concepts. It should be noted here that the total weight of an F1 vehicle without driver and ballast was at times well below 500 kg.

Torsional stiffness values of various vehicles are compiled in Fig. 9.38. The torsional stiffness of the Lotus 25 was 3250 Nm/°. This was a considerable increase over the values of a conventional steel tubular frame, whose torsional stiffnesses were only about 1400 Nm/° [4, 9]. The safety cell of the Lotus 79 was of monocoque construction, using mainly simple aluminum sheets which were bent over the transverse bulkheads. Thus, a torsional stiffness of 6780 Nm/° was achieved [4]. The increase in torsional stiffness was a necessity resulting from the permanent increase in aerodynamic downforce and the concomitant increase in body spring stiffnesses. With the use of aluminum sandwich structures, stiffnesses up to 8140 Nm/° could be achieved [9]. A quantum leap in torsional stiffness was achieved with the introduction of the first fiber-reinforced safety cell in the

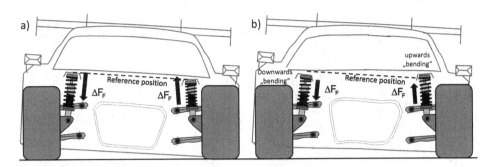

Fig. 9.39 The significance of torsional stiffness for driving dynamics with (**a**) stiff body, (**b**) torsionally soft front end

McLaren MP4/1. At 19,660 Nm/°, it was possible to more than double the values that had been common until then [9]. Modern F1 cars and closed LMP1 cars have torsional stiffnesses between 31,400 and 40,000 Nm/°. A VW Polo WRC even achieves a torsional stiffness of about 45,000 Nm/° by integrating a roll bar. The torsional stiffness of series-production cars can also be as high as 40,000 Nm/°. The torsional stiffness of a Bugatti Chiron is even said to be around 50,000 Nm/°. Although this means that production cars achieve the same order of magnitude as the structures of high-performance racing cars, it must be taken into account that high-performance racing cars are significantly lighter and have narrower cross-sections for aerodynamic reasons.

Figure 9.39 illustrates the significance of torsional stiffness in terms of vehicle dynamics. The theoretical case of an infinitely torsionally stiff structure is shown on the left. The centrifugal forces that occur during cornering generate a roll moment, which is supported by the spring force changes, creating a roll angle. The spring deflection changes generate a torsional moment acting on the structure. Since the structure does not deform, the motion of the structure is fully translated into a change in spring deflection. Figure 9.39b shows the case where the front end of the vehicle is torsionally elastic. The torsional moment causes deformation of the structure so that the spring mounts yield in the direction of load application. Thus, the vehicle structure and body springs represent a series connection of springs. The yielding of the body results in reduced spring deflection at the front axle. The effective roll stiffness of the front axle is therefore reduced by the deformation.

▶ According to Chap. 5 this effect increases the understeer tendency of the vehicle. So the basic rule applies:

The more torsionally soft the vehicle structure in front of the center of gravity is compared to the vehicle structure behind the center of gravity, the more the vehicle understeers.

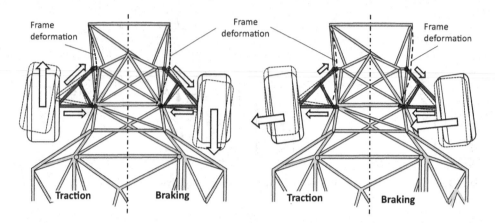

Fig. 9.40 Influence of frame stiffness on wheel alignment

Precise tuning of the chassis is therefore only possible with a sufficiently stiff vehicle structure, as otherwise changes to springs, dampers and stabilizers may not result in the expected vehicle reactions.

In Fig. 7.52 it was shown that compliance of the frame favors the development of positive camber at the outer wheel of the curve. In addition, the toe of the vehicle is influenced by deformations of the frame and the chassis components. The former case is exemplified in Fig. 9.40. Accordingly, a stiff frame is also advantageous for precise wheel tracking. Detailed guidance for the design of the frame and the suspension mounting points are given in [10], where the different structural designs are also discussed. The elastic behavior of the wheel position under the action of longitudinal and lateral forces superimposes the kinematic behavior described by the wheel lift curves. For the general effect of a changing wheel position due to elasticities, Fig. 7.63 can be referred to. In addition to the frame, the elastic behavior is determined by suspension components such as wheel bearings, control arms or mounts, among others. Appropriate attention must be paid to the design of these elements. By means of a targeted spatial arrangement of the control arms, undesired elastic wheel position changes can be partially compensated for or, especially in series-produced vehicles, can be used to specifically influence the driving behavior. Such a design is the task of elastokinematics. Basic examples of this are described in [11].

References

1. Friedrich, H.E.: Leichtbau in der Fahrzeugtechnik. Springer, Wiesbaden (2013)
2. Klein, B.: Leichtbau-Konstruktion – Berechnungsgrundlagen und Gestaltung. Springer, Wiesbaden (2013)
3. Weber, W.: Fahrdynamik in Perfektion – Der Weg zum optimalen Fahrwerk-Setup. Motorbuch, Stuttgart (2011)

4. Aird, F.: The Race Car Chassis. HPBooks, New York (2008)
5. Morgan, P.: Porsche 956/962 – The Enduring Champions. Haynes Publishing, Sparkford (2003)
6. Savage, G.: Formula 1 composites engineering. Formula 1. Compos. Eng. **17**(1), 92 (2010). https://doi.org/10.1016/j.engfailanal.2009.04.014
7. Schürmann, H.: Konstruieren mit Faser-Kunststoff-Verbunden. Springer, Berlin (2007)
8. Theissen, M., Duesmann, M., Hartmann, J., Klietz, M., Schulz, U.: Ten Years of BMW F1 Engines. 31. Internationales Wiener Motorensymposium (2010)
9. Matchett, S.: The Chariot Makers – Assembling the Perfect Formula 1 Car. Orion Books, London (2005)
10. Trzesniowski, M.: Handbuch Rennwagentechnik – Gesamtfahrzeug. Springer, Wiesbaden (2017)
11. Matschinsky, W.: Radführungen der Straßenfahrzeuge. Springer, Berlin (2007)

10.1 Historical Review

In the first decades of motorsport, the concept of safety was a foreign word for everyone involved, and death was a more or less accepted concomitant of motorsport. Even the deaths of spectators actually uninvolved in the race usually only led to a brief flare-up of indignation. Even the tragic events during the 24 h of Le Mans in 1955, in which according to official information the Mercedes driver Pierre Levegh and 83 spectators lost their lives, did not lead to a rethinking of the subject of safety. People were used to worse suffering from the Second World War, which was just 10 years ago. Only the Mille Miglia, which was held on public roads, was banned after two passengers and 11 spectators died in an accident in 1957. A similar tragedy took place at the 1962 Italian Grand Prix in Monza, where Ferrari driver Wolfgang Berghe von Trips and 15 spectators were killed. Such accidents were favored by the fact that on the racetracks of the time there were no structural separations between the track and the pit areas or adjacent wooded areas and spectator stands.

Sustained improvements did not begin until the late 1960s. At that time, the British racing driver Jackie Stewart led a campaign to improve safety in motorsport. A key event for this was the accident Jackie Stewart suffered during the 1966 Belgian Grand Prix. Jackie Stewart was trapped in his deformed BRM, soaked in petrol, for 25 min before fellow drivers Graham Hill and Bob Bondurant managed to free him. With the impression of having only narrowly escaped death, Stewart became aware that something fundamental had to change in the safety thinking in motorsport. So in 1967 Jackie Stewart was the first driver to insist that his BRM H16 was fitted with seatbelts. Jackie Stewart's campaign received a tragic boost in 1968 with the death of two-time F1 world champion Jim Clark, who was killed in a Formula 2 race at the Hockenheimring on 7 April. Clark lost control of

L. Frömmig, *Basic Course in Race Car Technology*, https://doi.org/10.1007/978-3-658-38470-8_10

his Lotus 48 at the exit of the first chicane, presumably due to a puncture on the right rear wheel, and went off the track at high speed, crashing sideways into a tree in the adjacent woodland. A delimitation of the race track by crash barriers would probably have saved Jim Clark's life. Guard rails were already used sporadically at the beginning of the 1960s, e.g. since 1962 at Monza, to separate the race track from the pit lane. However, the high costs for their installation made the large-scale use of crash barriers, also known as Armco barriers, unattractive for the track operators. The fact that Jim Clark, an absolute idol of the sport, lost his life and that the drivers Mike Spencer, Codovico Scarfiotti and Jo Schlesser also died within a month of each other led to critical questioning of the way safety was handled in motorsport.

Making progress, however, remained a tough process. It was not until the GPDA drivers' union threatened to boycott races on tracks that were not adequately secured by crash barriers, and the mandatory safety inspection by the CSI, that the track operators came to their senses. Armco barriers dominated the scene at race tracks from then on. However, there was a painful learning curve before the Armco barriers could deliver their full protective effect. The wedge-shaped F1 cars of the 1970s drove under the barriers or got caught in them.[1] Guardrails that deformed too much unleashed a catapult-like effect. The two- to three-stage design of crash barriers and optimum anchoring in the ground solved these problems. Safety fences were also used on some routes at this time, but they fell into disrepute because drivers were fatally injured by the fence posts in some accidents. Today, crash barriers and safety fences are central components of a modern track safety system. Gravel beds were introduced at Grand Prix circuits from 1977 onwards. Tire stacks, the positioning of which has been continuously optimized since then, became compulsory from 1981 onwards.

The drivers' safety equipment also changed dramatically. In 1968, for example, the first "full-face" helmet and fireproof racing suits made of Nomex were introduced. By the mid-1970s, full protective gear consisting of fireproof overalls, underwear, gloves and boots were standard. However, the use of seatbelts was not made compulsory in Formula One until 1972. On the vehicle side, the use of on-board fire extinguishing systems and safety fuel tanks became mandatory in 1969 and 1970. From 1972, the fuel tank also had to be lined with safety foam. From 1973, the protection of the fuel tank by lateral crash structures became mandatory. Nevertheless, fires caused by leaking fuel remained a deadly threat until well into the 1970s.

At the 1967 Monaco Grand Prix, Lorenzo Bandini lost control of his Ferrari and overturned, causing the roll bar to fail and the fuel tank to rupture. The vehicle immediately caught fire. Headfirst, Bandini's arm was pinned between the vehicle and the track. It took the helpless track marshals an eternity to extricate Bandini from the wreckage. Bandini suffered severe burns and lung injuries. Three days later he died in hospital [1]. At the 1968 French Grand Prix, Jo Schlesser lost control of his Honda, which instantly burst into flames

---

[1] Such problems still occur today in public road traffic in connection with motorcycle accidents.

on impact. The car was largely made of magnesium, so the fire could not be fought by water-based extinguishing agents. The marshals were powerless. Jo Schlesser burned to death in his vehicle. Piers Courage suffered a comparable fate at Zaandvoort in 1970. At Brands Hatch, Jo Siffert lost his life in a race that was not part of the World Championship. Due to suspension damage, Siffert lost control of his BRM and rolled over. Jo Siffert was trapped in the burning vehicle with a broken leg. He eventually died of asphyxiation, as once again insufficient rescue measures were initiated.

Another fatal racing accident, which was related to an emergence of fire, occurred at the 1973 Dutch Grand Prix in Zaandvoort. 25-year-old Roger Williamson lost control of his March and shot into the crash barrier at a 45° angle. The yielding guardrail catapulted the car into the air. Upside down, the vehicle landed back on the ground and slid down the track for another 100 yards or so. Leaking fuel ignited. Purley, who was following, stopped his vehicle and rushed to Williamson's aid. Williamson was trapped in his vehicle screaming for help. Once again, track officials were left behind. Purley tried desperately for minutes to tip the March onto its wheels and fight the fire with a fire extinguisher. All efforts were in vain. As the fire intensified, Purley, who could still hear Williamson's cries for help, was removed from the wreck by the marshals. Enraged and desperate, Purley surrendered. Otherwise uninjured, Williamson also suffered asphyxiation. This accident was another tragic example of the lack of availability, training and equipment of the marshals. This was also shown by the accident of Niki Lauda at the Nürburgring in 1976, who could also only be rescued from his burning vehicle by the help of other drivers. The regulations for the condition and location of the fuel tank have been continuously adapted. Since 1984, the fuel tank has had to be located between the driver and the engine so that it is better protected in the event of a collision. Later, the regulation was further tightened, so that today the fuel tank must be integrated into the safety cell. Another symbol for the lack of professionalism of the marshals was the terrible accident that took place at the 1977 South African Grand Prix in Kyalami. Renzo Zorzi's car came to a halt on the opposite side of the pit area with the engine on fire. From the pit wall, two marshals ran down the start and finish straight to the damaged car. The second marshal was 19-year-old Jansen van Vuuren, who was carrying a heavy fire extinguisher. Hans-Joachim Stuck was able to avoid him reflexively. The following Pyrce caught van Vuuren and hurled him through the air. Pyrce was hit in the head by the fire extinguisher and died instantly. Training and equipment of the marshals are crucial elements of modern concepts for safety in motorsport.

Many safety improvements, especially in the area of medical care, are associated with the commitment of Sid Watkins, who was initially engaged by Bernie Ecclestone as a race doctor for Formula One in 1978 [2] and in 1981 took over as head of the newly founded FISA Medical Commission.[2] However, the year 1978 is primarily associated with the death of Swedish Formula One driver Ronnie Peterson. Peterson was involved in a pile-up

---

[2] Fédération Internationale du Sport Automobile (Former World Motor Sport Organisation).

shortly after the start in which his Lotus was rammed by Brambilla's Surtees. The ruptured fuel tank immediately set the car ablaze. James Hunt, Patrick Depaillier and Clay Reggazoni extricated Peterson, who was trapped between the vehicle and the steering wheel, from the wreckage. Unlike other cars, the Lotus did not yet have a removable steering wheel [3]. Watkins was denied access to the accident site by Italian police officers. He could only take care of Peterson after he had been transported to the medical center. The ambulance needed for this did not reach the scene of the accident for 10 min. With serious injuries to both legs and minor burns to his shoulders and chest, Peterson was flown to hospital by helicopter 40 min after the accident. His injuries were initially determined to be non-life threatening. However, complications arose that night, and X-rays revealed a pulmonary embolism discovered too late, which eventually caused Peterson's brain death. Peterson was the victim of poor initial care and inadequate intensive care, which is now another critical component in the chain of care. Peterson's fatal accident prompted a significant expansion of Watkins' expertise, which he used to establish new standards for medical care at the race track. At Watkins' instigation, permanent medical centers and a chase car manned by doctors became mandatory for the first lap in 1980. From then on, the medical safety concept of an F1 circuit had to be approved by the FIA.

During qualifying for the 1982 Belgian Grand Prix, the charismatic Gilles Villeneuve suffered a fatal accident. On a fast lap Villeneuve hit the slow driving Jochen Maas. Due to a momentous misunderstanding, Villeneuve's Ferrari collided with the March of Jochen Mass. The Ferrari rolled over several times and Villeneuve, strapped to his seat, was thrown out of the car. Villeneuve succumbed to his severe spinal injuries. His cervical spine was fractured at the point where it connected to his head [2]. In the same year, Riccardo Paletti also died as a result of a starting accident at the Canadian Grand Prix in Montreal. Paletti crashed into Didier Pironi's stalled Ferrari at the start, hitting the steering wheel with his chest. Although medical assistance was immediately on the scene and the fire that had broken out in the meantime was quickly brought under control, Paletti succumbed to his severe chest and internal injuries at the hospital to which he was taken by helicopter [2]. The safety cell of a modern racing vehicle and its restraint systems have the task of avoiding such injury patterns.

The downforce values achieved by the ground effect and the search for lighter and stiffer construction methods, which led to the introduction of carbon-fiber reinforced safety cells in 1981, also produced increased safety as a positive effect, which, however, lulled motorsport into a deceptive sense of security for many years. Between 1982 and 1994 there were no fatal accidents during official F1 events. Many spectacular accidents had a mild outcome and the drivers were able to leave the accident scenes without significant injuries. Formula One's safety standards had not fully kept pace with the performance development of the cars, which, in retrospect, is easy to analyze. This statement also applied to endurance racing with sports car prototypes. In 1985, German racing drivers Manfred Winkelhock and Stefan Bellof had fatal accidents within a few weeks of each other. Winkelhock died in a high-speed accident with a Porsche 962 in Mosport, Canada. Bellof was killed at Spa Franchorchamps when he attempted to overtake Jackie Ickx in the

Eau Rouge in his Porsche 956 in a maneuver that could be classified as impossible. Bellof's Porsche 956 crashed head-on into a concrete pillar behind the crash barriers at almost unabated speed. Ickx was only slightly injured in the accident. With Bellof, German motorsport lost one of its most promising talents. Austrian Jo Gartner suffered a fatal accident in a Porsche 962 during the 1986 Le Mans 24 Hours.

In 1994, however, the illusion of absolutely safe high-performance racing came to an abrupt end in Formula 1 as well, as described in detail in Sect. 2.1.10. If one wants to take something positive away from the events of Imola, it is the fact that this created the willingness to fundamentally improve the safety standards in racing. For the 1995 season, Formula 1 initially reduced the engine capacity from 3.5 l to 3.0 l and increased the width of the safety belts to 7.5 cm. Around the then FIA President Max Mosley, the racing doctor Sid Watkins and the technical advisor Harvey Postlethwaite, the Expert Advisory Group (later "Safety Commission") was founded, which analyzed the safety in motorsport according to strict scientific standards from all essential points of view, for example with regard to the vehicle, the driver and his equipment, the track design and the rescue chain. Numerous improvements have resulted from the work of this institution. To protect the head, raised side walls made of energy-absorbing Confor foam have been mandatory in Formula 1 since 1996 and are now standard on all open-wheel racing cars. Since 1997, an accident data recorder ADR (Accident Data Recorder) standardized by the FIA has to be used. Since then, Mercedes-Benz has used both a high-performance safety car and a medical car equipped to the latest medical standards. In 1999, a special recovery seat was introduced and the wheels are attached to the safety cell by ropes. Wearing the HANS system ("Head and Neck Support") became mandatory in Formula 1 from 2003. In NASCAR and DTM, the wearing of the HANS system had already been mandatory since 2001 and 2002 respectively. Furthermore, the standards for conducting crash tests, training track personnel, medical care and the use of the safety car have been continuously improved. A tabular overview of the safety improvement measures introduced in the formula since its existence until 2000 can be found in [4]. The FIA Institute for Motor Sport Safety was founded in 2004 and since 2017 has continued its work as the Global Institute for Motor Sport Safety, which is dedicated to the sustainable improvement of safety in motor sport.

Other racing series have also made lasting improvements to their safety. In NASCAR's top racing series, which has been called Monster Energy Cup since 2017 (Winston Cup 1972–2003, Nextel Cup 2004–2007, Sprint Cup 2008–2016), there has not been another fatal accident since the legendary Dale Earnhardt's 2001 crash at Daytona Beach. Earnhardt had crashed head-on into the boundary wall while in third position on the final lap of the Daytona 500 after spinning at high speed, sustaining a fractured skull base and severe brain injuries. Ironically, Earnhardt was a big skeptic of the HANS system. His son, Dale Earnhardt Jr., finished second in the race and later became a NASCAR legend in his own right. After Earnhardt Sr.'s death, NASCAR mandated the use of the HANS system and introduced deformable safety barriers at its tracks. It also initiated the development of the "Car of Tomorrow," which was a new generation of stock car vehicles that was also designed with safety improvements in mind.

In the 1990s, the US CART series had already experienced an accumulation of fatal accidents. In 1996, drivers Scott Brayton and Jeff Krossnoff and a track marshal died. Gonzalo Rodriguez and Greg Moore suffered fatal accidents in 1999. Paul Dana suffered a fatal accident in 2006. Racing on U.S. ovals is particularly challenging from a safety standpoint due to high vehicle density and virtually non-existent run-off zones. The increased risk resulting from CART and IRL events can be seen in the statistics (cf. Fig. 1.4). In no other circuit racing series have fatal accidents occurred with such regularity.

Despite the intensive efforts to improve safety in international high-performance motorsport and the impressive successes achieved along the way, not every residual risk can be ruled out. Twenty years after Ayrton Senna, Frenchman Jules Bianchi crashed so badly at the Japanese Grand Prix in Suzuka in October 2014 that he succumbed to the consequences of his severe head injury in July 2015. Bianchi had gone off the track under yellow flags on a wet track and crashed into a recovery vehicle which was in action at the same location due to a previous accident involving Adrian Sutil. In 2011, Briton Dan Wheldon suffered a fatal accident in an IndyCar race in Las Vegas. After being catapulted into the air while hitting a car in front of him, his helmet hit a mounting post as he slammed into the catch fence. Wheldon left behind his wife and two children. On August 23, 2015, IndyCar driver Justin Wilson died after a crash at Pocono Raceway. The nose, which came loose from a vehicle ahead, struck Wilson in the head. Wilson was survived by his wife and two daughters.

All accidents have the common feature that the driver's head, which is exposed in a monoposto, came into contact with an obstacle or solid object. This is one of the major remaining weaknesses in motorsport. This is why Formula 1 introduced the so-called halo system for the 2018 season. The quote by Peter Wright "Anything to do with safety soon makes you aware that you are aiming for a moving target" is emblematic of the fact that efforts to optimize safety in motorsport must continue with sustainability in the future. The other sections of this chapter show the interaction of the essential components that determine safety in motorsport, with the focus on the technical aspects.

10.2 Accident Sequence and Rescue Chain

The main causes of racing accidents are the unexpected failure of components and the loss of control over the vehicle due to a driving error or a collision with another vehicle. Figure 10.1 shows an example of a typical racing accident and divides it into the following phases:

- Phase I: Loss of control
- Phase II: Impact of the vehicle on an obstacle

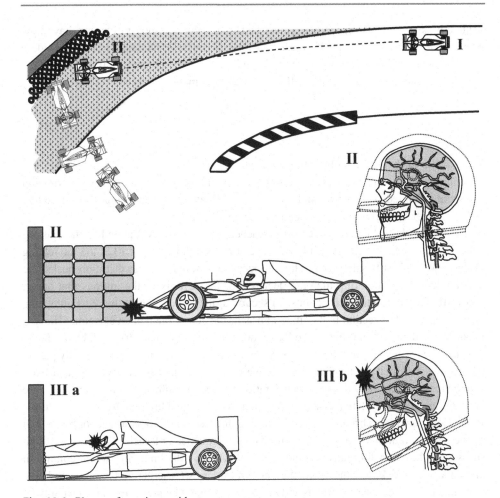

Fig. 10.1 Phases of a racing accident

- Phase IIIa: Impact of body parts on the vehicle
- Phase IIIb: Impact of internal organs on the skeleton
- Phase IV: Rebound

At the time of loss of control, the vehicle has the speed v_I, its kinetic energy at this time is:

$$E_{kin,I} = \frac{m \cdot v_I^2}{2}. \tag{10.1}$$

Until the vehicle comes to a standstill, this kinetic energy must be dissipated without exposing the driver to dangerously high decelerations. Modern race track concepts are designed in such a way that a high proportion of this energy can still be dissipated on the track and in the run-off zone. The most efficient means of doing this is by applying the vehicle brakes. For this reason, the gravel beds that were common in the past have now been replaced in many places by asphalted run-off zones. Asphalted surfaces provide a defined contact of the wheels to the ground. Until impact, the vehicle is braked down to the speed v_{II}. At the beginning of phase II, the remaining kinetic energy is:

$$E_{kin,II} = \frac{m \cdot v_{II}^2}{2}.$$

(10.2)

This kinetic energy ultimately acts on the structure of the vehicle and determines the resulting accident consequences. The aim is to convert this kinetic energy into deformation energy at a defined acceleration and force level and without serious damage to the safety cell. These relationships are explained in more detail in Sect. 10.3.

The end of the vehicle deformation is reached in phase III. At the end of the existing deformation zone, the vehicle still has the speed v_{III} and the vehicle strikes the unyielding obstacle with the stiff safety cell. This produces high forces and acceleration peaks, the magnitude and duration of which are decisive for the severity of the injuries that occur. Phase III is divided into two sections: In section a, the driver's body first encounters the vehicle parts, and in section b, the internal organs encounter the human skeleton. The aim in phase II and phase IIIa is to avoid relative movements between the driver and the vehicle. This is done by the belts and the HANS system. In the best case, the belt is the only part of the vehicle on which the head and body have contact. In this case, the vehicle and body experience approximately the same deceleration values. In contrast, belts that are too loose or a missing HANS system cause the body parts to build up a high relative velocity to the vehicle and then to be abruptly braked by hitting the dashboard, for example, or by the end of the spinal column's range of motion. The consequences of such a process are high stresses on the internal organs and the skeleton, whereby the brain and spinal column in particular are at risk, together with the vertebral arteries (Arteriae vertebrales). Relative movements between internal organs and the skeleton cannot be prevented; they can only be kept within a window of high survival probability by limiting the accelerations that occur.

At first, only the driver and the vehicle are involved in the actual accident. Then the rescue chain (Fig. 10.2) is activated.

▶ Taken as a whole, the following aspects contribute to safety in motorsport:

- Technical regulations: performance, standards, crash tests
- Vehicle design or "crashworthiness": survival space, restraint systems, crash structures, cockpit openings
- Track design: run-off zones, crash barriers, catch fences
- Drivers: equipment, physical fitness
- Sporting regulations and race control: safety car procedures, speed limit in the pit lane
- Trackside staff: training, equipment, staff density, signaling

- First aid: Recovery, medical car, emergency medical assistance, route hospital ...
- Evacuation: helicopter transport, accessible special hospitals
- Inpatient care: Special clinics, specialists for accident and trauma surgery
- Accident research: biomechanics, load values, accident data collection

The technical aspects are dealt with in detail in the following sections, while only a basic description of the rescue chain is given here. This includes all the elements that ensure optimum medical care after the accident and thus guarantee the highest possible probability of survival and possibility of rehabilitation. First of all, it includes the track personnel and their training. During a Grand Prix, around 250 marshals are deployed at the track, who specialize in radio contact with race control, warning signaling by flags or firefighting, and securing the accident site. The track personnel must also ensure a quick and professional recovery or access for medical first aid without causing further injuries.

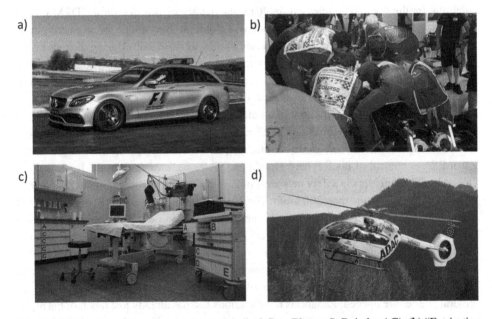

Fig. 10.2 Elements of the rescue chain: (**a**) Medical-Car. (Photo: © Daimler AG), (**b**) "Extrication Team Hockenheimring". (Photo: © Dr. Maria Mogg), (**c**) Shock room in the Medical Center of the Nürburgring. (Photo: © Nürburgring 1927 GmbH & Co. KG), (**d**) Rescue helicopter type Airbus Helicopters H145. (Photo: © ADAC)

Immediate medical first aid is provided by the Medical Car crew, which is manned by a specialist trauma surgeon and the FIA Safety Coordinator. In addition, there are about five vehicles with a medical intervention team consisting of a race doctor and a paramedic, who can reach any location on the track in a maximum of 30 s. The main task of the intervention team is to diagnose the severity of the injury and to initiate further necessary measures. In an emergency, the patient is stabilized at the accident site, which includes immediately securing the oxygen supply and stopping bleeding. An immediate oxygen supply reduces the risk of permanent brain damage, as otherwise the first brain cells begin to die after about 3-4 min. Because of the immediate first aid, racing drivers have a significantly higher chance of fully recovering even from serious accidents than people who have accidents in public road traffic.

Three extrication teams, each with six helpers, are stationed along the route for the professional recovery of the driver. The patient can be transferred to the Medical Center by ambulance for further treatment. There, two experienced trauma surgeons, two anesthesiologists and a burn expert are on alert. Two rescue helicopters, each manned by a doctor and two paramedics, are on standby for transfer to clinics specializing in the relevant injury pattern. The readiness of the helicopters for take-off is now a mandatory condition for the track to be released. In 2013, the qualifying practice for the DTM race in Moscow was cancelled because the airspace was closed at short notice for an overflight by Russian President Vladimir Putin. The selected clinics are on standby during the race event. The medical teams there are also familiar with the injuries to be expected, so that not only the obvious injuries are treated, but also various preventive measures are carried out to check the patient for further injuries typical of motorsport.

The importance of the instrument of an effective rescue chain is shown by the accident of Alex Zanardi (Fig. 10.3) during a CART race at the Lausitzring in 2001. Shortly before the end of the race, Alex Zanardi drives into the pits for a fuel stop. Upon exiting pit lane, Zanardi loses control of his Reynard Honda and slides backwards onto the track in Turn One. Alex Tagliani hits the safety cell level of Zanardi's vehicle with an impact angle 90°. The impact speed is more than 300 km/h. The TV pictures make us fear the worst. The nose of Tagliani's vehicle hits Zanardi's pelvis and lower extremities. Wreckage flings across the entire track. Various fragments penetrate Zanardi's legs, destroying tissue and causing multiple bone fractures. The first recovery vehicle arrives at the accident site after 19 s. Within seconds, the second vehicle follows with the race doctor. Zanardi's neck is stabilized and his helmet is removed. Zanardi is intubated to ensure oxygen supply. Both femoral arteries are opened. Zanardi is not conscious and has already lost much of his blood volume. The medical team slows the bleeding and extricates Zanardi from the vehicle to take him by ambulance to the Medical Center [5]. There, the rescue helicopter is already put on alert. Due to the severity of the injuries, it is decided to transport him immediately to the Trauma Center in Berlin. Further measures to stabilize Zanardi are to be taken directly in the helicopter in order to avoid unnecessary loss of time. It is already clear at this point that Zanardi will lose both legs. When the ambulance arrives at the landing zone, Zanardi's pulse is barely perceptible and he has to be artificially ventilated. Zanardi is wrestling with

Fig. 10.3 Alessandro "Alex" Zanardi. (Courtesy of © BMW AG 2018. All Rights Reserved)

death. The blood loss is compensated by a saline solution. Heart rate and oxygen levels stabilize. After less than 30 min, the rescue helicopter takes off for its approximately half-hour flight to Berlin. From impact to arrival at the emergency room takes only 59 min. Nevertheless, Zanardi has lost 75% of his blood volume at this point. He is immediately transferred to the operating room. In a 6-h operation, both of Zanardi's legs are amputated and he is placed in an induced coma. His condition is critical but stable. Without immediate medical attention, he would not have been able to survive such an injury. Tagliani miraculously sustained only minor injuries to his back in the accident. After 3 days, Zanardi wakes up from the artificial coma without permanent brain damage. In [5] 2011, Zanardi expresses his gratitude for the rescue team and says he still possesses more in life than he lost in the accident. Zanardi has never lost his optimism and courage to face life, which makes him one of the most impressive personalities in motorsport in this author's view. He later returns to racing and in 2012 becomes a two-time Paralympic champion in the handbike discipline. In 2016, he wins another gold and silver medal each.

Accident research is another important safety aspect in motorsport. Its task is to clarify the causes of a racing accident and of the injuries that have occurred, and to derive measures for improving safety from this. One instrument for this is the recording of the driving condition by an Accident [or Incident] Data Recorder (ADR). Accident data recorders (Fig. 10.4a) are now mandatory in almost all high-class racing series (in all FIA championships, in the DTM, in various NASCAR racing series, etc.), although the quality of the data recorded can vary due to the measurement technology used. These systems are often too cost-intensive for use in mass sports. The spearhead here is once again Formula 1, which has been using accelerometers in the drivers' earpieces since 2015 (Fig. 10.4b). These "in-ear accelerometers" are based on sensors such as those used in commercially available smartphones. However, they are capable of detecting peak acceleration values of up to 400 g. Accelerations of the head are an important measure for assessing the severity of injuries. They can sometimes differ significantly from the

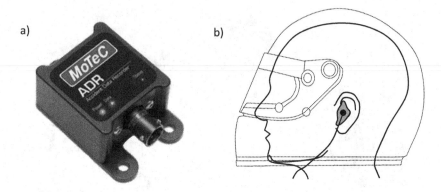

Fig. 10.4 (a) Accident data recorder, (b) In-ear accelerometer

acceleration of the vehicle (Fig. 10.1). In addition to the analysis of driving condition data (e.g. accelerations in all three spatial directions, vehicle speed, yaw rate, steering wheel angle as well as pedal values), accident research also includes the examination of the helmet for damage and contact traces in order to obtain information about the origin of the injury pattern.

The Abbreviated Injury Scale (Fig. 10.5b) is used for the basic classification of injury severity, but is only valid for single injuries. Extended methods have been developed for the assessment of multiple injuries. An essential task of accident research is to find a systematic relationship between the loads acting on the human body in the form of forces or accelerations and the injury severity. This work is summarized under the term biomechanics.

A typical example of the research results of biomechanics is the so-called Wayne State Tolerance Curve (Wayne load curve, Fig. 10.5a), which establishes a relationship between the acceleration acting on the head and its duration of action and the risk of fatal head injuries. These relationships can be further substantiated by the Head Injury Criterion or HIC value. This is defined as:

$$HIC = \left(\frac{1}{t_2 - t_1} \cdot \int_{t_1}^{t_2} a_S \cdot dt \right)^{2.5} \cdot (t_2 - t_1). \tag{10.3}$$

In Eq. (10.3), the acceleration a_S acting on the head must be entered in g and the time interval in seconds. The time interval is to be selected in such a way that the maximum HIC value can be derived from the respective measurement record. Figure 10.6 shows the assignment of the HIC value to the expected injury severity as well as two calculation examples. The HIC value only takes into account the linear acceleration of the head. The real movement of the head during a racing accident is usually superimposed by other head movements. In [6] it is shown that a very high correlation to the severity of head injuries is

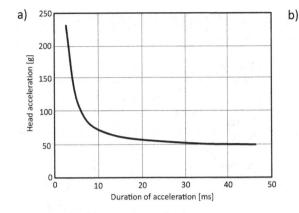

a) [graph with y-axis "Head acceleration [g]" ranging 0 to 250, x-axis "Duration of acceleration [ms]" ranging 0 to 50]

b)

AIS	Severity	Fatality Range [%]
0	None	0,00
1	Minor	0,00
2	Moderate	0,07
3	Severe life threatening	2,91
4	Critical	6,88
5	Critical Survivival uncertain	32,32
6	Maximum fatal	100,00

Fig. 10.5 Elements of accident research: (**a**) Wayne State Tolerance Curve, (**b**) Abbreviated Injury Scale (AIS)

obtained from the superposition of the peak values of linear and rotational head accelera-tion. The HIC value is also used as a threshold for crash tests. Similarly, the effectiveness of the HANS system (Sect. 10.4) has been scientifically researched and proven through various crash tests and the calculation of such characteristic values. Among other things, it has been shown that better results can be achieved with a HANS system than with an airbag [4]. Corresponding characteristic load values have also been defined for other parts of the body, in particular the cervical spine, thoracic spine or pelvis, which are also used as limit values for crash tests.

The Global Institute for Motorsport Safety (which took over from the FIA Institute for Safety and Sustainability in 2016), the Australian Institute for Motor Sport Safety (AIMSS) and the NASCAR Research and Development Center are among those dedicated to researching racing accidents and optimizing safety in racing. Today, all the data available on a racing accident, such as the ADR data, TV footage, on-board recordings (in F1 cars, the use of a high-speed camera is mandatory for this purpose), photos of the accident site, eyewitness reports, medical diagnoses and the final analysis results are stored in a database there. This facilitates the search for systematic weaknesses in the safety concept of the vehicles and restraint systems used as well as the track design.

10.3 Crashworthiness

"Crashworthiness" is the ability of a vehicle to protect its occupants from serious injury in the event of a crash. It is determined by the components

- Safety Cell,
- Deformation or crash structures and
- Restraint systems

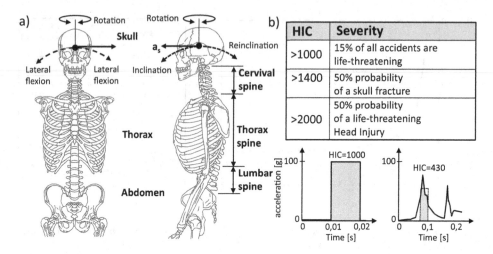

Fig. 10.6 (**a**) Human skeleton and stress variables, (**b**) Head Injury Criterion (HIC) with severity levels [4]

ensured. The safety cell is a component of the vehicle structure, which was dealt with in Chap. 9, and has the task of preventing mechanical effects on the driver caused by external obstacles, penetrating foreign bodies or severe deformations to the structure of the driver's own vehicle. By preventing deformations to the vehicle, it is ensured that the driver is not trapped and can therefore either leave the vehicle under his own power as quickly as possible or be recovered from the vehicle without unnecessary complications. Another task of the safety cell is to protect the driver from fire, whereby the term "crashworthiness" already includes the prevention of fire on the driver's own vehicle.

The area of the vehicle in which these tasks are performed is referred to as the survival space. The vehicle's crash structures convert the vehicle's kinetic energy into deformation energy. The force level required for the deformation determines the acceleration behavior of the vehicle. In conjunction with the restraint systems, this results in the acceleration values acting on the individual body parts, which in turn are a measure of the expected injury severity. The restraint systems prevent relative accelerations from building up between the vehicle and the body or sensitive parts of the body or from being reduced by abrupt decelerations. The technical regulations ensure that the safety aspects are not compromised by performance-oriented requirements.

Figure 10.7 uses the example of a Formula 1 vehicle to show essential characteristics of the safety cell specified by the technical regulations. The front and rear roll bars must be designed in such a way that the driver's helmet has a prescribed clearance below the line connecting the two roll bars. This prevents contact of the head with the ground during a rollover. The resistance of the roll bars is guaranteed by prescribed load tests, the fulfilment of which is a prerequisite for participation in official racing events.

Accidents from the past show that the partially unprotected head of the driver continues to be a weak point in the safety concept of open racing cars. The accident of Felipe Massa at

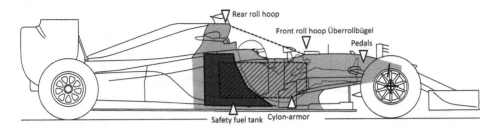

Fig. 10.7 Influence of the regulations on the design of the safety cell of an F1 car

the Hungarian Grand Prix and the accident of Henry Surtees during a round of the English Formula 2 Championship at Brands Hatch, both of which occurred in 2009, can be taken as examples. Felipe Masse was hit on the helmet during qualifying by a coil spring weighing about 800 g that had broken from the rear suspension of another car. His speed at the time was about 190 km/h. Massa lost consciousness and drove almost unbraked into the tire barrier. He suffered a broken bone in his skull as well as a concussion, but this did not result in any permanent damage. A week earlier, Henry Surtees was fatally struck in the head by a wheel that had broken loose from a crashed vehicle. As a result, to increase the survival space, Formula One cars were required to use the so-called halo system from the 2018 season. In parallel with its testing, trials were carried out with the more visually appealing shield system, but its implementation proved problematic due to reflections. Both variants are shown in Fig. 10.8.

Since 1987, the pedals have had to be positioned in front of the wheel center line of the front axle, so that the driver's lower extremities are better protected from possible penetrating chassis parts. Various racing accidents in the US racing series for monoposti as well as various crash tests of the FIA have shown that the driver is particularly endangered in the event of a lateral impact ("T-bone crash") at the height of the passenger compartment due to the penetrating nose of the other party involved in the accident. For this reason, the FIA prescribes a lateral armoring of the monocoque made of Zylon (Fig. 10.7). Further regulations on the nature of the safety cell are discussed in Sect. 10.4 in connection with the restraint systems.

The wheels also pose a particular danger if they become detached from the vehicle during an accident and are catapulted either into the spectator stands or the area of other vehicles. To avoid this scenario, many racing series now stipulate the use of wheel tethers, which are attached to the wheel carrier and monocoque or gearbox. The nature of the tethers, which are shown in Fig. 10.9, has been standardized by the FIA (FIA Standard 8664).

The fuel tanks of early racing cars were constructed of metal, which made them susceptible to cracking in the event of an accident. The escaping fuel could then easily ignite on the hot engine and transmission parts. Flexible fuel tanks have been mandatory for several decades. There has not been a serious fire accident in F1 since Gerhard Berger's 1989 accident at Imola. Berger had gone off the track at Tamburello and smashed into the

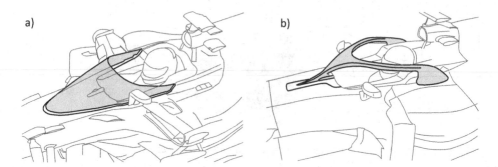

Fig. 10.8 Protection of the head by (**a**) The Shield system, (**b**) The Halo system

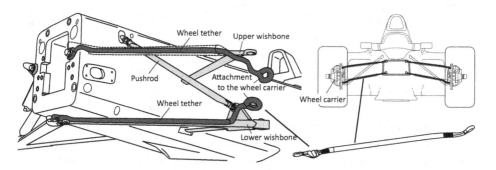

Fig. 10.9 Securing the front wheels with wheel tethers

track barrier. The monocoque and fuel tank were so badly damaged that leaking petrol ignited. However, track marshals were able to extinguish the fire within about 25 s. The chase car with race doctor Sid Watkins was on the scene after about 1 min. Berger suffered only second degree burns to his hands and a broken rib and collarbone. The high standard of equipment used by the track marshals made all the difference to Elio den Angelis' accident in 1986: during testing at Le Castellet, Elio de Angelis lost control of his car at high speed. The car overturned and leaking fuel ignited on the hot turbocharger. It took 8 min to bring the fire under control. A rescue helicopter was also unavailable. In hospital, de Angelis succumbed to the consequences of asphyxiation. It was the last fatal accident in Formula 1 that was caused by a fire.

Since 1984, the fuel tank has had to be installed inside the safety cell so that it is protected from external mechanical impact in a similar way to the driver. Figure 10.10a shows the fuel tank of an F1 car from the 2000s, which is located between the engine and the driver and takes on the contour of the seat at its front. The outer shell of a modern safety fuel tank is made of a flexible but highly rigid material that ensures tightness even in the event of a severe impact. The materials used are rubber-covered Kevlar or nylon, depending on the requirements of the racing series, so that a high resistance to penetrating foreign bodies is achieved. The fuel lines must be self-sealing. As further protection against

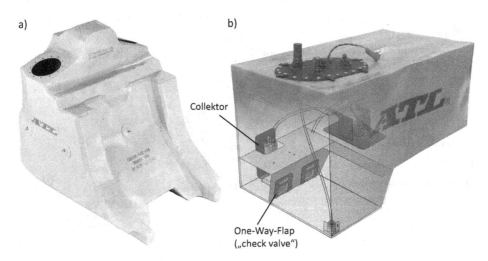

a)   b)

Collektor

One-Way-Flap
("check valve")

Fig. 10.10 Safety fuel tank: (**a**) of a Formula 1 car, (**b**) of a BTCC car. (Courtesy of © ATL Ltd. 2018. All Rights Reserved)

fire, a onboard fire extinguishing system must be installed in high-performance racing vehicles.

The tank and fuel system also has the demanding task of supplying the engine with fuel even under maximum longitudinal and lateral accelerations and minimizing variations in the center of gravity position due to fuel sloshing. Within the fuel tank of a high-performance racing vehicle, therefore, is a complex network of partitions and one-way valves (Fig. 10.10b) that prevent fuel from sloshing and backing up into the upper portions of the tank. Inside the fuel tank is a collector from which the high-pressure fuel pump is supplied and which can provide fuel for about 30 s even at maximum accelerations. In high-performance racing vehicles, fuel is supplied to the collector by up to four low-pressure pumps, ensuring fuel supply even at low levels.

The second central element of the crash safety of a racing vehicle is its deformation or crash structures. Figure 10.11 shows the various crash structures of an F1 vehicle. The nose of the vehicle represents the frontal crash structure. Behind the transmission is the rear impact structure. To the side of the safety cell are the side impact structures, which are integrated into the sidepods. The task of the crash structures is to convert the kinetic energy of the vehicle into deformation energy and to keep the deceleration values acting on the driver as low as possible. Figure 10.12 explains this process in a simplified way using the example of a frontal crash structure. The frontal crash structure is pushed against an unyielding barrier by a hydraulic press. The force required to deform the crash structure is measured.

In the example shown, the deformation force F_{Defo} remains constant over the entire deformation path $s*$. The area below the force curve is the deformation energy absorbed by the crash structure E_{Defo}. In this simple example:

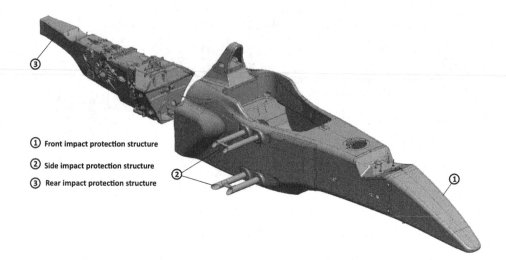

Fig. 10.11 Crash structures of an F1 car. (Courtesy of © Dallara 2018. All Rights Reserved)

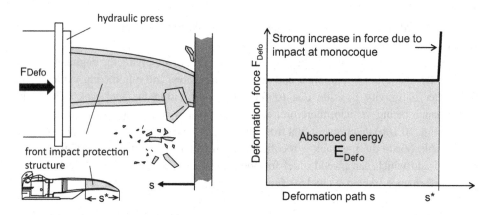

Fig. 10.12 Deformation force and deformation energy of a crash structure

$$E_{Defo} = F_{Defo} \cdot s^*. \tag{10.4}$$

The acting deformation force determines the decelerations occurring on the vehicle in the event of a collision. Simplified:

$$\ddot{x} = \frac{F_{Defo}}{m}. \tag{10.5}$$

The deformation force must therefore be sufficiently high to absorb as much kinetic energy as possible over the available deformation path, although it must not lead to unbearably

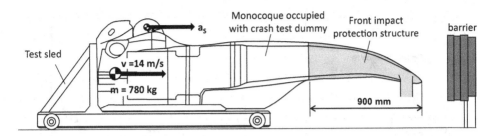

Fig. 10.13 FIA frontal crash test

high deceleration values. A prerequisite for this is that the crash structures do not fail during the impact, as only then the available deformation path canbe used efficiently. The kinetic energy remaining after deformation determines the residual speed at which the safety cell impacts against the barrier. At this point, high acceleration values occur because the safety cell can resist high forces without deforming significantly (Fig. 10.13). The magnitude and duration of the acceleration values that occur depend on the residual speed and determine the risk of injury to the driver. Figure 10.14 shows the measured values for speed and acceleration of the monocoque recorded during an FIA frontal crash test (Fig. 10.13).

These measurements can be used to predict the severity of the injuries. Examples of this are the Wayne load curve shown in Figs. 10.5 and 10.6 or the Head Injury Criterion (HIC). Comparable characteristic values have been defined for numerous other body parts. In order to pass the crash tests, it is necessary to comply with fixed limit values for these characteristic values. The level of the limit values and the regulations for conducting the crash tests are continuously adapted to the current state of the art.

10.4   Restraint Systems and Driver Equipment

An important component in reducing the consequences of accidents is the prevention of relative movements of the occupants to the vehicle. In racing vehicles, a five-point or six-point belt system is used for this purpose instead of the three-point belt system commonly used in passenger cars. Figure 10.15 shows such a six-point belt system. The six partial belts are brought together in a single belt buckle with a quick-release buckle, which is fully opened by a single movement of the hand. The advantage of a six-point belt system (Fig. 10.15) is on the one hand the better fixation of the body in the seat, and on the other hand the higher belt surface reduces the local surface pressures on the torso.

However, the belt system only restricts the movement of the upper body. The dangerous relative movements between head and upper body are only prevented by the use of a "head-and-neck support" (HANS) system (Fig. 10.16). This consists of a shoulder corset which is fixed to the driver's body by the belt system. The shoulder corset has a brace which is connected to the rider's helmet via two retaining straps. In the event of a frontal impact, the retaining straps prevent overstretching of the cervical vertebrae and impact of the helmet on

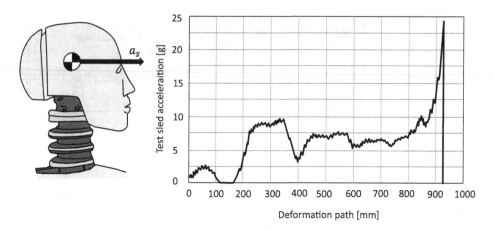

Fig. 10.14 Exemplary acceleration curve in an FIA frontal crash test

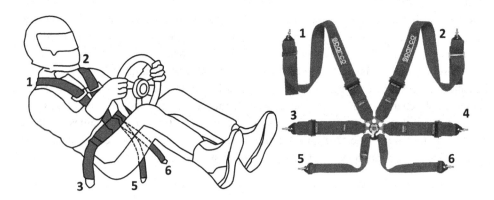

Fig. 10.15 Six-point belt system

the instrument panel, since the retaining straps do not permit any forward rotational movement of the head. The inertial forces acting on the head are now absorbed by the retaining straps, which reduces the load on the head and the spinal system. The values of tensile, shear and total load on the cervical vertebral system fall below the threshold values above which serious injuries are to be expected.

Lateral movement of the head is restricted by the raised cockpit walls, which were introduced in 1996 in response to the fatal accidents of 1994. Figure 10.17 shows the U-shaped cockpit insert used in Formula 1. This consists of an energy absorbing Confor foam covered with two layers of aramid fibers in an epoxy resin matrix. Analogous to the HANS system, the insert reduces the loads acting on the side of the head and the cervical vertebrae. The rear part of the insert has a damping function in the event of rear impacts. The cockpit walls also prevent the head from hitting the barrier in the event of side impacts, as was the case in Wendlinger's accident in Monaco in 1994. The main dimensions of the U-shaped insert are specified by the regulations.

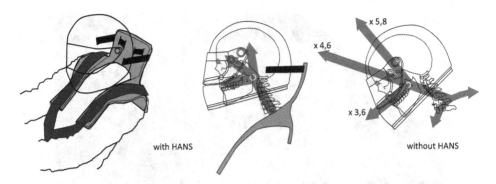

Fig. 10.16 Function of the HANS system

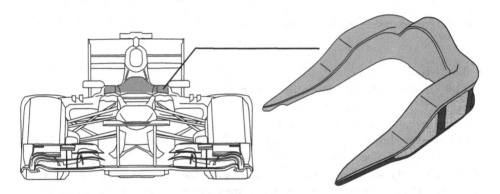

Fig. 10.17 Cockpit walls of a modern Formula 1 car

In closed racing cars, lateral movement of the head is usually prevented by U-shaped projections or ears on the racing seat, as can be seen in the example of the Audi R8 LMS ultra in Fig. 10.18.

Racing helmets have been standard equipment for racers since the late 1960s. Put vividly, they form a separate survival space for a racer's head. In order to meet the high requirements for modern racing helmets, which are formulated in the EA2016 of the Snell Memorial Foundation or the FIA Standard 8859-2015, among others, the following requirements must be fulfilled:

- Wide field of view
- High position stability
- Good ventilation and cooling
- High noise insulation
- Compatibility with HANS system
- High resistance against intrusion of foreign bodies
- Wide-area load application and high energy absorption capacity
- High refractoriness

Fig. 10.18 Race seat of an Audi R8 LMS ultra. (Courtesy of © Audi AG 2018. All Rights Reserved)

- Low weight
- Low aerodynamic disturbance effects (open racing vehicles)

The total weight of an F1 driver's racing helmet, as shown in Fig. 10.19, is only about 1.2 kg. Its outer shell consists of up to 20 layers of a carbon fiber reinforced plastic, Kevlar or Zylon, and glass fibers. Polyethylene foams are used for the inner padding, ensuring the helmet's snug fit and high energy absorption capacity. The inner padding is covered with a fire-resistant Nomex material. When designing the ventilation, care must be taken to ensure adequate cooling of the head and a good supply of breathing air. The breathing air is cleaned by a particle filter. At high vehicle speeds, about ten liters of fresh air per second flow through the helmet. Outer shell and helmet have a high resistance against the intrusion of stones, wreckage or other foreign bodies. For this purpose, the visor consists of a 3 mm thick polycarbonate layer. In the event of an impact, the shell and padding distribute the forces acting over a large area of the skull. The avoidance of local stress peaks significantly reduces the risk of injury. The performance of modern racing helmets was impressively demonstrated during the Hungarian Grand Prix in 2009, when Felipe Massa was hit directly on the helmet by a coil spring weighing around 0.8 kg while driving at full speed. Massa suffered severe but not fatal skull injuries.

10.5   Track Design and Safety

There are various objectives and requirements for the safety aspects of the track design. The top priority is the protection of spectators and track personnel. At this point, however, only those elements are considered which serve to mitigate the consequences of a racing accident. These elements include run-off zones and track boundaries.

1 Außenschale Kohlefasern mit bis zu 20 Lagen, Aramid, Polyäthylen

2 Aufnahme für HANS-System

3 Energieabsorbierende Schaumstoffpolsterung basierend auf Polystyrenen, ummantelt mit feuerfestem Nomex

4 Partikelfilter

5 Helmvisier aus ca. 3mm dickem Polycarbonat

6 Belüftung

7 Schloss zur Helmanbindung

1 Outer shell carbon fiber with up to 20 layers, aramid, Polyethylene

2 HANS tether anchorage

3 Energy absorbing Foam padding based on polystyrenes, coated with fire-resistant Nomex (in helmet)

4 Particle filter

5 Helmet visor made of approx. 3 mm thick polycarbonate

6 Ventilation

7 Tether with connector

8 HANS- collar

Fig. 10.19 Racing helmet for high-performance motorsport. (Courtesy of © BMW AG 2018. All Rights Reserved)

It has been shown in Sects. 10.2 and 10.3 that the kinetic energy of the vehicle and its ability to convert this into deformation energy are the decisive factors for the severity of the accident to be expected. The potential of a vehicle to convert its kinetic energy into deformation energy is limited by the available deformation paths. The purpose of a run-off zone is to provide a path in which the vehicle's kinetic energy can be reduced through a decrease in speed. In addition, the run-off zone serves as an avoidance space to avoid the collision of several vehicles. The highest deceleration is achieved when the vehicle has all four wheels in contact with an asphalt surface and the driver applies the brakes. For this reason, asphalt run-off zones have replaced the gravel beds that were common until then in many areas.

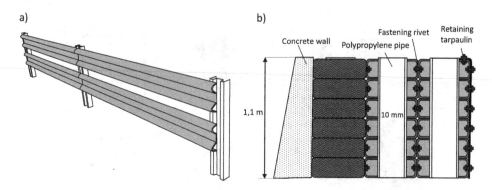

Fig. 10.20 (**a**) Guard rail or Armco barrier, (**b**) Tire pile

The first task of the track barrier is to prevent the vehicle from hitting areas where it could endanger the lives of spectators and track personnel. It also prevents the vehicle from hitting obstacles that reduce the effectiveness of the deformation structures. This is the case if the obstacle is a massive one with a small surface area, such as a tree. This is then referred to as a "small overlap" between the vehicle and the obstacle. In such a case, the structure of the vehicle has to absorb a lot of energy locally. An effective track boundary helps to introduce the forces acting in a crash into the vehicle structure over as large an area as possible, so that the deformation elements are used optimally and the function of the safety cell as a survival space is guaranteed.

The simplest means of track delimitation are crash barriers or Armco barriers, as sketched in Fig. 10.20a. Long straight sections in particular are still secured by crash barriers as standard today. As a rule, these are designed with at least three rails in order to prevent both underriding and rolling over the track boundary. Armco barriers deform on impact and thus serve as a secondary additional deformation zone. In the fatal accident of Allen Simonsen during the 24 h of Le Mans in 2013, this mechanism was overridden because there was a tree directly behind the guard rail.

The effect of energy absorption is also achieved by placing stacks of tires in front of a concrete wall. The concrete walls of a track boundary must be 1.10 m high, which allows six tires to be stacked (Fig. 10.20b). A plastic tube ensures dimensional stability and increases the ability to absorb energy. The tires are placed in rows of three and connected by rivets to prevent them from being driven apart by an intruding vehicle. In front of the tires is a fabric-reinforced plastic sheet which ensures contact over a large area.

Another standard element are safety fences, which are used in front of spectator stands, on road courses and in the US ovals (Fig. 10.21). They protect the rear areas from wreckage and rising vehicles. A critical scenario is a direct impact with one of the required mounting piers, which was Dan Wheldon's undoing in 2011. However, there is no alternative to this design in the areas mentioned.

Fig. 10.21   Safety fence at the Lausitzring. (Courtesy of © Audi AG 2018. All Rights Reserved)

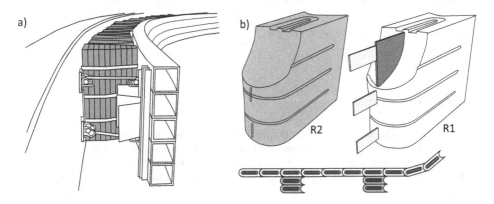

Fig. 10.22   (**a**) "SAFER barrier", (**b**) TECPRO barrier

Basically, the US ovals do not provide any run-off zones. In the event of a loss of control, there is therefore hardly any possibility of dissipating the kinetic energy on the track. For this reason, the "SAFER barrier (Steel And Foam Energy Reduction)" is now used in the US ovals. The "SAFER barrier", as illustrated in Fig. 10.22a, is installed in front of the original concrete wall. It consists of a guard rail formed by steel pipes and deformation elements. The deformation elements located between the guard rail and the concrete wall are capable of absorbing a considerable amount of energy.

The French company TECPRO Barriers developed the track barrier of the same name, which has the same objective as the SAFER barrier. A TECPRO barrier consists of two elements which can be retrofitted in front of a solid roadway barrier. Figure 10.22b shows the corresponding R1 and R2 elements. The R1 elements are lined up like a chain. The purpose of this chain is to provide a large area of attack for an impacting vehicle. The R1 element is made of high-strength polyethylene with a metal plate in the middle. The R1 elements are connected by means of continuous nylon belts. Their shape is designed in such a way that the chain can also follow tight curves. The R2 elements are made entirely of energy-absorbing polyethylene. They are placed between the R1 chain and the actual track boundary. The combination of R1 and R2 elements can vary depending on the track section. In the appropriate configuration, TECPRO barriers are also capable of absorbing the energy of a high-speed accident. Since the compact elements weigh only 110 or 45 kg each, they are also suitable for protecting non-permanent race tracks.

Today, great importance is attached to the proper condition of the track boundaries, which can be seen from the fact that after an accident, these areas are driven through under a yellow flag until their repair has been completed.

The topic of safety will continue to play a central role in motorsport in the future, as more than 30 people lost their lives at registered racing events worldwide in 2017 (www. motorsportmemorial.org). The high priority of this issue can also be seen in the controversial introduction of the Halo system (Fig. 10.22b): Although the majority of fans, teams and drivers were against its introduction, the FIA pushed through its implementation in the name of safety.

References

1. Tremayne, D.: The Science of Safety. Haynes Publishing, Sparkford (2000)
2. Watkins, S.: Triumph und Tragödien in der Formel 1. HEEL, Königswinter (1997)
3. Prüller, H.: Grand Prix Story 1978 – Andretti, Nummer 1. ORAC, Wien (1978)
4. Watkins, S.: Über dem Limit – Formel-1-Stars zwischen Leben und Tod. HEEL, Königswinter (2001)
5. Olvey, S.: Rapid Response – My Inside Story as Motor Racing Life-Saver. Haynes Publishing, Sparkford (2011)
6. Mellor, A.: Formula One Accident Investigations. SAE-technical paper 2000-01-3552. (2000). https://doi.org/10.4271/2000-01-3552

11.1 Questionnaire

Chapter 2	What technological milestones have had a lasting impact on the development of racing cars?
	Which key elements have found their way into various technical regulations in the course of motorsport history?
	What influence does the discussion about CO_2 and pollutant emissions have on motorsport?
	What are the main items in the budget of a professional racing team?
Chapter 3	State any significant differences between the technical regulations of Formula 1 and the LMP1 category.
	What measures can be taken in the technical regulations to reduce costs within a racing series?
	Which components are assigned to the body by the regulations?
	What are the conflicting goals when creating technical regulations for a high performance racing series?
Chapter 4	Which mechanisms determine the force transmission behavior of a racing tire?
	How does viscoelasticity affect the force transfer characteristics of a racing tire?
	What is meant by the phenomenon of wheel load sensitivity?
	What are the advantages and disadvantages of radial tires?
Chapter 5	How do dynamic wheel load changes affect the handling of a race car?
	How does the weight distribution of a vehicle affect its handling?
	What are the advantages of mixed tires?
	From a performance perspective, what are the conceptual advantages of a rear axle drive?

(continued)

L. Frömmig, *Basic Course in Race Car Technology*, https://doi.org/10.1007/978-3-658-38470-8_11

Chapter 6	What circumstance favors the phenomenon of flow separation?
	What is the relationship between ground effect and the handling of a racing car?
	What measures can be taken in a technical regulation to limit the downforce of a racing vehicle?
	What is the relationship between the safety of a racing vehicle and its aerodynamics?
Chapter 7	What are the causes of positive wheel camber, and how can it be avoided or compensated for?
	How can the driving behavior be influenced by creating a rising rate geometry for the springs?
	How do the aerodynamics of a race car affect its chassis design?
	What is a brake-by-wire system needed for in racing?
Chapter 8	What are the essential tasks and driving dynamic properties of a conventional differential? What disadvantages do they have?
	What influence does a limited slip differential have on braking in a curve?
	What adjustment options does a Salisbury differential offer?
	In addition to limited-slip differentials, there are other technical solutions for achieving asymmetrical drive force distribution. Why are these not used in modern high-performance motorsport?
Chapter 9	How are the driving dynamic properties influenced by a mid-engine concept?
	What are the characteristics of a monocoque design in high performance racing?
	How does the engine concept influence the center of gravity of a vehicle?
	How do the stiffness properties of a vehicle structure affect its handling characteristics?
Chapter 10	Which aspects are mainly responsible for safety in motorsport?
	At which points do the technical regulations determine the design of a racing vehicle in order to optimize "crashworthiness"?
	What restraint systems does a racing vehicle have, and what are their tasks?
	How does track design contribute to safety in motorsport?

11.2 Collection of Tasks and Solutions

11.2.1 Tasks

Task 11.1
The following data were determined at the Silverstone Circuit. At what speed does a Golf VII GTI pass through the "Copse" bend? (Fig. 11.1).

Task 11.2
For a vehicle with a 50–50 weight distribution, the steering wheel angle curve is to be determined via the lateral acceleration, whereby the vehicle is equipped in one case with identical tires on the front and rear axles (combination V1, H1) and once with mixed tires

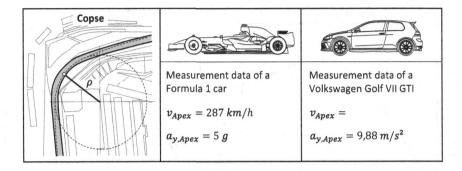

	Measurement data of a Formula 1 car	Measurement data of a Volkswagen Golf VII GTI
	$v_{Apex} = 287\ km/h$	$v_{Apex} =$
	$a_{y,Apex} = 5\ g$	$a_{y,Apex} = 9{,}88\ m/s^2$

Fig. 11.1 Measured values in "Copse"

(combination V1, H2). A stationary circular drive is performed on a 75 m radius. The associated tire characteristics can be seen in Fig. 11.2. The vehicle mass is 1400 kg, the wheelbase 2.8 m and the steering ratio 15.

Task 11.3
A near-series racing vehicle, which does not generate any downforce, negotiates a curve with a lateral acceleration of 10 m/s². The mass of the vehicle is 1200 kg, with 45% of the weight on the front axle. The front and rear axles have coefficients of adhesion of $\mu_V = 1.2$ and $\mu_H = 1.4$ respectively. What longitudinal acceleration can the rear axle driven vehicle achieve in this driving situation? The aerodynamic drag is negligible. What is the steering tendency of the vehicle when it reaches its limit lateral acceleration?

Task 11.4
A racing vehicle negotiates a curve at 15 m/s². The distance between the roll axis and the center of gravity of the body is 0.35 m and the body mass is 1150 kg. What is the roll angle and roll angle gradient of the vehicle if the body springs and stabilizers have the following data. What is the percentage distribution of roll stiffness between the front and rear axles?

Front axle: $c_{FV} = 80{,}000$ N/m, $s_{FV} = 1.5$ m, $c_{StV} = 75{,}000$ Nm/wheel.
Rear axle: $c_{FH} = 92{,}000$ N/m, $s_{FH} = 1.4$ m, $c_{StH} = 25{,}000$ Nm/wheel.

Task 11.5
At the 1998 Belgian Grand Prix, Michael Schumacher drove his Ferrari into the leading McLaren of David Coulthard. At about 260 km/h Coulthard abruptly went off the throttle. Schumacher, following in a spray of rain, crashed into the rear of the McLaren. In the pits, Schumacher later stormed up to Coulthard and yelled, "You tried to kill me!"

What deceleration occurred in Coulthard's maneuver, and how does it compare to a production car?

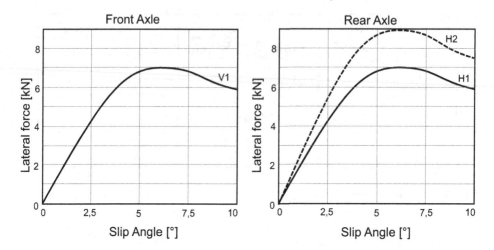

Fig. 11.2 Tire characteristics

Coulthard's McLaren weighed 600 kg and, with a frontal area of 1.6 m², had a drag coefficient $c_x = 1$.

Task 11.6

The Volkswagen I. D. R Pikes Peak generates an downforc of 10,000 N at a speed of 130 km/h in the high-speed oval of the proving ground in Ehra-Lessien (0 m above sea level).

At what speed can this vehicle negotiate the right-hand bend of the 'Bottomless Pit' section (3890 m above sea level) if it has a radius of curvature of 66 m?

The vehicle has a 50:50 weight distribution and an aerodynamic balance of 50%. The frontal area is 2.05 m². Front and rear tires have a coefficient of adhesion $\mu_V = \mu_H = 1.5$. The mass of the VW I. D. R Pikes Peak including the driver is given as 1100 kg.

Task 11.7

Please determine from Fig. 6.17 the drag coefficient c_x as well as the downforce coefficients c_{zV} and c_{zH}, which the Williams-Renault FW14B needs at least to drive through the curve "Bridge" with the measured transverse acceleration of 4.5 g. The frontal area of the vehicle is $A = 1.55$ m², and it has a power output of 525 kW. The coefficients of adhesion of the tires are $\mu_V = 1.2$ at the front axle and $\mu_H = 1.4$ at the rear axle. The weight of the Williams FW14B, including driver and fuel, is given as 620 kg, of which 55% is on the rear axle.

Task 11.8

A vehicle with rear axle drive is accelerated on a μ-split track with $\mu_{low} = 0.4$ and $\mu_{High} = 1$. The vehicle mass is 1400 kg, of which 55% is on the rear axle. The vehicle does not

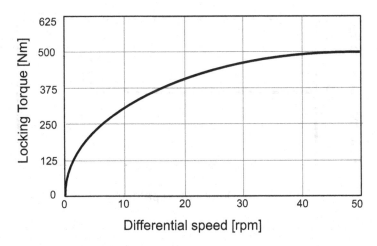

Fig. 11.3 Characteristic curve of the Visco coupling

generate aerodynamic downforce. Dynamic wheel load changes are neglected. The dynamic tire radius is given as $r = 0.34$ m. What is the maximum longitudinal acceleration generated by the vehicle when

(a) the locking value of the differential is $S = 0.3$,
(b) the locking value of the differential is $S = 0.7$,
(c) a Visco coupling according to Fig. 11.3 is used?

Task 11.9
A rear-axle-driven racing vehicle negotiates a curve at 2.0 g. The wheel loads on the inside and outside rear wheels are 3000 N and 4500 N respectively. The vehicle with 50:50-weight distribution weighs 1000 kg. Front and rear tires have coefficients of adhesion $\mu_V = \mu_H = 1.5$. A lateral force of 5800 N is generated at the outside wheel of the curve. What longitudinal acceleration can the vehicle still achieve in this driving situation, and what is the minimum locking value required for this? The air resistance is neglected.

Task 11.10
Figure 11.4 shows the course of the deformation force over the deformation path during a racing accident. Determine the maximum deceleration during the impact and the energy absorbed by deformation. (There are no braking forces or drag forces acting.) The mass of the vehicle is 950 kg. What is the maximum impact velocity that will dissipate all of the vehicle's kinetic energy during deformation?

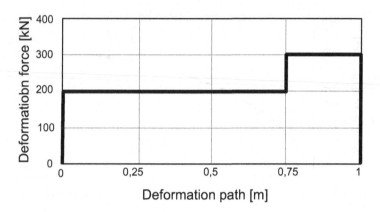

Fig. 11.4 Deformation force curve

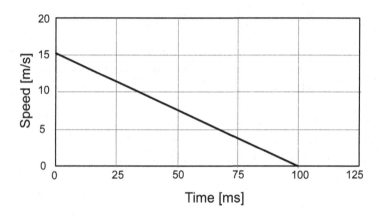

Fig. 11.5 Velocity curve

Task 11.11

Figure 11.5 shows the time course of the sled speed during a frontal crash test of an F1 nose. The mass of the sled is 600 kg. Determine the time course of the deformation force acting on the sled, the deformation path of the nose and the kinetic energy absorbed during the crash test.

11.2.2 Solutions

Solution 11.1

Determine radius of curvature from F1 vehicle data: $a_y = \frac{v^2}{\rho} \rightarrow \rho = 130\ m$.

From this follows for the VW Golf VII GTI: $v = 129\ km/h$.

Solution 11.2

The following applies to the steering wheel angle: $\delta_L = i_L \cdot \left(\frac{l}{\rho} + \alpha_V - \alpha_H \right)$.

The slip angle is calculated from: $F_{yV} = \frac{l_H}{l} \cdot m \cdot \ddot{y} \rightarrow \alpha_V \, (F_{yV})$ read from tire diagram. Repeat this calculation for α_H and calculate the steering wheel angle. The steering wheel angles obtained for different lateral accelerations are then plotted against the lateral acceleration. In this way, the curves shown below are obtained.

Combination V1-H1

a_y [m/ s^2]		1	2	3	4	5	6	7	8	9	$a_{y,}$ $_{max}$ [m/ s^2]	10
δ_L [°]	32.1	32.1	32.1	32.1	32.1	32.1	32.1	32.1	32.1	32.1	δ_L [°]	32.1

The vehicle has neutral steering.

Combination V1-H2

a_y [m/ s^2]		1	2	3	4	5	6	7	8	9	$a_{y,}$ $_{max}$ [m/ s^2]	10
δ_L [°]	32.1	33.6	35.1	36.6	38.1	39.6	41.1	42.6	44.1	50.1	δ_L [°]	72.6

The vehicle is understeering.

Solution 11.3

Friction circle on the rear axle: $\mu_H \cdot F_{zH} = \sqrt{F_{xH}^2 + F_{yH}^2} \rightarrow F_{xH}$.

The following applies: $F_{zH} = m \cdot \frac{l_V}{l} \cdot g$ and $F_{yH} = m \cdot \frac{l_V}{l} \cdot \ddot{y}$ with $\frac{l_V}{l} = 0.55$.

The acceleration is obtained from: $m \cdot \ddot{x} = F_{xH} \rightarrow \ddot{x} = 5.18 \ m/s^2$.

Limiting lateral acceleration of the front axle: $m \cdot \frac{l_H}{l} \cdot a_{y,gr,V} = \mu_V \cdot F_{zV} \rightarrow a_{y, \, gr,}$ $v = \mu_V g = 11.77 \ m/s^2$.

Limiting lateral acceleration of the rear axle: $m \cdot \frac{l_V}{l} \cdot a_{y,gr,H} = \mu_{VH} \cdot F_{zH} \rightarrow a_{y, \, gr,}$ $v = \mu_H g = 13.73 \ m/s^2$.

The vehicle is understeering at the limit because $a_{y, \, gr, \, v} < a_{y, \, gr, \, H}$.

Solution 11.4

Front axle roll resistance: $M_V = c_V \cdot \kappa = \left(\frac{c_{FV} \cdot s_{FV}^2}{2} + c_{StV} \right) \cdot \kappa$.

Roll resistance of the rear axle: $M_H = c_H \cdot \kappa = \left(\frac{c_{FH} \cdot s_{FH}^2}{2} + c_{StH} \right) \cdot \kappa$.

Equilibrium about the roll axis: $m_A \cdot \ddot{y} \cdot h' + m_A \cdot g \cdot h' \cdot \kappa = M_V + M_H \rightarrow \kappa = 1.25$ $^\circ \rightarrow WG = 0.82\ °/g$.

Rolling stiffness ratio of the front axle $= \frac{c_V}{c_V + c_H} = 58.9\%$.

Solution 11.5

When pressing the accelerator pedal from high speeds, it can be simplistically postulated that only drag brakes the vehicle. It follows from this:

$$m \cdot \ddot{x} = c_x \cdot A \cdot \frac{\rho_L}{2} \cdot v^2 \rightarrow \ddot{x} = 8.52 m/s^2.$$

The braking distance of a typical production car from 100 m/h is about 36 m. This corresponds to a deceleration of 10.72 m/s^2.

At high speeds, the deceleration of an F1 car due to air resistance is comparable to the emergency braking of a passenger car.

Solution 11.6

It follows from the terms of reference: $F_{Abrieb,Front\ axle} = 5000\ N$ at $v = 130\frac{km}{h}$..

Due to the center load, the following applies: $\frac{l_H}{l} = 0.5$.

The air density can be determined from the literature or an Internet search: $\rho_L(0m) = 1.225\frac{kg}{m^3}$.

At Pikes Peak, the following applies to the "Bottomless Pit" section: $\rho_L(3890\ m) = 0.832\frac{kg}{m^3}$.

In Ehra-Lessien: $F_{Abrieb,Front\ axle} = - c_{zV} \cdot A \cdot \frac{\rho_L(0\ m)}{2} \cdot v^2 \rightarrow c_{zV} = -3.05$.

At Pikes Peak: $F_{yV} = m \cdot \frac{l_H}{l} \cdot \frac{v^2}{\rho} = \mu_V \cdot \left(m \cdot \frac{l_H}{l} \cdot g - c_{zV} \cdot A \cdot \frac{\rho_L(3890m)}{2} \cdot v^2 \right) \rightarrow$

$v = 153.92\ km/h$.

Solution 11.7

Top speed shortly before Stowe (after approx. 32 s) $\rightarrow$ v$_{max} \approx 182$ mph ≈ 293 km/h

$$P_{max} = c_x \cdot A \cdot \frac{\rho_L}{2} \cdot v_{max}^3 \rightarrow c_x \approx 1.03.$$

Values for bridge: a$_y$ = 4.5 g with a$_x$ = 0 and v = 174 mph $\approx$ 280 km/h with $\frac{l_H}{l} = 0.45$ (according to the task text).

$$F_{yV} = m \cdot \frac{l_H}{l} \cdot a_y = \mu_V \cdot \left(m \cdot \frac{l_H}{l} \cdot g + c_{zV} \cdot A \cdot \frac{\rho_L}{2} \cdot v(Bridge)^2 \right) \rightarrow c_{zV} = 1.31.$$

The output coefficient for the rear axle is calculated analogously: $c_{zH} = 1.29$.

Solution 11.8

At maximum acceleration, the low-friction wheel is always at the adhesion limit. Furthermore, it must be examined whether the limited slip differential locks completely. In the case of complete locking on μ-split, the adhesion potential is completely utilized on both drive wheels, so that the following applies:

$$S_{sperr} = \frac{M_4 - M_3}{M_4 + M_3} = \frac{F_{x4} \cdot r - F_{x3} \cdot r}{F_{x4} \cdot r + F_{x3} \cdot r} = \frac{(\mu_{High} - \mu_{low}) \cdot m \cdot \frac{l_V}{l \cdot 2} \cdot g}{(\mu_{High} + \mu_{low}) \cdot m \cdot \frac{l_V}{l \cdot 2} \cdot g} = \frac{\mu_{High} - \mu_{low}}{\mu_{High} + \mu_{low}}$$

$$\rightarrow S_{sperr} = 0.43.$$

The locking torque required for this is: $M_{Sperr} = (F_{x4} - F_{x3}) \cdot r$, see case (c).

In case (a), the following applies: $S_{Sperr} > S = 0.3$. The locking value is not sufficient to completely lock the differential. Therefore, the following must apply:

$$S = 0.3 = \frac{F_{x4} - F_{x3}}{F_{x4} + F_{x3}} \rightarrow F_{x4} = \frac{S+1}{1-S} \cdot F_{x3}.$$

The low friction wheel operates at the adhesion limit: $F_{x3} = \mu_{Low} \cdot m \cdot g \cdot \frac{l_V}{2 \cdot l}$.

The following applies to the acceleration: $m \cdot \ddot{x} = F_{x3} + F_{x4} \rightarrow \ddot{x} = \frac{F_{x3} \cdot \left(1 + \frac{S+1}{1-S}\right)}{m} = 3.1 \ m/s^2$.

In case (b), $S_{Sperr} < S = 0.7$. The differential is fully locked and both wheels operate at the friction limit: $F_{x3} = \mu_{Low} \cdot m \cdot g \cdot \frac{l_V}{2 \cdot l}$ and $F_{x4} = \mu_{High} \cdot m \cdot g \cdot \frac{l_V}{2 \cdot l}$.

The following applies to the acceleration: $m \cdot \ddot{x} = F_{x3} + F_{x4} \rightarrow \ddot{x} = 3.8 \ m/s^2$.

In case (c), $M_{Sperr} = (\mu_{High} + \mu_{Low}) \cdot m \cdot g \cdot \frac{l_{HV}}{2 \cdot l} \cdot r = 770 \ Nm > M_{Visco, max} = 500 \ Nm$.

The differential is not fully locked, and therefore: $F_{x4} = F_{x3} + \frac{M_{Visco, max}}{r}$.

The following applies to the acceleration: $m \cdot \ddot{x} = 2 \cdot F_{x3} + \frac{M_{Visco, max}}{r} \rightarrow \ddot{x} = 3.2 \ m/s^2$.

Solution 11.9

The following applies to the total lateral force of the driven axle: $F_{yH} = m \cdot \frac{l_V}{l} \cdot \ddot{y}$.

The following applies to the lateral force on the inside wheel: $F_{y3} = F_{yH} - F_{y4}$.

The acceleration that can still be achieved is reached when both wheels are working at the frictional limit:

$$\mu_H \cdot F_{z3} = \sqrt{F_{x3}^2 + F_{y3}^2} \rightarrow F_{x3} \ and \ \mu_H \cdot F_{z4} = \sqrt{F_{x4}^2 + F_{y4}^2} \rightarrow F_{x4}.$$

The following applies to the acceleration: $m \cdot \ddot{x} = F_{x3} + F_{x4} \rightarrow \ddot{x} = 5.5 \ m/s^2$.

For the locking value, $S = \frac{F_{x4} - F_{x3}}{F_{x4} + F_{x3}} \rightarrow S \geq 0.26$.

Solution 11.10

The maximum deformation force causes the maximum deceleration. This can be taken from the diagram (Fig. 11.4).

It is true: $m \cdot \ddot{x}_{max} = F_{Defo, max} = 300 \ kN \rightarrow \ddot{x}_{max} = 316\frac{m}{s^2} = 32 \ g$.

The deformation energy corresponds to the area under the deformation force curve:

$$E_{Defo} = 2,00,000N \cdot 0.75m + 3,00,000N \cdot 0.25m \rightarrow E_{Defo} = 2,25,000Nm.$$

The deformation energy must completely absorb the kinetic energy.

It follows: $E_{Defo} = E_{kin} = \frac{1}{2} \cdot m \cdot v^2 \rightarrow v = 78.4$ km/h.

Solution 11.11

Due to the linear velocity progression, the following applies: $\ddot{x} = \frac{\Delta v}{\Delta t} = -150 \ \frac{m}{s^2} = -15.3 \ g$.

For the deformation force, it follows: $F_{Defo} = m \cdot \ddot{x} = 90 \ kN$.

Because of the constant deceleration, the deformation force is also constant over time.

The following applies to the deformation path: $s_{Defo} = \frac{1}{2} \cdot \ddot{x} \cdot t^2 = -0.75 \ m$.

The deformation energy is obtained from: $E_{Defo} = F_{Defo} \cdot s_{Defo} = 67,500 \ Nm$.

Printed in the United States
by Baker & Taylor Publisher Services